Nonlinear Dynamics
in Solids

H. Thomas (Ed.)

Nonlinear Dynamics in Solids

Edited in Cooperation with the
Deutsche Physikalische Gesellschaft

With 148 Figures

Springer-Verlag

Berlin Heidelberg New York
London Paris Tokyo
Hong Kong Barcelona
Budapest

Professor Dr. Harry Thomas

Institut für Physik, Universität Basel, Klingelbergstrasse 82,
CH-4056 Basel, Switzerland

ISBN-13: 978-3-642-95652-2   e-ISBN-13: 978-3-642-95650-8
DOI: 10.1007/978-3-642-95650-8

Library of Congress Cataloging-in-Publication Data. Nonlinear dynamics in solid state physics / H. Thomas
(ed.), edited in cooperation with the Deutsche Physikalische Gesellschaft. (p. cm.) Includes bibliographical re-
ferences.ISBN-13: 978-3-642-95652-2
1. Solid state physics. 2. Dynamics. 3. Nonlinear theories. I. Thomas, H. (Harry)
II. Deutsche Physikalische Gesellschaft (1963–) QC176.2.N66   1992   530.4'1 – dc20   91-45673

© Springer-Verlag Berlin Heidelberg 1992
Softcover reprint of the hardcover 1st edition 1992

The use of general descriptive names, registered names, trademarks, etc. in uns publication does not imply even
in the absence of a specific statement, that such names are exempt from the relevant protective laws and regula-
tions and therefore free for general use.

Typesetting: Camera ready by author

56/3140 - 5 4 3 2 1 0 – Printed on acid-free paper

Preface

This volume contains the notes of lectures given at the school on "Nonlinear Dynamics in Solids" held at the Physikzentrum Bad Honnef, 2–6 October 1989 under the patronage of the Deutsche Physikalische Gesellschaft.

Nonlinear dynamics has become a highly active research area, owing to many interesting developments during the last three decades in the theoretical analysis of dynamical processes in both Hamiltonian and dissipative systems. Research has been focused on a variety of problems, such as the characteristics of regular and chaotic motion in Hamiltonian dynamics, the problem of quantum chaos, the formation and properties of solitary spatio-temporal structures, the occurrence of strange attractors in dissipative systems, and the bifurcation scenarios leading to complex time behaviour.

Until recently, predictions of the theory have been tested predominantly on instabilities in hydrodynamic systems, where many interesting experiments have provided valuable input and have led to a fruitful interaction between experiment and theory. Fluid systems are certainly good candidates for performing clean experiments free from disturbing influences: with fluids, compared to solids, it is simpler to prepare good samples, the relevant length and time scales are in easily accessible ranges, and it is possible to do measurements "inside" the fluid, because it can be filled in after the construction of the apparatus. Further, the theory describing the macroscopic dynamics of fluids is well established and contains only very few parameters, all of which have well-known values.

The dynamics of solids, on the other hand, is much richer and has a higher degree of physical interest. Moreover, for several solid-state systems, the methods of sample preparation and measuring techniques have reached a level where experiments on nonlinear dynamic behaviour of comparable quality and detail have become possible. It therefore appeared to be the right time to bring together a number of leading experts working on nonlinear dynamic properties of various solid-state systems to give introductory reports on the state of knowledge, problems incurred, and future prospects.

The volume starts with a presentation of the basic concepts of formation, symmetry and stability of dynamic structures, and a discussion of the analogies and differences to phase transitions in equilibrium systems. This is followed by a brief introduction to deterministic chaos and strange attractors, and an outline of methods for the characterization of chaotic motion and the reconstruction of attractors from experimental time series.

The next group of topics is concerned with nonlinear oscillations and chaos occurring in various solid-state systems: current instabilities and optical instabilities in semiconductors, driven Josephson junctions, and spin-wave instabilities in ferromagnets.

First, an introduction is given to various mechanisms of current instabilities in semiconductors, their theoretical description, and the methods used for their analysis. The current instability occurring in p-germanium due to avalanche breakdown gives rise to a rich variety of self-generated dynamical structures, as discussed in a contribution containing a detailed experimental analysis of oscillatory and chaotic states due to breathing of current filaments. Formation and dynamics of current filaments in semiconductor devices such as pin diodes is reviewed in a further contribution.

Next follows a discussion of optical bistabilities in passive semiconductors based on photo-thermal nonlinearities, and of self-oscillations in optical ring resonators with bistable elements.

A variety of nonlinear dynamical phenomena and chaos is expected to occur in Josephson junctions and devices driven by ac or dc currents. Such systems are the subject of a contribution reviewing experimental and numerical investigations of the characteristics of chaotic dynamics in various parts of parameter space.

Another class of solid-state systems exhibiting chaotic dynamics is ferromagnetic samples excited by strong microwave fields giving rise to spin-wave instabilities. The last contribution of this group reports and analyses complex multistable behaviour, self-oscillations, and bifurcation sequences leading to chaos found in magnetic resonance experiments on YIG (yttrium iron garnet) spheres.

A theme discussed in several of these contributions concerns the interplay between nonlinear dynamics and solid-state physics, and in particular the question to what extent it is possible to relate the details of the observed nonlinear dynamical effects – bifurcation scenarios, onset of chaos, characteristics of chaotic dynamics, etc. – to typical solid-state properties. Owing to the extreme sensitivity of the nonlinear dynamical properties to the detailed sample structure, this presents a serious problem: even samples cut from the same carefully grown material show differences in their nonlinear dynamic behaviour. On the other hand, this sensitivity is an important aspect for future development. It may well turn out that observation of nonlinear dynamical effects can be developed to become a most sensitive tool for the study of solid-state properties.

An interesting aspect of nonlinear dynamics, the formation of solitary spatiotemporal structures (kinks, domain walls), had actually already found experimental attention in solid-state physics at an early stage. Under certain conditions, these structures are expected to occur in thermodynamic equilibrium as a gas of nonlinear excitations; detailed investigations have been carried out in particular for quasi-one-dimensional magnetic systems, which are the subject of the next two contributions. The first of these reviews the theory of the formation, propagation and stability of such nonlinear excitations as well as their statistical mechanics, and the second reports results of inelastic neutron scattering, and NMR and ESR experiments for $CsNiF_3$ and TMMC (Tetramethylammonium Manganese Trichloride).

Macroscopic dynamical processes in solids are usually dissipative, and are therefore properly described in terms of the concepts of the dynamics of dissipative systems such as attractors and transients, with well-known exceptions of certain (nuclear and electron) spin systems with extremely small damping. However, semiconductor physics has reached a stage where mesoscopic semiconductor structures are now available in which electron transport is essentially dissipation-free, and which are therefore expected to exhibit nonlinear phenomena based on Hamiltonian dynamics. The relevance of Hamiltonian chaos and KAM theory for phenomena occurring in such systems is discussed in a contribution focusing on lateral surface superlattices.

It is an interesting question whether the concept of chaos may be extended to describe irregular behaviour in spatial dimensions, and whether spatial chaos is a useful concept for characterizing disordered or glassy structures. These questions are addressed in the last contribution of this volume, where simple models are introduced which have a multitude of spatially chaotic metastable states showing at least qualitatively some of the typical properties of glassy materials.

Basel
January 1992 *H. Thomas*

Contents

Dynamical Structures: Formation, Symmetry, Stability

H. Thomas

Institut für Physik der Universität Basel,
Klingelbergstrasse 82, CH-4056 Basel, Switzerland

1. Introduction

The subject of this lecture is the formation of dynamic structures in physical systems under the influence of external forces which drive currents through the system. It is a common feature of such "driven" systems that the driving force keeps them far from thermodynamic equilibrium, and that dissipation gives rise to the production of heat which has to be carried away by coupling the system to a heat sink (Fig. 1).

Our main interest is in systems under the influence of a *stationary* driving force. Here, one may distinguish two cases:

- Systems in which the force always gives rise to a current, for example a semiconductor in an electric field.
- Systems in which the occurrence of a current depends on the boundary conditions, for example a magnetic system in a static magnetic field: Here, thermodynamic equilibrium is possible because of the absence of magnetic charges, but continuous flow of magnetic flux may still occur if the field drives a domain wall between two oppositely magnetized domains.

Examples of driven systems in solid-state physics are treated in the other chapters of this book.

We shall concentrate on the behaviour of the system after the decay of transients. Of particular importance is the fact that, in contrast to thermodynamic systems which always approach a state of thermodynamic equilibrium, driven systems may either approach a stationary nonequilibrium state, or remain permanently time-dependent. It is the possiblity of the occurrence of such "dynamic structures" with a spontaneously broken time translation symmetry which makes the study of driven systems especially fascinating.

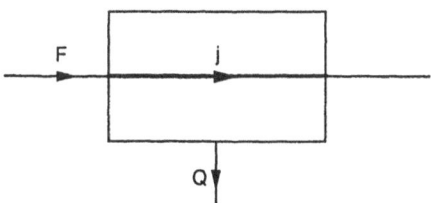

Fig. 1. The external force F drives a current j through the system; the dissipated energy Q is absorbed by a heat sink.

The formation of dynamic structures in driven systems bears a certain resemblance to the occurrence of phase transitions in thermodynamic equilibrium systems. In the latter case, symmetry aspects have proved to be of central importance. It appears therefore appropriate to extend the symmetry concept for application to driven systems, and to determine its significance for the prediction and classification of the types of dynamic structure which may occur in a given situation.

In this lecture, I shall give a simple introduction to the basic features of structure formation and its relation to symmetry. For more detailed presentations and discussions of other aspects, I refer to some standard texts [1–4] and a number of recent discussions of the subject [5–13].

2. Description of Driven Systems

2.1 Equation of Motion

We describe the dynamics of driven systems on a macroscopic level, similar to that used in Landau's theory of phase transitions: The *state* of the system is represented by a point $\boldsymbol{\theta}$ in a multidimensional state space (in general a differentiable manifold); the relevant external parameter is called the *control parameter* μ.

The state change in the course of time is described in terms of a velocity field $\boldsymbol{B}(\boldsymbol{\theta}, \mu)$ in state space, giving rise to an evolution equation

$$\frac{\mathrm{d}\boldsymbol{\theta}}{\mathrm{d}t} = \boldsymbol{B}(\boldsymbol{\theta}, \mu) \tag{2.1}$$

which is nonlinear and local in time (no memory!). In the case of a stationary control parameter μ, (2.1) is invariant with respect to time translations.

In dissipative systems, the vector field $\boldsymbol{B}$ is contracting, i.e. in the course of time any trajectory $\boldsymbol{\theta}(t)$ approaches an *attractor* in state space. The type of attractor may be characterized by a set of *Lyapunov exponents* (LE) which describe the asymptotic behaviour of the distance between two initially adjacent state points for $t \to \infty$. A trajectory $\gamma : \boldsymbol{\theta} = \boldsymbol{\theta}(t)$ has an LE λ if the distance $|\boldsymbol{\theta}(t) - \boldsymbol{\theta}_1(t)|$ between a point $\boldsymbol{\theta}(t)$ on γ and a neighboring point $\boldsymbol{\theta}_1(t)$ varies for $t \to \infty$ asymptotically as $\exp(\lambda t)$. In a state space of N dimensions, a trajectory has in general N LEs λ_i which satisfy $\sum_i \lambda_i < 0$ in dissipative systems. One distinguishes the following types of attractor:

- *Fixed points* in state space corresponding to stationary states $\boldsymbol{\theta}_s(\mu)$, for which *all* Lyapunov exponents are negative, and which are found as stable solutions of

$$\boldsymbol{B}(\boldsymbol{\theta}_s, \mu) = 0 \ . \tag{2.2}$$

- *Limit cycles* (1-*tori*) with a periodic time dependence, $\boldsymbol{\theta}_c(t) = \boldsymbol{\theta}_c(t + T)$, having 1 vanishing LE.

- Multiply periodic structures (*n-tori*) having n vanishing LEs.
- *Strange attractors* exhibiting *chaotic motion* characterized by 1 or more positive LEs.

Equation (2.1) represents a *deterministic* evolution equation. Fluctuations may be taken into account by adding a stochastic force $f(\theta, t)$ with $\langle f(\theta, t) \rangle = 0$ to the r.h.s of (2.1).

2.2 Symmetry

The set of transformations $g : \theta \mapsto g\theta$ of state space which leave the velocity field invariant forms the symmetry group G of the system:

$$G := \{ g \,|\, gB(\theta) = B(g\theta) \; \forall \theta \} \; . \tag{2.3}$$

The symmetry group G_s of the stationary state θ_s consists of all transformations $g \in G$ which leave θ_s invariant:

$$G_s := \{ g \in G \,|\, g\theta_s = \theta_s \} \; . \tag{2.4}$$

The stationary state θ_s may be fully symmetric (G_s coincides with G), or it may already have a broken symmetry (G_s is a genuine subgroup of G). In the latter case, there exists a set of symmetry-related stationary states corresponding to the left cosets of G_s in G.

Further, a stationary state is invariant under all time translations $\mathbf{T}(\tau)$: $t \mapsto t + \tau$ forming the full-time-translation group

$$\mathcal{I} := \{ \mathbf{T}(\tau) \,|\, \tau \in \mathbb{R} \} \; . \tag{2.5}$$

The *extended symmetry group* $\mathcal{G}_s$ of the stationary state θ_s is defined as

$$\mathcal{G}_s := G_s \times \mathcal{I} \; . \tag{2.6}$$

It corresponds to the "H-group" in Landau's theory of phase transitions. Structure formation is associated with a breaking of symmetry: A structure bifurcating from θ_s will be characterized by an "L-group", which is a subgroup of $\mathcal{G}_s$ (see Sect. 3.2).

3. Self-oscillations (Limit Cycles)

3.1 Bifurcation of Self-oscillations (Hopf Bifurcation)

3.1.1 Destabilization of a Stationary State. In order to obtain conditions for the formation of self-oscillations, we study the linear stability of the stationary state $\theta_s(\mu)$ with respect to small perturbations $\vartheta(t)$: $\theta(t) = \theta_s + \vartheta(t)$. Linearization of (2.1) about θ_s yields the equation of motion for ϑ

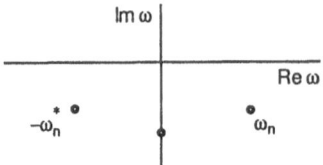

Im ω

Re ω

$-\omega_n^*$ ω_n

Fig. 2. Poles in the complex frequency plane with $\mathrm{Re}\,\omega_n \neq 0$ occur in pairs $(\omega_n, -\omega_n^*)$

$$\frac{d\boldsymbol{\vartheta}}{dt} = \mathbf{L}(\mu) \cdot \boldsymbol{\vartheta} \ , \tag{3.1}$$

where $\mathbf{L}(\mu)$ is a real time-independent matrix defined by

$$\mathbf{L} := \partial B/\partial\theta\big|_{\theta \,=\, \theta_s} \ . \tag{3.2}$$

The *normal modes* of the stationary state $\boldsymbol{\theta}_s$ are solutions of (3.1) of the form

$$\boldsymbol{\vartheta}_n(t) = \boldsymbol{p}_n e^{-i\omega_n t} \ ; \tag{3.3}$$

their polarization amplitudes $\boldsymbol{p}_n$ and frequencies ω_n are determined by the linear eigenvalue problem

$$\mathbf{L}(\mu) \cdot \boldsymbol{p}_n = -i\,\omega_n \boldsymbol{p}_n \ . \tag{3.4}$$

Because of the reality of $\mathbf{L}(\mu)$, normal modes with $\mathrm{Re}\,\omega_n \neq 0$ occur in pairs $\{(\boldsymbol{p}_n, \omega_n), (\boldsymbol{p}_n^*, -\omega_n^*)\}$ (see Fig. 2).

We assume that the stationary state $\boldsymbol{\theta}_s$ is stable in a control-parameter range $\mu < \mu_c$:

$$\mathrm{Im}\,\omega_n(\mu) < 0 \ \text{ for } \ \mu < \mu_c \qquad \forall n \tag{3.5}$$

(except for zero-frequency modes),

and that at the stability threshold $\mu = \mu_c$ there occurs an instability with respect to a single, possibly degenerate mode (single mode pair if $\mathrm{Re}\,\omega_1 \neq 0$):

$$\mathrm{Im}\,\omega_1(\mu_c) = 0, \qquad \frac{d\,\mathrm{Im}\,\omega_1}{d\mu}\bigg|_{\mu \,=\, \mu_c} > 0 \ . \tag{3.6}$$

Such an instability marks the bifurcation of a new structure.

3.1.2. The Order Parameter. Types of Bifurcating Structures. Close to the bifurcation threshold, the bifurcating structure is uniquely described by the projection ϕ of the state vector $\boldsymbol{\theta}$ onto the space Ω spanned by the undamped normal modes. In analogy to Landau's theory of phase transitions, the projection ϕ is called the "order parameter" (OP) associated with the instability. By adiabatic elimination of the other components of $\boldsymbol{\theta}$, one obtains from (2.1) a reduced equation of motion in OP space

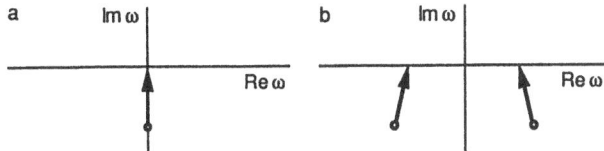

Fig. 3. (a) Soft-mode instability. (b) Hard-mode instability

$$\frac{d\phi}{dt} = -i\,\omega_1(\mu)\,\phi + N(\phi) \ , \tag{3.7}$$

where $N(\phi)$ contains all nonlinear terms. Symmetry requires that the terms in (3.7) transform equivariant with the OP. Therefore, the form of this equation may be constructed without actually carrying out the adiabatic elimimation, by writing $N(\phi)$ as a sum of polynomial equivariants of increasing degree, with coefficients considered as model parameters.

The type of instability is determined by the real part $\omega_c := \mathrm{Re}\,\omega_1(\mu_c)$ of the mode frequency at threshold (see Fig. 3 and Table 1):

- $\omega_c = 0$: If a purely *relaxational* mode becomes undamped ("soft-mode instability"), one expects a bifurcation of a new *stationary state*.
- $\omega_c \neq 0$: If an *oscillating* mode becomes undamped ("hard-mode instability"), one expects a bifurcation of a *limit cycle* with period $T_c = 2\pi/\omega_c$ at threshold.

Table 1. Unstable modes and bifurcating structures

Type of Instability	Bifurcating Structure
Soft Mode: $\omega_c = 0$	Stationary: $\phi = \text{const}$
Hard Mode: $\omega_c \neq 0$	Oscillating: $\phi(t+T) = \phi(t)$ $T_c = 2\pi/\omega_c$

These expectations are indeed confirmed by the formal bifurcation analysis based on an expansion of the solution of (3.7) (or even of the original equation (2.1)) in powers of the amplitude $\varepsilon = |\phi|$ of the OP. From such an analysis, one obtains the components of the OP, the control parameter μ, and in the case of a limit cycle its period T expressed in powers of ε,

$$\phi = \phi(\varepsilon) \quad \mu = \mu(\varepsilon), \quad T = T(\varepsilon) \ . \tag{3.8}$$

The form of the function $\mu(\varepsilon)$ determines the existence range of the new structure (see Fig. 4):

- $\mu < \mu_c$ (subcritical bifurcation),
- $\mu > \mu_c$ (supercritical bifurcation),
- $\mu <> \mu_c$ (transcritical bifurcation).

The dependence of the amplitude of the OP, its orientation in OP space, and (if applicable) its period T on the control parameter μ are found from (3.8) by

5

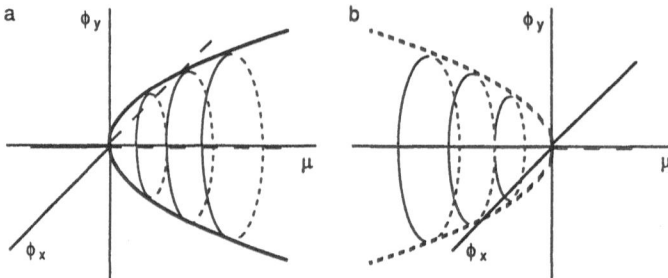

Fig. 4a,b. Hopf bifurcation. (a) supercritical, (b) subcritical

elimination of the amplitude parameter ε. Finally, the structures identified in this way have to be tested for stability in order to determine which of them can actually occur. It is found that close to the bifurcation threshold, subcritical branches are always unstable; only the supercritical branches are candidates for stable structures. If there is no symmetry at all, there either occurs a transcritical bifurcation with exchange of stability, or a Hopf bifurcation.

The bifurcation problem depends in an essential way on the degeneracy of the destabilized mode, i.e. on the dimension of the OP space Ω (assuming that the representation of the OP occurs with multiplicity 1) (Table 2):

– Nondegenerate soft mode, dim $\Omega = 1$: The direction of ϕ is fixed by symmetry; only its amplitude has to be determined from bifurcation analysis.
– Degenerate soft mode, dim $\Omega > 1$: Both the amplitude and the direction of ϕ have to be determined from bifurcation analysis.

Correspondingly:

– Nondegenerate hard mode, dim $\Omega = 2$ (Hopf bifurcation): The orientation of the limit cycle is fixed by symmetry, only its amplitude has to be determined from bifurcation analysis.
– Degenerate hard-mode, dim $\Omega > 2$ (degenerate Hopf bifurcation): Both the orientation of the limit cycle and its amplitude have to be determined from bifurcation analysis.

Note that in the case of a hard-mode instability dim Ω is even, i.e. the smallest dimension for a degenerate Hopf bifurcation is 4. Degenerate Hopf bifurcations with dim $\Omega = 4$ are studied in [6,7,9]; other examples are provided by wave bifurcations associated with degenerate modes (see Sects. 4, 5).

Table 2. OP repesentation and bifurcating structure

Instability type	dim Ω	Bifurcating Structure
Soft mode	dim $\Omega = 1$	uniquely determined by symmetry
	dim $\Omega \geq 2$	depending on higher-order terms
Hard Mode	dim $\Omega = 2$	uniquely determined by symmetry
	dim $\Omega \geq 4$	depending on higher-order terms

In most cases it is found that the direction of the OP satisfies a rule of "minimally broken symmetry":

The symmetry group of the bifurcating structure is a maximal subgroup of the symmetry group of the state $\boldsymbol{\theta}_s$,

which may serve as a first indication of the structures to be expected.

Other types of bifurcation which occur frequently but which cannot be found from the stability analysis of an "unperturbed" stationary state $\boldsymbol{\theta}_s(\mu)$ include:

– "Saddle-node bifurcation": At a critical value of the control parameter there appears a pair of fixed points, one of them stable, the other unstable, but unconnected to any other fixed point.
– "Homoclinic bifurcation" of a limit cycle: At a critical value of the control parameter there appears a fixed point $\boldsymbol{\theta}_h$ with a homoclinic orbit starting at $t = -\infty$ at $\boldsymbol{\theta}_h$ and returning at $t = \infty$ to $\boldsymbol{\theta}_h$ which develops into a limit cycle bifurcating with zero frequency but finite amplitude.

3.1.3 Symmetry Properties of the Order Parameter. The OP space Ω carries a representation of the symmetry group G_s of the parent state $\boldsymbol{\theta}_s$. In analogy to Landau's theory of phase transitions, one expects that this representation contains important information on the bifurcation behaviour. Of particular interest is the question whether it can be predicted on the basis of symmetry alone if the instability gives rise to a stationary or an oscillating structure. The following result shows the extent to which this is the case [8]:

The OP associated with the bifurcation of a *stationary state* transforms as a real irreducible representation of G_s

The OP associated with a *limit cycle* bifurcation transforms
– either as a physically irreducible representation consisting of two complex conjugate irreducible representations ("symmetry-induced limit-cycle bifurcation")
– or as a reducible real representation consisting of two equivalent real irreducible representations of G_s ("coupling-induced limit-cycle bifurcation").

The two types of limit-cycle bifurcation may be illustrated for the case of a two-dimensional OP space by considering viscous motion in a potential $V(\phi_x, \phi_y) = -\frac{1}{2}\mu(\phi_x^2 + \phi_y^2) + V_0(\phi_x, \phi_y)$ under the action of an azimuthal force $\omega_0\{-\phi_y, \phi_x\}$,

$$\frac{d\phi_x}{dt} = \mu\phi_x - \omega_0\phi_y - \frac{\partial V_0}{\partial \phi_x}, \qquad \frac{d\phi_y}{dt} = \mu\phi_y + \omega_0\phi_x - \frac{\partial V_0}{\partial \phi_y} \ . \qquad (3.9)$$

If $V(\phi_x, \phi_y)$ has n-fold symmetry, it will for sufficiently large values of the control parameter μ develop n minima separated by saddles with heights of $O(|\phi_0|^n)$, where $|\phi_0|$ is the amplitude of the OP at minimum.
– For $n \geq 3$, the representation spanned by (ϕ_x, ϕ_x) is irreducible, and the azimuthal force can in fact drive the system across the saddles, independent of the value of the coupling constant ω_0 (symmetry-induced limit cycle).

– For $n = 2$, on the other hand, the representation decomposes into two equivalent real representations spanned by ϕ_x and ϕ_y. In this case, the coupling constant w_0 has to exceed a critical value in order to drive the system across the saddle (coupling-induced limit cycle); for values of w_0 smaller than the critical value, the system becomes trapped in a stationary state in one of the two potential troughs.

3.2 Symmetry of Cycles

A cycle breaks the symmetry of the stationary state which is described by its symmetry group G_s and the extended symmetry group $\mathcal{G}_s$ defined in (2.4,6). The symmetry group of the cycle $\boldsymbol{\theta}_c$ consists fo all transormations $g \in G_s$ which leave the *orbit* $[\boldsymbol{\theta}_c]$ of the cycle invariant:

$$G_c := \{ g \in G_s \,|\, [g\boldsymbol{\theta}_c] = [\boldsymbol{\theta}_c] \} \ . \tag{3.10}$$

Further, the cycle $\boldsymbol{\theta}_c(t) = \boldsymbol{\theta}_c(t + T)$ is left invariant under *discrete* time translations $\mathbf{T}(nT)$ forming a *time lattice*

$$\mathcal{I}_T := \{ \mathbf{T}(nT) \,|\, n = 0, \pm 1, \ldots \} \ . \tag{3.11}$$

The *extended symmetry group* $\mathcal{G}_c$ of the cycle consists of all transformations $g \in \mathcal{G}_s$ which leave the cycle invariant:

$$\mathcal{G}_c := \{ g \in \mathcal{G}_s \,|\, g\boldsymbol{\theta}_c(t) = \boldsymbol{\theta}_c(t) \ \forall t \} \ . \tag{3.12}$$

It consists of a product of the time lattice $\mathcal{I}_T$ and a group L which is isomorphic to G_c:

$$\mathcal{G}_c = L \times \mathcal{I}_T \quad \text{with} \quad L \cong G_c \ . \tag{3.13}$$

According to the rule of minimally broken symmetry, G_c is a maximal subgroup of G_s. The extended symmetry group $\mathcal{G}_c$ corresponds to the space group of a spatial structure, and the group G_s to its point group.

The time lattice $\mathcal{I}_T$ gives rise to a *Brillouin-zone* (BZ) structure on the $\text{Re}\,\omega$ axis. In terms of the reciprocal lattice vector $\Omega := 2\pi/T$, the first BZ is given by

$$-\Omega < \text{Re}\,\omega \le \Omega \tag{3.14}$$

(see Fig. 5). The analog of Bragg scattering by spatial lattices is the occurrence

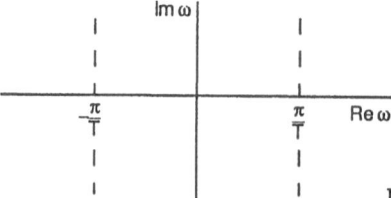

Fig. 5. "Brillouin zone" on the real frequency axis

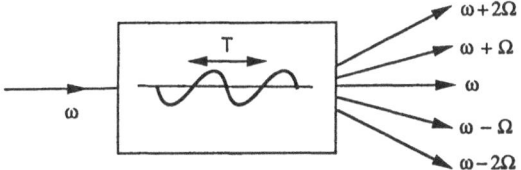

Fig. 6. "Bragg scattering" at a time lattice: Linear response to a force at frequency ω gives rise to side bands at frequencies $\omega \pm n\Omega$ where $\Omega = 2\pi/T$

of *side-bands* at frequencies $\omega \pm n\Omega$ in the linear response of the cycle to an external force of frequency ω (Fig. 6).

3.3 Destabilization of a Cycle

The linear stability of the cycle $\boldsymbol{\theta}_c(t)$ is determined by the time evolution of small perturbations $\boldsymbol{\vartheta}(t)$: $\boldsymbol{\theta}_c(t) + \boldsymbol{\vartheta}(t)$. Linearization of (2.1) about $\boldsymbol{\theta}_c(t)$ yields the equation of motion

$$\frac{d\boldsymbol{\vartheta}}{dt} = \mathbf{L}(t, \mu) \cdot \boldsymbol{\vartheta} \ , \tag{3.15}$$

where $\mathbf{L}(t)$ is a periodically time-dependent matrix of period T,

$$\mathbf{L}(t) = \partial B/\partial\theta\big|_{\boldsymbol{\theta} = \boldsymbol{\theta}_c(t)} = \mathbf{L}(t+T) \ . \tag{3.16}$$

According to the Floquet–Bloch theorem, the normal modes of the cycle are solutions of (3.15) of the form

$$\boldsymbol{\vartheta}_n(t) = \boldsymbol{\psi}_n(t) \, e^{-i\omega_n t} \ , \tag{3.17}$$

where $\boldsymbol{\psi}_n(t)$ is T-periodic, and $\mathrm{Re}\,\omega_n$ lies in the first BZ,

$$\boldsymbol{\psi}_n(t+T) = \boldsymbol{\psi}_n(t), \qquad -\frac{\pi}{T} < \mathrm{Re}\,\omega_n \le \frac{\pi}{T} \ . \tag{3.18}$$

The eigenvalues ω_n and eigenfunction $\boldsymbol{\psi}_n(t)$ have to be found as solutions of the eigenvalue problem

$$\frac{d\boldsymbol{\psi}_n}{dt} - \mathbf{L}(t, \mu) \cdot \boldsymbol{\psi}_n = i\,\omega_n(\mu)\,\boldsymbol{\psi}_n \ . \tag{3.19}$$

We assume that the cycle $\boldsymbol{\theta}_c(t)$ is stable up to a second threshold μ_{c2}, where an instability occurs with respect to a single mode:

$$\mathrm{Im}\,\omega_1(\mu_{c2}) = 0, \qquad \frac{d\,\mathrm{Im}\,\omega_1}{d\mu}\bigg|_{\mu = \mu_{c2}} > 0 \ . \tag{3.20}$$

The type of instability is again determined by the real part of the eigenvalue ω_1 at threshold (see Fig. 7):

9

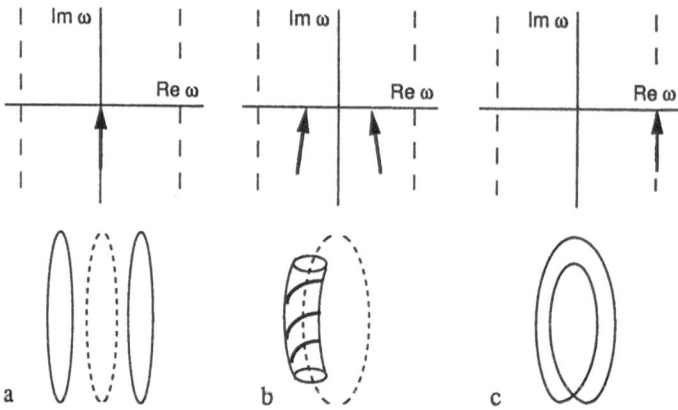

Fig. 7a–c. Destabilization of a cycle. (a) Bifurcation of another cycle with the same frequency. (b) Torus bifurcation. (c) Period doubling

- $\mathrm{Re}\,\omega_1(\mu_{c2}) = 0$: Bifurcation of another cycle with the same period at threshold.
- $\mathrm{Re}\,\omega_1(\mu_{c2}) = \pi/T(\mu_{c2})$: Bifurcation of another cycle with twice the original period (*period doubling*).
- $\mathrm{Re}\,\omega_1(\mu_{c2}) \neq 0, \pi/T(\mu_{c2})$: Bifurcation of a quasiperiodic structure (*torus bifurcation*).

4. Spatio-temporal Structures

4.1 Description of Spatio-temporal Systems

The state of a spatio-temporal system is described in terms of an in general multi-component field $\boldsymbol{\theta}(z;t)$ satisfying an equation of motion

$$\frac{\partial \boldsymbol{\theta}}{\partial t} = \boldsymbol{B}(\boldsymbol{\theta}, \mu) \ , \tag{4.1}$$

where $\boldsymbol{B}(\boldsymbol{\theta}, \mu)$ is now a nonlinear (partial-differential or integral) operator acting on the field $\boldsymbol{\theta}(z,t)$. I shall concentrate here on the case of *one* space dimension z.

The attractors of such a system, i.e. the structures which the solutions of (4.1) approach after the decay of transients, are characterized by the form of the space and time dependence of the field $\boldsymbol{\theta}(z,t)$. One may distinguish purely temporal structures $\boldsymbol{\theta}(t)$, purely spatial structures $\boldsymbol{\theta}(z)$, travelling-wave structures $\boldsymbol{\theta}(z-vt)$, and truly spatio-temporal structures with $\boldsymbol{\theta}(z,t)$ depending separately on z and t. Possible structures are obtained by combining a space dependence of the form

| spatially uniform | $\boldsymbol{\theta}$ independent of z |
| solitary | $\boldsymbol{\theta}(z) \to \boldsymbol{\theta}_{+,-}$ for $z \to \pm\infty$ |

10

spatially periodic	$\theta(z) = \theta(z + \Lambda)$
spatially multiply periodic	
spatially chaotic	

with a time dependence of the form

stationary	θ independent of t
oscillatory	$\theta(t) = \theta(t + T)$
temporally multiply periodic	
temporally chaotic	

In this lecture, I shall concentrate on the bifurcation of waves from a stationary uniform state.

There are two types of tranformation which may occur as basic symmetry operations of a spatio-temporal system:

– *External symmetries* with respect to transformations of the independent variables z and t:

spatial translations	$z \mapsto z + \zeta$
mirror reflection	$z \mapsto -z$
time translations	$t \mapsto t + \tau$

– *Internal symmetries* with respect to transformation of the field components:

$$\theta \mapsto g\theta \ .$$

Other symmetry operations may be formed by combining these basic transformations. The symmetry group of the system consists of all transformations which leave the equation of motion (4.1) invariant, and the symmetry group of a spatio-temporal structure is the subgroup containing those elements which transform the structure into itself.

4.2 Bifurcation of Waves

We assume that the equation of motion (4.1) has a solution $\theta_s(\mu)$ describing a spatially uniform stationary state and consider the bifurcation of wave-like structures occurring at the instability threshold $\mu = \mu_c$ of this state.

Because of the symmetry of $\theta_s(\mu)$ with respect to space and time translations, its normal modes are of the form

$$\vartheta_k(z,t) = p_k e^{i(kz - \omega_k t)} \tag{4.2}$$

with polarization amplitudes p_k and frequencies ω_k determined by the linearized equation of motion for the deviation $\vartheta(z,t) = \theta(z,t) - \theta_s(\mu)$. Reality of the field requires

$$\omega_{-k} = -\omega_k^* \ . \tag{4.3}$$

We assume that $\theta_s(\mu)$ is stable for control parameters in the range $\mu < \mu_c$:

$$\text{Im}\,\omega_k(\mu) < 0 \quad \text{for} \quad \mu < \mu_c \; \forall k \; \forall \text{ branches} \tag{4.4}$$

(except for zero-frequency modes),

and that at the stability threshold $\mu = \mu_c$, an instability occurs with respect to a single, possibly degenerate mode with $k = k_c$:

$$\text{Im}\,\omega(k_c, \mu_c) = 0, \qquad \left.\frac{d\,\text{Im}\,\omega(k_c, \mu_c)}{d\mu}\right|_{\mu = \mu_c} > 0 \;. \tag{4.5}$$

Close to bifurcation, the bifurcating structure is uniquely described by the projection of the field $\theta(z, t)$ onto the space Ω spanned by the undamped normal modes the amplitudes of which form the components of the order parameter (OP) ϕ associated with the instability. In the case of a spatially periodic structure with period Λ, this amounts to taking the Fourier amplitudes with $k = \pm 2\pi/\Lambda$ as components of the OP. By this projection, the treatment of wave bifurcation is formally reduced to the case studied in Sect. 3: One obtains the same type of reduced equation of motion (3.7) in OP space, and the formal bifurcation analysis works exactly as described by (3.8) and the associated text. In particular, the bifurcation of waves with a fixed spatial period is reduced to a Hopf bifurcation problem.

The type of instability is determined by the critical wavenumber k_c and the real part $\omega_c := \text{Re}\,\omega_{kc}(\mu_c)$ of the critical mode frequency at threshold (Table 3):

Table 3. Unstable Modes and bifurcating structures

Type of Instability	Bifurcating Structure
Uniform Soft Mode $\omega_c = 0, k_c = 0$	Static, Uniform $\phi = \text{const}$
Non-uniform Soft Mode $\omega_c = 0, k_c \neq 0$	Static, Spatially Periodic $\phi(z) = \phi(z + \Lambda), \Lambda_c = 2\pi/k_c$
Uniform Hard Mode $\omega_c \neq 0, k_c = 0$	Oscillating, Uniform $\phi(t) = \phi(t + T), T_c = 2\pi/\omega_c$
Non-uniform Hard Mode $\omega_c \neq 0, k_c \neq 0$	Periodic in Space and Time $\phi(z, t) = \phi(z + \Lambda, t) = \phi(z, t + T)$ $\Lambda_c = 2\pi/k_c, T_c = 2\pi/\omega_c$

- $\omega_c = 0, k_c = 0$: Instability with respect to a uniform soft mode gives rise to the bifurcation of another spatially uniform stationary state.
- $\omega_c = 0, k_c \neq 0$: Instability agains a nonuniform soft mode gives rise to the bifurcation of a spatially modulated stationary structure with wavenumber k_c at threshold.
- $\omega_c \neq 0, k_c = 0$: Instability with respect to a uniform hard mode gives rise to the bifurcation of a spatially uniform oscillating structure with frequency ω_c at threshold.
- $\omega_c \neq 0, k_c \neq 0$: Instability with respect to a nonuniform hard mode gives rise to the bifurcation of a wave-like structure with frequency ω_c and wave number k_c at threshold.

The bifurcation problem depends again in an essential way on the degeneracy of the destabilized mode, i.e. on the dimension of OP space Ω (assuming that the representation of the OP occurs with multiplicity 1):

- In the nondegenerate case, the bifurcating pattern is – up to a phase – uniquely determined by symmetry, and only its amplitude has to be determined from bifurcation analysis.
- In the degenerate case, any linear combination of the degenerate modes is a possible candidate for giving rise to a bifurcating pattern, and it has to be determined from bifurcation analysis which of these patterns is in fact selected by the nonlinear terms in the equation of motion (3.7).

In the wave bifurcation problem ($\omega_c \neq 0$, $k_c \neq 0$), two types of degeneracy are of practical importance: degeneracy of modes belonging to wave vectors k_c and $-k_c$ (mirror-symmetric medium), and degeneracy of modes with different polarization amplitudes p (polarization-symmetric medium).

The pattern selection problem in a mirror-symmetric medium carrying a single-component field $\theta(z,t)$ has been studied in [15–18]. The bifurcation analysis shows that generically only two types of pattern can occur:

- Travelling waves $\phi(z+vt)$ and $\phi(z-vt)$ corresponding to the modes ϑ_{k_c} and ϑ_{-k_c}, respectively.
- Standing waves corresponding to linear combinations of ϑ_{k_c} and ϑ_{-k_c} with equal amplitudes.

Both patterns satisfy the rule of minimally broken symmetry.

Pattern selection in a polarization-symmetric medium carrying a two-component field $\theta(z,t) = \{\theta_x(z,t), \theta_y(z,t)\}$ is studied in [19], and is discussed as an example for wave bifurcation in Sect. 5. The OP equation for this case turns out to be equivalent to that for the mirror-symmetric case, and the results of the bifurcation analysis are therefore easily translated. Generically, only two types of pattern can occur:

- Linearly polarized waves corresponding to linear combinations $\vartheta_x \cos \chi + \vartheta_y \sin \chi$ with arbitrary χ.
- Circularly polarized waves corresponding to linear combinations $\vartheta_x \pm i\vartheta_y$.

Again, both patterns satisfy the rule of minimally broken symmetry.

When the control parameter is increased beyond its critical value, a whole branch of normal modes with wavenumbers differing from the critical wavenumber becomes successively destabilized. This gives rise to the bifurcation of a whole family of new structures with broken time-translation symmetry [14].

5. Wave Bifurcation in a Polarization-Symmetric Medium

5.1 Wave Instability of the Quiescent State

As an application, we discuss wave bifurcation in a one-dimensional dissipative medium carrying a two-component field

$$\boldsymbol{\theta}(z,t) = \{\theta_x(z,t),\ \theta_y(z,t)\}\ . \tag{5.1}$$

The system is assumed to be invariant with respect to spatial translations $z \mapsto z + \zeta$ and to arbitrary orthogonal transformations $\boldsymbol{\theta} \mapsto R\boldsymbol{\theta}$ of the field components (polarization symmetry) [19].

Interest in this problem was partly stimulated by experiments on an instability of a cylindrical solid–liquid phase boundary under the influence of convection [20]: The central part of a long cylindrical sample of succino-nitrile is melted by an electrically heated coaxial wire. The observations show that above a critical value of heat input the originally cylindrical interface between melt and solid assumes the form of a moving helical wave. The mechanism for the occurrence of a wave instability was identified by a linear stability analysis carried out in [20], which yielded the critical value of the Grashof number in satisfactory agreement with experiment. But no explanation was given for the occurrence of a helical wave (corresponding to circular polarization) rather than a linearly polarized buckling wave which would be equally consistent with the form of the destabilized modes. The results of the following analysis provide a possible explanation of this observation.

Another system which corresponds even closer to the situation considered here is the polarization-symmetric laser, which has been proposed as a potential light source with a sharp photon linewidth [21-24]. The results of our analysis show that the nonlinear terms stabilize definite states of polarization even in the absence of external polarization symmetry-breaking effects.

In the quiescent state $\boldsymbol{\theta}_s = 0$ of such a system, the normal modes

$$\vartheta_k^{(1,2)}(z,t) = \boldsymbol{p}_{1,2}e^{i(kz-\omega_k t)}\ , \tag{5.2}$$

with $\boldsymbol{p}_1 = \boldsymbol{e}_x$ and $\boldsymbol{p}_2 = \boldsymbol{e}_y$, are degenerate. Further, in the absence of reflection symmetry, $\omega_{-k} \neq \omega_k$, there is a unique propagation direction. If $\operatorname{Re}\omega_k > 0$ for $k > 0$, all normal modes propagate in the $+z$ direction.

The dispersion of the normal-mode frequency is of the form shown in Fig. 8. At a critical value $\mu = \mu_c$ of the control parameter, an instability (4.5) occurs with respect to a nonuniform hard mode:

$$\operatorname{Im}\omega(k_c,\mu_c) = 0, \qquad \left.\frac{d\operatorname{Im}\omega(k_c,\mu_c)}{d\mu}\right|_{\mu=\mu_c} > 0\ , \tag{5.3a}$$

$$\operatorname{Re}\omega(k_c,\mu_c) = \omega_c \neq 0\ , \tag{5.3b}$$

$$k_c \neq 0\ . \tag{5.3c}$$

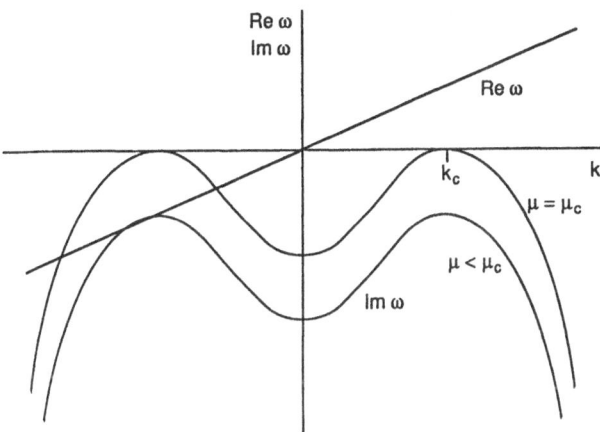

Fig. 8. Dispersion of the real and imaginary parts of the normal-mode frequency for a uni-directional medium

At the critical wavenumber k_c, we assume a linear dependence of $\operatorname{Im}\omega_k$ on μ and neglect a possible μ dependence of $\operatorname{Re}\omega$, such that

$$\omega(k_c, \mu) = \omega_c + i(\mu - \mu_c) \tag{5.4}$$

for an appropriate choice of the scale of μ.

The simplest model field equation showing such a wave instability in a polarization-symmetric medium is of the form

$$\partial_t\theta = \mu\,\theta - c\,\partial_z\theta - D_1\partial_z^2\theta - D_2\partial_z^4\theta + N(\theta) \;, \tag{5.5}$$

where $c > 0$, $D_{1,2} > 0$, and $N(\theta)$ contains all terms nonlinear in the field θ. Under rotations in θ space, $N(\theta)$ must transform equivariant to θ. From the field components (θ_x, θ_y), one can form only the trivial equivariant

$$N^{(0)}(\theta) = -U|\theta|^2\theta \;, \tag{5.6}$$

but there may also occur terms containing derivatives such as

$$N^{(1)}_{lmn}(\theta) = -U_{lmn}(\partial_z^l\theta \cdot \partial_z^m\theta)\,\partial_z^n\theta \;, \tag{5.7a}$$

$$N^{(2)}_{lmn}(\theta) = -V_{lmn}(\partial_z^l\theta \times \partial_z^m\theta) \times \partial_z^n\theta \;. \tag{5.7b}$$

The normal-mode frequency is given by

$$\omega_k(\mu) = c\,k + i(\mu + D_1k^2 - D_2k^4) \;, \tag{5.8}$$

from which one finds the critical values

$$k_c = \sqrt{D_1/2D_2}\,, \qquad \mu_c = -D_1^2/4D_2 \;. \tag{5.9}$$

15

We shall use this simple model only for demonstration, and base our conclusions on general properties of the wave equation: Translation symmetry in space and in time, $O(2)$ symmetry in θ space, and the occurrence of a hard-mode instability (5.3,4).

5.2 Bifurcation of Nonlinear Waves

The destabilitzation of a nonuniform hard mode with wavenumber $k = k_c$ gives rise to the bifurcation of nonlinear waves $\theta_w(z, t)$ with wavelength $\Lambda = 2\pi/k_c$, which may be expanded into a Fourier series

$$\theta_w(z, t) = \sum_{-\infty}^{+\infty} \theta_n(t) e^{ink_c z} ; \qquad \theta^*_{-n} = \theta_n . \tag{5.10}$$

The components of the order parameter are given by the Fourier amplitudes $\theta_1(t)$, $\theta_{-1}(t)$,

$$\phi(t) = \theta_1(t), \qquad \phi^*(t) = \theta_{-1}(t) . \tag{5.11}$$

They completely describe the bifurcation behaviour; all other Fourier amplitudes $\theta_n(t)$, $n \neq \pm 1$, follow the motion of power products of the components of ϕ and ϕ^* adiabatically.

We construct the equation of motion for $\phi(t)$ by symmetry considerations. Since the translations $z \mapsto z + \zeta$ are represented by $\phi \mapsto \phi \exp(ik_c\zeta)$, the symmetry group of the system is represented in OP space by the group $O(2) \times U(1)$. Under this symmetry, all terms are of odd order in ϕ. The linear term is proportional to ϕ, and its coefficient is determined by the normal-mode analysis (5.4). In third order, there occur two terms, $(\phi^* \cdot \phi)\phi$ and $(\phi^* \times \phi) \times \phi$. We therefore have, to leading nonlinear order, the equation of motion

$$\frac{d\phi}{dt} = (\mu - \mu_c - i\omega_c)\phi - u(\phi^* \cdot \phi)\phi - v(\phi^* \times \phi) \times \phi \tag{5.12}$$

whith in general complex coefficients u and v which may in principle be calculated from the nonlinear terms in the full wave equation by carrying out the adiabatic elimination procedure. For the model equation (5.5) with a non-derivative third-order term (5.6), u and v are real and are given by

$$u = 3U, \qquad v = -U . \tag{5.13}$$

In the basis $\{e_+ = (e_x - ie_y)/\sqrt{2}, e_- = (e_x + ie_y)/\sqrt{2}\}$, (5.12) coincides with the equation of motion for the order parameter describing wave bifurcation in a mirror-symmetric medium derived in [18]. Correspondingly, there exists a one-to-one correspondence between the results for the two cases.

Ordinary travelling waves with a single oscillation period $T = 2\pi/\omega$ are decribed by solutions of the form

$$\phi(t) = Ae^{-i\omega t} . \tag{5.14}$$

The equation of motion (5.12) requires

$$[\mu - \mu_c + i(\omega - \omega_c)]A - u(A^* \cdot A)A - v(A^* \times A) \times A = 0 . \qquad (5.15)$$

If $v \neq 0$, vectorial multiplication by A yields an algebraic condition for the complex amplitudes $A = \{A_x, A_y\}$

$$(A_x^* A_y - A_x A_y^*)(A_x^2 + A_y^2) = 0 . \qquad (5.16)$$

Inspection of this equation shows that there are two types of solution:
1. $A_x^2 + A_y^2 = 0$:

$$A_x = \frac{|A|}{\sqrt{2}} e^{i\alpha} , \qquad A_y = \mp i \frac{|A|}{\sqrt{2}} e^{i\alpha} . \qquad (5.17)$$

This solution represents a family of *circularly* polarized waves

$$\theta_w^{cp}(z,t) = \frac{|A|}{\sqrt{2}} (e_x \mp i e_y) e^{i(k_c z - \omega t + \alpha)} + \text{c.c.}$$
$$+ \text{ terms of higher order} \qquad (5.18)$$

with arbitrary phase α. The amplitude $|A|$ and the frequency ω are to leading order determined by (5.15),

$$\mu - \mu_c + i(\omega - \omega_c) = |A|^2 (u + v) , \qquad (5.19)$$

which yields

$$|A| = |A^{cp}| = \sqrt{\frac{\mu - \mu_c}{\mathrm{Re}\,(u + v)}} , \qquad \omega = \omega^{cp} = \omega_c + |A|^2 \,\mathrm{Im}\,(u + v) . \qquad (5.20)$$

The bifurcation is supercritical (the solution exists for $\mu > \mu_c$) if $\mathrm{Re}\,(u+v) > 0$.
2. $A_x^* A_y = A_x A_y^*$:

$$A_x = |A| e^{i\alpha} \cos \chi , \qquad A_y = \pm |A| e^{i\alpha} \sin \chi . \qquad (5.21)$$

This solution represents a family of *linearly* polarized waves

$$\theta_w^{lp}(z,t) = |A| (e_x \cos \chi \pm e_y \sin \chi) e^{i(k_c z - \omega t + \alpha)} + \text{c.c.}$$
$$+ \text{ terms of higher order} \qquad (5.22)$$

with arbitrary phase α and arbitrary polarization angle χ ($0 \leq \chi \leq \pi/2$). The amplitude $|A|$ and the frequency ω are to leading order determined by

$$\mu - \mu_c + i(\omega - \omega_c) = |A|^2 u , \qquad (5.23)$$

which yields

$$|A| = |A^{lp}| = \sqrt{\frac{\mu - \mu_c}{\mathrm{Re}\,u}} , \qquad \omega = \omega^{lp} = \omega_c + |A|^2 \,\mathrm{Im}\,u . \qquad (5.24)$$

The bifurcation is supercritical if $\mathrm{Re}\,u > 0$.

17

The symmetry group of a circularly polarized wave is that of a helix which is isomorphic to SO(2). Thus, the bifurcation breaks only one continuous symmetry of O(2)×U(1). Correspondingly, we find a family of solutions (5.18) depending on a single parameter, the phase α, and we expect a single Goldstone mode.

The symmetry group of a linearly polarized wave, on the other hand, is the mirror group generated by a reflection at the plane of polarization, i.e. the bifurcation breaks both continuous symmetries of O(2)×U(1). Correspondingly, there is a two-parameter family of solutions (5.22) depending on the phase α and the polarization angle χ, and we expect the occurrence of two Goldstone modes.

Both these groups are maximal subgroups of O(2)×U(1) in OP space, i.e. the rule of minimal symmetry breaking is satisfied.

5.3 Stability of the Nonlinear Waves

Since the nonlinear travelling wave θ_w is a periodic function of the variable $k_c z - \omega t$ with period 2π, its normal modes are of the form

$$\vartheta(z,t) = g(k_c z - \omega t)\, e^{iqz + \lambda_q t}, \qquad g(\zeta + 2\pi) = g(\zeta) . \tag{5.25}$$

The wave θ_w is stable if $\mathrm{Re}\,\lambda_q < 0$ $\forall q$ except for Goldstone modes. Because of the downward curvature of $\mathrm{Im}\,\omega_k$ at $k = k_c$ in the quiescent state, in the region close to bifurcation considered here, $\mathrm{Re}\,\lambda_q$ assumes its maximum at $q = 0$. Therefore, the stability analysis may be restricted to the mode $q = 0$. We set in leading order

$$\vartheta(z,t) = (\delta A\, e^{i(k_c z - \omega t)} + \delta B\, e^{-i(k_c z - \omega t)})\, e^{\lambda t} , \tag{5.26}$$

where $\delta B(\lambda^*) = \delta A^*(\lambda)$, and obtain for the determination of the Floquet exponent λ the eigenvalue problem

$$\lambda \delta A = \{[\mu - \mu_c + i(\omega - \omega_c) - (u + v)|A|^2]\mathbf{1} - (u + v)AA^* + 2v A^* A\} \cdot \delta A$$
$$- \{(u + v)AA - v(A \cdot A)\mathbf{1}\} \cdot \delta B \tag{5.27a}$$

$$\lambda \delta B = -\{(u + v)^* A^* A^* - v^*(A^* \cdot A^*)\mathbf{1}\} \cdot \delta A + \{[\mu - \mu_c - i(\omega - \omega_c)$$
$$- (u + v)^*|A|^2]\mathbf{1} - (u + v)^* A^* A + 2v^* A A^*\} \cdot \delta B . \tag{5.27b}$$

For both types of wave θ_w there exists a phase mode (Goldstone mode)

$$\delta A = \frac{\partial A}{\partial \alpha} = iA , \qquad \delta B = -iA^*, \qquad \lambda_1 = 0 \tag{5.28}$$

and an amplitude mode

$$\delta A = [\mu - \mu_c + i(\omega - \omega_c)]A , \qquad \delta B = \delta A^*, \qquad \lambda_2 = -2(\mu - \mu_c) . \tag{5.29}$$

18

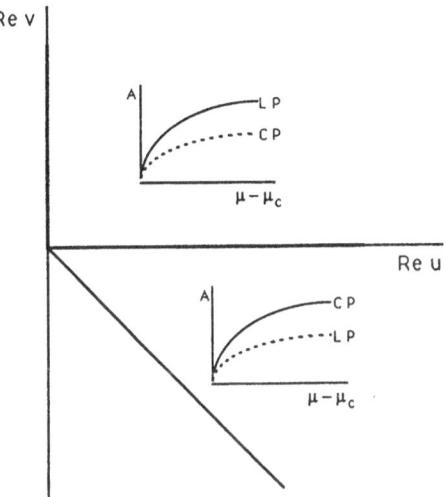

Fig. 9. Stability regions of two types of wave in the (Re u, Re v)-plane. Insets show the amplitudes (stable branches —, unstable branches - - - -) for circular polarization (CP) and linear polarization (LP) as functions of the control parameter

For the circularly polarized wave (5.18), the other two modes have complex conjugate eigenvalues:

$$\delta A = A^* , \qquad \delta B = 0 , \qquad \lambda_3^{\mathrm{cp}} = 2\,|A^{\mathrm{cp}}|^2\, v \quad , \tag{5.30a}$$

$$\delta A = 0 \;, \qquad \delta B = A , \qquad \lambda_4^{\mathrm{cp}} = 2\,|A^{\mathrm{cp}}|^2\, v^* \;. \tag{5.30b}$$

For the linearly polarized waves (5.22), there exists a second Goldstone mode

$$\delta A = \frac{\partial A}{\partial \chi} = e_z A , \qquad \delta B = \delta A^* \qquad \lambda_3^{\mathrm{lp}} = 0 \tag{5.31}$$

and another mode with real eigenvalue

$$\delta A = \mathrm{i}\, v\, e_z \times A , \qquad \delta B = \delta A^* , \qquad \lambda_4^{\mathrm{lp}} = -2\,|A^{\mathrm{lp}}|^2\, \mathrm{Re}\, v \;. \tag{5.32}$$

These results show that we have the following stability regions (see Fig. 9):

circularly polarized waves: $-\mathrm{Re}\,u < \mathrm{Re}\,v < 0$;
linearly polarized waves: $\mathrm{Re}\,u > 0$, $\mathrm{Re}\,v > 0$.

In the case of a wave equation with a nonderivative third-order term (2.8), (5.7a,b) shows that only circularly polarized waves are stable, provided that $U > 0$. This represents a possible explanation for the observation that only helical waves occur in the experiments of [20].

5.4. Bifurcation of a Branch of Modulated Waves

The fact that the circularly polarized waves become unstable at the common stability boundary $\operatorname{Re} v = 0$ with respect to a pair of modes with complex conjugate Floquet exponents $\lambda_{3,4}^{cp}$ suggests that at this boundary there bifurcates a branch of *modulated waves* described by two independent frequencies ω_a, ω_b (we assume in this Section that $\operatorname{Im} v \neq 0$). Such waves are described by solutions of the equation of motion (5.12) of the form

$$\phi(t) = a\,e^{i\omega_a t} + b\,e^{-i\omega_b t} \,. \tag{5.33}$$

By following a procedure analogous to that in Sect. 5.3, we obtain in leading order the following results.

The two parts represent circularly polarized waves with opposite polarization $e_+ = (e_x - ie_y)/\sqrt{2}$, $e_- = (e_x + ie_y)/\sqrt{2}$,

$$a = |a|\,e^{i\alpha}e_+, \qquad b = |b|\,e^{i\beta}e_- \,, \tag{5.34}$$

with amplitudes satisfying

$$|a|^2 + |b|^2 = \frac{\mu - \mu_c}{\operatorname{Re} u} \,, \tag{5.35}$$

and frequencies determined by

$$\omega_a = \omega_c + |a|^2 \operatorname{Im}(u + v) + |b|^2 \operatorname{Im}(u - v) \,, \tag{5.36a}$$
$$\omega_b = \omega_c + |a|^2 \operatorname{Im}(u - v) + |b|^2 \operatorname{Im}(u + v) \,. \tag{5.36b}$$

We have thus found a branch of modulated waves

$$\begin{aligned} \theta_{mw}(z,t) = (|a|\,e_+\,e^{i(-\omega_a t + \alpha)} + |b|\,e_-\,e^{i(-\omega_b t + \beta)})\,e^{ik_c z} \\ + \text{ terms of higher order} \end{aligned} \tag{5.37}$$

representing states with completely broken symmetry in OP space (only the discrete translations $z \mapsto z + \Lambda$ remain).

With increasing values of the control parameter μ, the stability limits of the two types of ordinary wave still coincide to order $\mu - \mu_c$ and shift according to

$$\operatorname{Re} v_c = \Delta \cdot (\mu - \mu_c) \,. \tag{5.38}$$

The relative amplitudes $|a|$, $|b|$ and the coefficient Δ are determined only by higher-order terms in the equation of motion (5.12).

References

1. Marsden, J., McCracken, M.: The Hopf Bifurcation and Its Applications. Applied Mathematical Sciences 19, Springer, Berlin, Heidelberg 1976
2. Guckenheimer, J., Holmes, P.: Nonlinear Oscillations, Dynamical Systems and Bifurcations of Vector Fields. Applied Mathematical Sciences 42, Springer, Berlin, Heidelberg 1980
3. Golubitsky, M., Schaeffer, D.G.: Singularities and Groups in Bifurcation Theory, Vol. I. Applied Mathematical Sciences 51, Springer, Berlin, Heidelberg 1985
4. Golubitsky, M., Stewart, I., Schaeffer, D.G.: Singularities and Groups in Bifurcation Theory, Vol. II. Applied Mathematical Sciences 69, Springer, Berlin, Heidelberg 1988
5. Ihrig, E., Golubitsky, M.: Pattern selection with O(3) symmetry. Physica 13D (1984) 1–33
6. Golubitsky, M., Stewart, I.N.: Hopf bifurcation in the presence of symmetry. Arch.Rat. Mech.Anal. 87 (1985) 107–165
7. Knobloch, E.: On the degenerate Hopf bifurcation with O(2) symmetry. In: Multiparameter Bifurcation Theory, Contemporary Mathematics Vol. 56, pp. 193–201. Am.Math.Soc. 1986
8. Höck, K.-H., Jordan, P., Thomas, H.: Symmetry aspects of instabilities in driven systems. J.Phys.C: Solid State Physics 20 (1987) 895–916
9. Höck, K.-H., Hackenbracht, D., Jordan, P., Thomas, H.: Degenerate Hopf bifurcation in the presence of symmetry. Physica 27D (1987) 338–356
10. Cigogna, G., Gaeta, G.: Bifurcations, symmetries and maximal isotropy subgroups. Nuova Cimento 102 (1988) 451–479
11. Crawford, J.D., Knobloch, E.: Classification and unfolding of degenerate Hopf bifurcations with O(2) symmetry: No distinguished parameter. Physica D31 (1988) 1–48
12. Crawford, J.D., Knobloch, E.: Period-doubling mode interactions with circular symmetry. Physica D44 (1990) 340–396
13. Crawford, J.D., Knobloch, E.: Symmetry and symmetry-breaking bifurcations in fluid dynamics. Annu.Rev.Fluid Mech. 23 (1991) 341–387
14. Büttiker, M., Thomas, H.: Bifurcation and stability of families of waves in uniformly driven spatially extended systems. Phys.Rev. A24 (1981) 2635–48
15. Renardy, M., Haken, H.: Bifurcations of solutions of the laser equation. Physica 8 (1983) 57–89
16. Livshits, M.A.: Chemical waves as a result of instability in reaction-diffusion systems. Z.Phys.B – Condensed Matter 53 (1983) 83–88
17. Malomed, B.A.: Nonlinear waves in nonequilibrium systems of the oscillatory type, Part I. Z.Phys.B – Condensed Matter 55 (1984) 241–248
18. Thiesen, S., Thomas, H.: Bifurcation, stability and symmetry of nonlinear waves. Z.Phy.B – Condensed Matter 65 (1987) 397–408
19. Thomas, H.: Bifurcation of transverse waves in a polarization-symmetric medium. Physicalia Magazine (Gent, Belgium) 12 Suppl (1999) 247–263
20. Fang, Q.T., Glicksman, M.E., Coriell, S.R., McFadden, G.B., Boisvert, R.F.: Convective influence on the stability of a cylindrical solid-liquid interface. J.Fluid Mech. 151 (1985) 121–140
21. Graham, R.: Order parameter fluctuations of a laser with polarization symmetry. Phys.Letters 103A (1984) 255–258
22. Grossmann, S., Krauth, W.: Laser with large correlation times. Phys.Rev. A35 (1987) 2523–2531
23. Krauth, W., Grossmann, S.: Linewidths of lasers with broken polarization symmetry. Phys.Rev. A35 (1987) 4192–4199
24. Grossmann, S., Krauth, W.: The polarization symmetric laser. In: Lasers and Synergetics, ed. by R. Graham, A. Wunderlich, pp. 1–10. Springer Proc. Phys., Springer, Berlin, Heidelberg 1987

Deterministic Chaos: Introduction and Recent Results

H.G. Schuster

Institut für Theoretische Physik, Universität Kiel,
Olshausenstrasse 40, W-2300 Kiel, Fed. Rep. of Germany

1. Introduction

The word *chaos*, from which our present word *gas* is derived, usually characterizes a state of disorder in a system with many degrees of freedom. We might imagine for example a gas whose molecules move around in a container in a completely disordered, i.e. chaotic, fashion.

Here the notion *deterministic chaos* describes the dynamical behavior of physical systems whose future can in principle be computed by solving differential of difference equations, but which become nevertheless unpredictable for long times, because the nonlinearities in the equations of motion tend to amplify enormously small, but practically inevitable, errors in the initial conditions [1]. It has been shown in recent years that many physical systems, even with only a few degrees of freedom, show this sensitive dependence on initial conditions, which leads to chaotic, i.e. unpredictable, behavior in time.

One of the simplest examples is the periodically driven pendulum shown in Fig. 1.

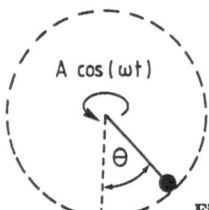

Fig. 1. Pendulum driven by an external torque $A \cos(\omega t)$ with drive frequency ω.

Its equation of motion

$$\ddot{\theta} + \gamma\dot{\theta} + \sin\theta = A \cos(\omega t) , \qquad (1)$$

where γ denotes the damping and the other variables are explained in Fig. 1, is nonlinear owing to the term $\sin\theta$.

2. Strange Attractors

The natual space in which the time behavior of dynamical systems may be visualized geometrically is phase space. Figure 2 shows qualitatively different kinds of motion

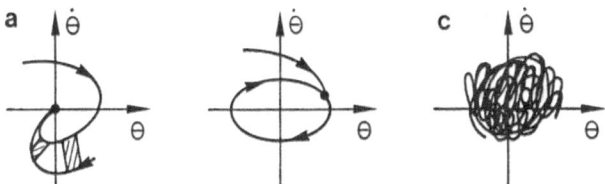

Fig. 2a–c. Schematic drawing of different attractors which occur in the phase space of a driven pendulum. (a) Fixed point. (b) Limit cycle. (c) Strange attractor.

Fig. 3. Pendulum in a critical position at $\theta = \frac{\pi}{2}$.

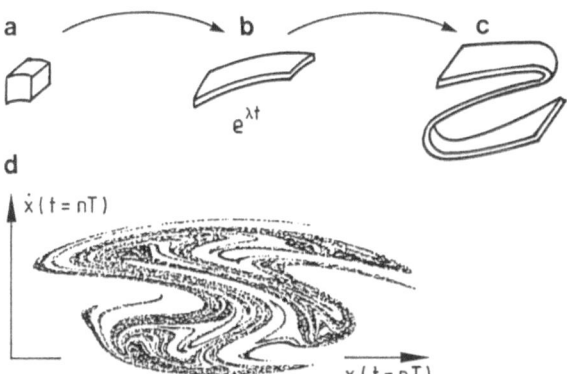

Fig. 4a–d. Deformation of a volume element on a strange attractor with increasing time. (a) Attracting region in phase space. (b) Stretching. (c) Folding. This leads to the foliated structure of a stange attractor as shown in (d), which shows a stroboscopic picture of a strange attractor which is generated by the differential equation of a periodically driven particle in an x^4 potential, $\ddot{x} + x\dot{x} + x^3 = A \cos(\omega_0 t)(T = 2\pi/\omega_0, n = 0, 1, 2, \ldots)$.

which occur for different values of the amplitude A of the external torque. We also indicate in Fig. 2a the fact that volume elements in phase space shrink to zero for dissipative systems.

Chaotic motion of dissipative systems (for conservative systems see eg. [2]) in phase space is characterized by the fact that the trajectory becomes attracted to a strange attractor, i.e. a bounded region in phase space where points on nearby trajectories become exponentially separated in time. This last property describes the sensitive dependence on initial conditions of chaotic systems, and it is easy to see, at least qualitatively, why a driven pendulum has this property.

23

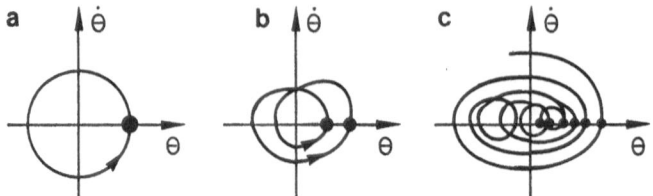

Fig. 5a–c. Poincaré section of trajectories with an axis. (a) Limit cycle. (b) Two-cycle. (c) Chaos. Successive crossing points are denoted by x_n, x_{n+1}, ... Note that there is only one crossing point for a circle.

Figure 3 shows that if the pendulum is just at the summit ($\theta = \frac{\pi}{2}$) a tiny touch is enough to deterimine whether it moves to the right or to the left, i.e. we have error amplification. Since the angular variable is confined to a bounded region $0 \le \theta \le 2\pi$, the pendulum passes the critical point $\theta = \pi/2$ again and again, i.e. we have "backfolding", and the error amplifies enormously.

We substantiate this qualitative description by performing in phase space a Poincaré section, which leads to a description of chaotic motion in terms of a simple one-dimensional map that models the properties of error amplification and backfolding. Figure 5 demonstrates qualitatively how a Poincaré section reduces the dimension of description by one.

3. Mechanism for Chaos

If the sequence of points x_0, x_1, x_2 in Fig. 5 is chaotic, we know that the system displays chaos (because these points belong to a trajectory).

Next we construct a nonlinear one-dimensional map

$$x_{n+1} = f(x_n) \qquad n = 0, 1, 2 \ldots \tag{2}$$

which has the properties of error amplification and backfolding that are needed to generate chaos.

Figure 6 shows that the map

$$x_{n+1} = (10 x_n) \bmod 10 \tag{3}$$

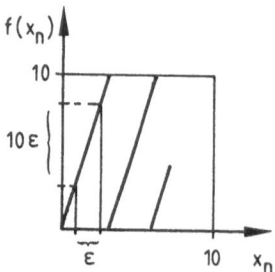

Fig. 6. The map $x_{n+1} = (10 x_n) \bmod 10$

amplifies errors by a factor of 10 and folds the iterates back to the interval [0, 10]. If we apply this map to an irrational number, say π, we obtain the following sequence of iterates:

$$x_0 = \pi \qquad\qquad = 3.14159 \ldots \; , \qquad (4)$$

$$x_1 = (31.4159\ldots) \mod 10 = 1.4159 \ldots \; ,$$

$$x_2 = (14.159\ldots) \quad \mod 10 = 4.159 \ldots \; ,$$

i.e. we see that the map shifts the sequence of decimals to the left and cuts off all but one digit before the decimal point. In this fashion it "pumps" the irregular sequence of decimals of π to the front of the decimal point and makes them *visible*. This is all that a chaotic system does: It makes the *chaos* which exists already in the data of the initial conditions (x_0 for our example) macroscopically visible. If we assume for example that x_0 is a measured variable which is known up to 3 digits after the decimal point, we could write

$$x_0 = a_0 \cdot a_1 \, a_2 \, a_3 \; ? ? ? \; , \qquad (5)$$

where the question marks indicate the unknown decimals. After 4 iterations with out map from Fig. 6 we obtain

$$x_4 = ? . \;\; ? ? ? \; , \qquad (6)$$

i.e. the fourth iterate is completely unknown, we cannot make any prediction about it, and the system displays chaos.

4. Characterization of Chaotic Motion

Let us now consider the question how we can characterize chaotic motion. First of all there can be different rates of error amplification in our map given by (3), i.e. instead of 10, the amplification factor can be $1 < a < \infty$; for $a \leq 1$ we do not have chaos. If we consider $x_0 = \varepsilon \ll 1$, we may iterate

$$x_{n+1} = (a x_n) \mod 10 \; . \qquad (7)$$

Ignoring the modulus we obtain

$$x_n = \varepsilon a^n = \varepsilon e^{\lambda n} \; , \qquad (8)$$

where we have introduced the Lyapunov exponent $\lambda = \log a$. More formally, we define λ for one-dimensional maps as

$$\lambda = \lim_{n \to \infty} \frac{1}{n} \log \left| \frac{d}{dx_o} f^n(x_o) \right| \; . \qquad (9)$$

A positive Lyapunov exponent always signals sensitive dependence on initial conditions and is in fact *the* quantity by which chaotic motion may be distinguished from regular motion ($\lambda = 0$) or white noise ($\lambda \to \infty$) [1].

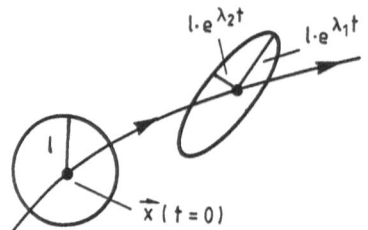

Fig. 7. Deformation of a sphere on a strange attractor in phase space as a function of time. The surface of the sphere consists of trajectory points.

Let us now extend these considerations to time-continuous systems, which are described by autonomous differential equations for the trajectory vector $x(t)$ in phase space:

$$\dot{x} = f[x(t)]; \qquad x(t) = [x_i(t), \ldots, x_m(t)] \ . \tag{10}$$

We can put a small sphere of radius l around a trajectory point $x(t)$ and watch how it changes in shape in the course of time.

As shown schematically in Fig. 7, the spere becomes an ellipsoid whose volume is smaller than that of the sphere because we consider dissipative systems, for which volumes in phase space shrink. But the ellipsoid may nevertheless have some major axes which are larger than the original radius of the sphere, because there is always stretching of nearby points on the strange attractor in some directions of phase space, which signals sensitive dependence on initial conditions. Generalizing from the two-dimensional picture shown in Fig. 7 to the m-dimensional case, it should be obvious that a strange attractor in m dimensions has m Lyapunov exponents. They may be computed from the geometrical picture in Fig. 7 by taking the limits $l \to 0$ and $t \to \infty$ and averaging the center point of the sphere $x(t = 0)$ over the whole attractor. There is also a formal expression for the Lyapunov exponents $\lambda_1, \ldots \lambda_m$ which allows for their computation in terms of the functions $f_1[x(t)] \ldots f_n[x(t)]$ that appear in the differential equations (10) for the trajectory:

$$\lambda_i = \text{real part of an eigenvalue of } \varLambda, \quad i = 1 \ldots m \ , \tag{11}$$

where the matrix $\varLambda$ is given by

$$\varLambda = \lim_{t \to \infty} \frac{1}{t} \log [\hat{T} \exp \int_0^t M(t')\, dt'] \ , \tag{12}$$

$M(t)$ is the Jacobian

$$M_{ij}(t) = \left. \frac{\partial f_i(x)}{\partial x_j} \right|_{x(t)} \ , \tag{13}$$

and $\hat{T}$ is the time-ordering operator which takes care of the fact that the matrices $M(t_1)$, $M(t_2)$ do not in general commute. Equations (11–13) may be derived by linearizing (10) and integrating the resulting equations of motion for a small deviation $\varepsilon(t)$ from the midpoint $x_0(t)$ of the sphere [3].

26

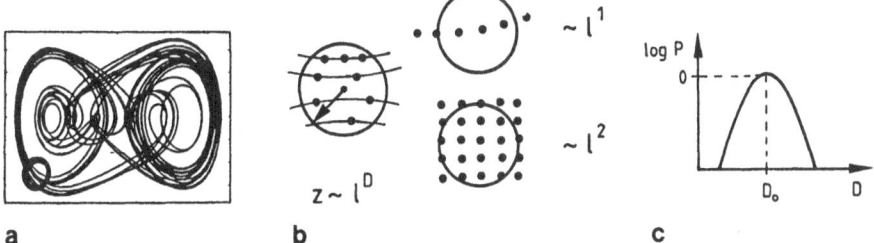

Fig. 8. (a) Definition of local dimensions α. Consider a small sphere of radius l around a point on the attractor. (b) The number of points Z which remain in the sphere for $l \to 0$ scales like $Z(l \to 0) \propto l^\alpha$. (c) The distribution of αs which is obtained by measuring α for a large set of points on the attractor is $p(\alpha) \propto l^{f(\alpha)}$.

Besides its Lyapunov exponents, a strange attractor is also characterized by its density in phase space, which in turn is encoded into local dimensions α and their distribution

$$p(\alpha) \propto l^{f(\alpha)} \ . \tag{14}$$

Figure 8 shows how local dimensions α on a strange attractor are defined, and it should be clear from their definition that α may also have noninteger values. If we take, in (14), the limit $l \to 0$, we obtain

$$f(\alpha) = \lim_{l \to 0} \frac{\log p(\alpha)}{\log l} \ . \tag{15}$$

The value $\alpha_{\max}$ where $f(\alpha)$ has its maximum gives just the usual Hausdorff dimension D_0 of the attractor, i.e. $D_0 = \alpha_{\max}$. Figure 9 shows the $f(\alpha)$ spectrum of the strange attractor which is generated by the Mackey–Glass equation [5]

$$\dot{x} = ax(t - \tau)/\{1 + [x(t - \tau)]^{10}\} - bx(t) \ , \tag{16}$$

which corresponds – because one needs an infinite number of initial conditions (i.e. $x(t)$ in the interval $[0, -\tau]$) – to an infinite-dimensional system. The interesting result is that this formally infinite-dimensional system generates a strange attractor with a bounded dimensionality spectrum [6].

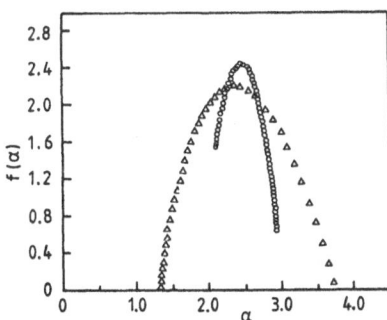

Fig. 9. $f(\alpha)$ spectrum for the Mackey–Glass equation (16) with parameter values $a = 0.2$, $b = 0.1$, $\tau = 30$.

5. Reconstruction of the Attractor from Time Series

After having introduced some basic concepts by which chaotic systems are characterized, we consider now the question what can be learned from an experimental time series of one variable $x(t)$ which looks chaotic. Before one can extract the relevant variables λ_i and $f(\alpha)$ mentioned above, one must first reconstruct in phase space the trajectory of the system which produces $x(t)$. This problem has been solved by Takens [7], and we illustrate his method by means of a simple example. Suppose that the true trajectory is a circle, i.e.

$$\boldsymbol{x}(t) = [x_1(t), x_2(t)] = [\sin t, \cos t] \ , \tag{17}$$

where the paramter t denotes time in dimensionless units. If we had only measured the variable $x_1(t) = \sin t$, we could reconstruct the second component $x_2(t)$ of the vector $\boldsymbol{x}(t)$ in phase space by performing in $x_1(t)$ a phase shift, i.e.

$$x_2(t) = \cos t = \sin\left(t + \frac{\pi}{2}\right) = x_1(t + \tau); \ \tau = \frac{\pi}{2} \ . \tag{18}$$

Takens has shown that this method also works for the reconstruction of vectors in higher-dimensional phase spaces, i.e.

$$x_1(t) \ldots x_m(t) = [x_1(t), x_1(t + \tau) \ldots x_1(t + (m - 1)\tau)] \ , \tag{19}$$

yielding a strange attractor which has the same topological features as the "true" attractor (but it need not "look" the same). However there remains the problem how to choose τ if the set of measured data is finite. For finite data sets the choice of τ should not matter, according to Takens (see also Fig. 10).

This problem, together with the question how large one has to choose the embedding dimension m (which is the smalles dimension of phase space that allows for a topologically correct reconstruction of the strange attractor) has been solved as follows [8]. We use the idea that an embedding via delay coordinates is a topological mapping which preserves neighborhood relations. Therefore, if the chosen embedding dimension is too small, one obtains a projection of the attractor which violates the condition of injectivity. For the proper values of m and τ one will obtain an optimal spanning of the attractor, and neighboring points will have their largest distances.

Figure 11 shows schematically how neighborhood relations may change for $m \rightarrow m+1$. In $\mathbf{R}^m$ the point closest to the point denoted by the black dot, which we could

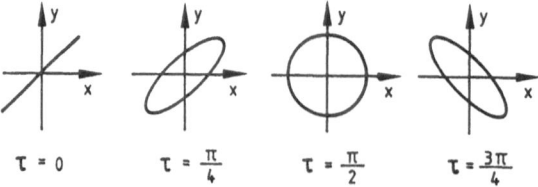

$$\tau = 0 \qquad \tau = \frac{\pi}{4} \qquad \tau = \frac{\pi}{2} \qquad \tau = \frac{3\pi}{4}$$

Fig. 10. Different reconstructions of the "circle attractor" for different values of τ. Note that for almost all τ values (with the exception of $\tau = 0$, π) the reconstructions are topologically equivalent to a circle.

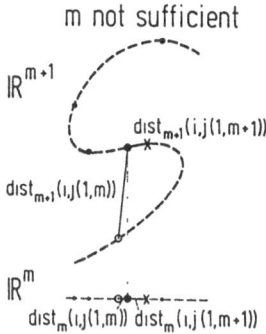

m not sufficient

$\mathbb{R}^{m+1}$

$dist_{m+1}(i,j(1,m+1))$

$dist_{m+1}(i,j(1,m))$

$\mathbb{R}^m$

$dist_m(i,j(1,m))$ $dist_m(i,j(1,m+1))$

Fig. 11. The delay coordinate mapping as a projection of an $(m-1)$-dimensional reconstruction space onto an m-dimensional one. The change of nearest-neighbor order near a reference point x_i by transition from embedding dimension m to $m+1$ is visualized for an insufficient embedding dimension. The crosses mark the nearest neighbor in $\mathbb{R}^{m+1}$ and the circles the nearest neighbor in $\mathbb{R}^m$.

call x_1, say, is denoted in Fig. 11 by a small circle. Its distance $dist_m(i, j \ (l, m))$ increases to $dist_{m+1}(1, j(l, m))$ if one passes to $\mathbb{R}^{m+1}$, i.e. the embedding dimension m was not sufficient to span the attractor to its full extent. The point closest to x_1 in $\mathbb{R}^{m+1}$, i.e. in the phase space with the correct dimension, is denoted by a cross, and its distance to x_1 is $dist_{m+1}(i, j \ (1, \ m+1))$. The ratio $Q(i, k)$ of distances,

$$Q(i, k) = \frac{dist_{m+1}(i, j(k, m))}{dist_{m+1}(i, j(k, m+1))} \geq 1 \ , \tag{20}$$

is larger than one, as follows from Fig. 11, and it becomes equal to one only if the nearest neighbor to a point in $\mathbb{R}^m$ also remains the nearest neighbor to this point in $\mathbb{R}^{m+1}$. This means that the equality sign in (20) holds only if the proper embedding is obtained. Since distances on strange attractors separate exponentially in time, we consider $\log Q(i, k)$ and take the average over p neighbors of N points x_1. The resulting quantity

$$\overline{W} = \frac{1}{\tau} \frac{1}{p} \frac{1}{N} \sum_{i,k} \log Q(i, k) \ , \tag{21}$$

in which a trivial τ dependence has been divided out, should approach zero for the proper choice of m and τ.

Figure 12 shows $\overline{W}$ versus τ for 9000 points from the Rössler attractor [9], which is generated by the differential equations

$$\dot{x} = -z - y \ , \tag{22}$$
$$\dot{y} = x + 0.15y \ ,$$
$$\dot{z} = 0.20 + z(x - 10) \ .$$

It can be seen that for $m = 3$ we have for the first time a value of $\overline{W}$ which is practically zero. The corresponding τ value $\tau^* = 0.143\tau_c$ is denoted by an arrow labeled 2.

In Fig. 13 we have plotted the Hausdorff dimension versus embedding dimension for various values of delay times τ as marked by the numbered arrows in Fig. 12. We see that D_0 converges for $m = 3$ and $\tau = \tau^*$. For $\tau < \tau^*$ (curve labeled 1) one still finds convergence only if the embedding is extended from $m = 3$ to $m = 4$.

29

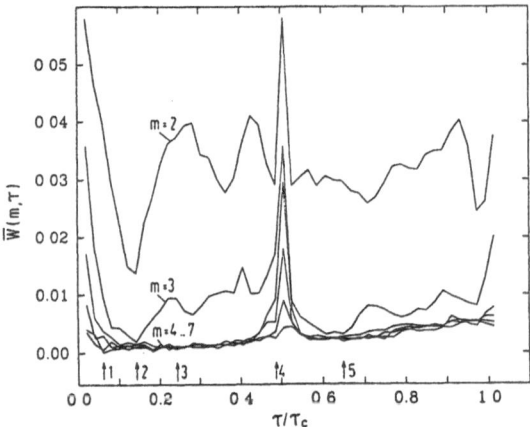

Fig. 12. $\overline{W}$ versus τ for various embedding dimensions m for the Rössler attractor defined by (19). τ_c denotes the first recurrence time of the Rössler system, and we have chosen $p = 10$.

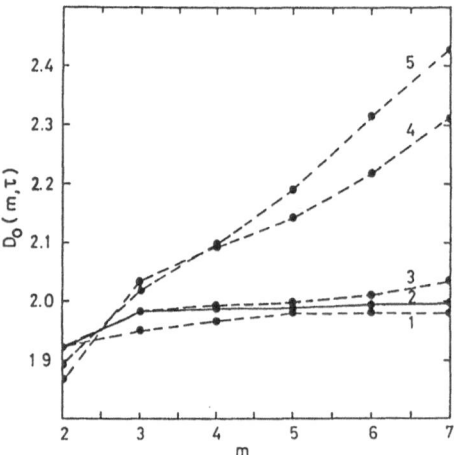

Fig. 13. Hausdorff dimension D_0 versus embedding dimension m for some values of delay time τ, as marked by the numbered arrows in Fig. 12.

However, there is no convergence for curves 4,5, which correspond to $\tau \gg \tau^*$. This means that finding correct values of τ and m simultaneously is a necessary condition for a proper data analysis of realistic, i.e. finite, sets of data.

References

1. Schuster, H.G.: *Deterministic Chaos*, second edition, Physik Verlag, Weinheim, 1988
2. Lichtenberg, A.J., Liebermann, M.A.: *Regular and Stochastic Motion*, Springer, Berlin, Heidelberg, 1982
3. Schuster, H.G., Martin, S., Martienssen, W.: Phys. Rev. **A33**, 3547 (1986)
4. Halsey, T.C., Jensen, M.H., Kadanoff, L.P., Procaccia, I., Shraiman, B.I.: Phys. Rev. **33A**, 1141 (1986)
5. Mackey, M.C., Glass, L.: Science **197**, 287 (1977)
6. Pawelzik, K., Schuster, H.G.: Phys. Rev. **35A**, 481 (1987)
7. Takens, F.: Lecture Notes in Mathematics 898, Springer, Berlin, Heidelberg, 1988
8. Liebert, W., Pawelzik, K., Schuster, H.G.: Europhys. Lett. **14**, 521 (1991)
9. Rössler, O.E.: Phys. Lett. **57A**, 397 (1976)

Current Instabilities in Semiconductors: Mechanisms and Self-organized Structures

E. Schöll

Institut für Theoretische Physik, Technische Universität Berlin,
Hardenbergstrasse 36, W-1000 Berlin 12, Fed. Rep. of Germany

An introduction to the theoretical description of current instabilities in semiconductors is given. Various physical mechanisms are surveyed, and the self-organized formation of spatio-temporal structures, in particular breathing current filaments, is discussed.

1. Introduction

Semiconductors are complex nonlinear dynamic systems which can in many cases exhibit electrical instabilities, like current runaway, switching between a nonconducting and a conducting state, or spontaneous oscillations of current or voltage, when they are driven far from thermodynamic equilibrium by strong external electric fields, optical irradiation, or current injection [1]. There is a variety of different physical mechanisms that can give rise to such current instabilities, but the observed phenomena are often similar, involving the self-organized formation of spatial and temporal structures. Such cooperative phenomena have been noted in a great number of very different dissipative systems occurring in physics, chemistry, and biology [2–4] when a state far from thermodynamic equilibrium is maintained by a continuous flux of energy or matter flowing through them. These instabilities bear a remarkable analogy with phase transitions [1].

Physical aspects of current instabilities in semiconductors have been treated in a number of older monographs and review articles, e.g. [5–9]. Recently, interest in this field has been greatly revived by the striking connections with the modern field of nonlinear dynamics, including bifurcations, chaos, and spatio-temporal pattern formation. The purpose of this article is to provide an introduction to a physical theory of current instabilities in semiconductors within this framework. More details and a large bibliography of the original literature may be found in [1] and in recent reviews on theoretical approaches to nonlinear and chaotic dynamics of generation–recombination processes [10], on a general dynamic-systems approach to nonlinear semiconductor transport and current instabilities [11], on the derivation of various levels of description for macroscopic nonlinear dynamic behavior from hot-electron transport theory [12] and on a microscopic theory of hot-carrier recombination and ionization processes [13]. For a guide to the recent experimental literature the reader is referred to [14–16] and to the articles by PEINKE [17] and JÄGER [18] in this volume.

2. Nonlinear Charge Transport in Semiconductors

The electrical transport properties of semiconductors are governed in a complex way by the macroscopic bulk properties, which determine the local current density j as a function of the local electric field $\vec{\mathcal{E}}$, by the contacts, and by the external circuit components. In the case of *linear* transport, Ohm's law holds:

$$j = \sigma \vec{\mathcal{E}} \tag{2.1}$$

with field-independent conductivity σ. Nonlinear transport may be described in the simplest case by a *field-dependent* local, static, scalar conductivity $\sigma(\mathcal{E})$. If the j–$\mathcal{E}$ characteristic has a regime of *negative differential conductivity* (NDC)

$$\sigma_{\text{diff}} := \frac{dj}{d\mathcal{E}} < 0 , \tag{2.2}$$

these states are often unstable against fluctuations, giving rise to the formation of spatial or temporal structures. Two important types of NDC are classified as NNDC (Fig. 1a) or SNDC (Fig. 1b) according to the respective shape ("N" or "S") of the j–$\mathcal{E}$ characteristic, but other more complicated shapes are also possible. NNDC and SNDC are in many cases associated with the formation of field domains (Fig. 2a) or current filaments (Fig. 2b). These self-organized spatial structures may be static or time-dependent, in which case current oscillations can arise due to domains moving

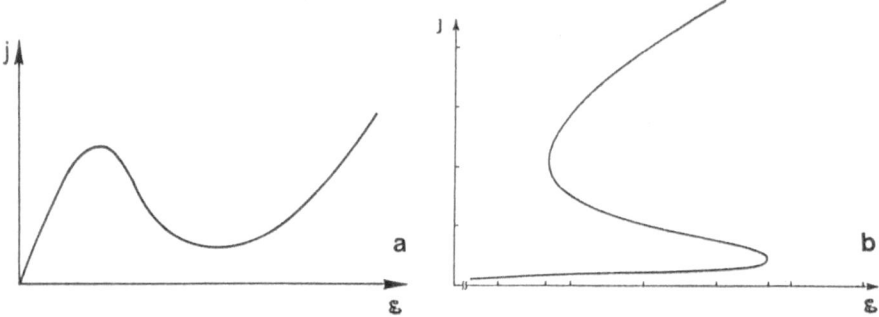

Fig. 1a,b. Current density j versus electric field $\mathcal{E}$ for two types of negative differential conductivity: (a) NNDC, (b) SNDC (schematic)

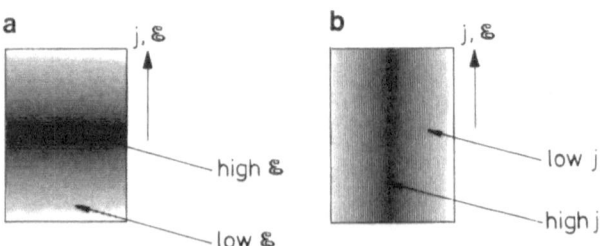

Fig. 2a,b. Sketch of (a) a field domain, (b) a current filament

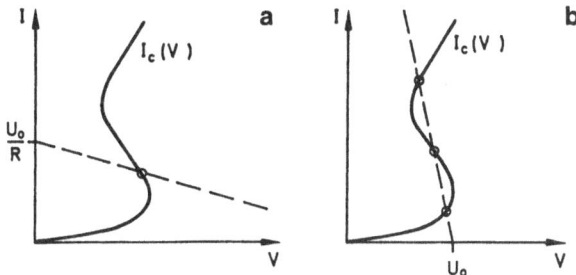

Fig. 3a,b. Stationary operating points of an SNDC element, given by the intersection of the current–voltage characteristic $I_C(V)$ with the load line (- - -). (a) large load resistence R, (b) small R

in the direction of the current flow [8], or filaments "breathing" transversally to the current flow [1], respectively.

At this point a word of warning is indicated. First, NDC does not always imply instability of the steady state, and vice versa. For example, SNDC states can be stabilized (and experimentally observed!) by a heavily loaded external circuit, and on the other hand, the Hopf bifurcation of a limit-cycle oscillation can occur on a $j–\mathcal{E}$ characteristic with positive differential conductivity [11]. Second, there is no one-to-one correspondence between SNDC and filaments, or between NNDC and domains [1]. Finally, it is important to distinguish between the local $j–\mathcal{E}$ characteristic and the global I–V characteristic which is determined by the total current

$$I = \int_A j \, d\boldsymbol{f} \tag{2.3}$$

and the total voltage drop

$$V = \int_0^L \mathcal{E}(z)dz \; , \tag{2.3}$$

where A is the cross-section of the current flow, and L is the contact distance. Only for spatially homogeneous states are the $j–\mathcal{E}$ and the I–V characteristics identical, up to rescaling.

The measured behavior also depends strongly upon the external circuit, which may contain resistive (load) or reactive (capacitance, inductance) components. The static operating point on the I–V characteristic is determined by the intersection of the load line

$$U_0 = I R_L + V \tag{2.5}$$

with the internal characteristic $I = I_c(V)$, where U_0 is the applied bias voltage, and R_L is the load resistance, giving one or three intersection points depending on the size of R_L (Fig. 3). For large R_L (Fig. 3a) the measured current is monotonic with increasing U_0, while for small R_L the current exhibits bistability and hysteretic switching between the lower and upper branch of the characteristic upon variation of U_0. Also, circuit-induced oscillations may be triggered by external capacitances or inductances [1].

3. Physical Mechanisms

In this section we shall review some microscopic mechanisms which give rise to negative differential conductivity and current instabilities. They are effective in a variety of semiconductor devices that are used as microwave oscillators (Gunn-diode, IMPATT diode) or electronic switches (thyristor, pin-diode). The NDC mechanism can be dominated by either junction or bulk properties. The mechanisms of the tunnel diode and the pnpn diode are based upon junction effects. For details the reader is referred to [19]. Bulk-dominated NDC includes three major classes of electronic instabilities: the nonmonotonic form of $j(\mathcal{E})$, which leads to $dj/d\mathcal{E} < 0$ in some range of $\mathcal{E}$, can be due to a nonlinearity of the mobility μ (*drift instability*), of the carrier-density n (generation–recombination or *g–r* instability), or of the electron temperature T_e (*electron overheating instability*).

3.1 Drift Instabilities

The best-known drift instability is the *Gunn effect* [8]. It is used in Gunn diodes to generate and amplify microwaves at frequencies typically beyond 1 GHz. The mechanism is based upon intervalley transfer of electrons from a state of high mobility to a state of low mobility by the influence of a strong electric field ($\mathcal{E} > 3$ kV/cm). The band structure of GaAs and other III-V compound semiconductors is shown schematically in Fig. 4. At low electric fields the electrons are essentially in the minimum of the central valley, which has a low effective mass m^* and hence a high mobility. As the field $\mathcal{E}$ is increased, the electrons are heated up and gain enough energy to be transferred to the satellite valley with a higher minimum energy, but larger effective mass, and hence lower mobility. As more and more electrons are transferred, the averaged mobility μ decreases strongly so that the current density $j = en\mu(\mathcal{E})\mathcal{E}$ decreases with increasing field, as a result of negative differential mobility (NDM). When most electrons are in the upper valley, j increases again. Thus an NNDC characteristic is produced. Depending upon the cathode boundary condition this can lead to the formation of moving domains which show up as transit time oscillations of the current [1].

Other mechanisms for negative differential mobility are due to the anisotropy of equivalent satellite valleys [20], for instance the *Erlbach* instability [6]. While the Gunn instability occurs for a current parallel to the applied field, the Erlbach instability involves off-diagonal elements of the differential conductivity tensor $dj_\alpha/d\mathcal{E}_\beta$. Consider in the simplest case two equivalent valleys 1 and 2 with anisotropic axes

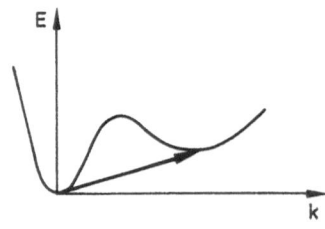

Fig. 4. Energy E versus wave vector k for the conduction band in GaAs (schematic)

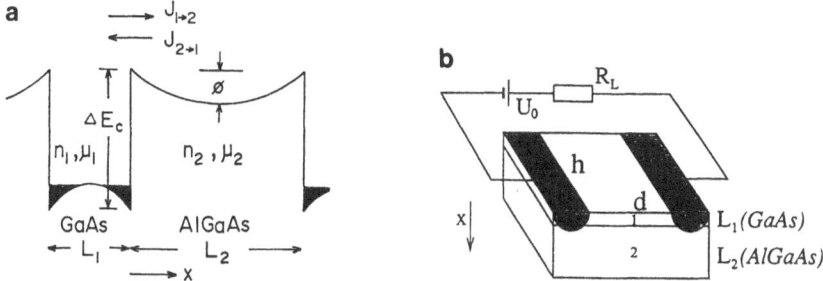

Fig. 5. (a) Conduction-band energy of a modulation-doped GaAs/Al$_x$Ga$_{1-x}$As heterostructure (schematic). (b) Sample configuration

pointing in different directions. If an electric field $\mathcal{E}$ is applied in the symmetry direction between these two valleys, the current j is also in the x-direction. Now suppose that the field is slightly off the x-direction, such that the conductivity effective mass in valley 1 is larger than the mass in valley 2. The rate at which electrons absorb energy from the electric field, i.e., the rate at which they are heated, is inversely proportional to the conductivity effective mass, so the electrons in valley 2 are heated more than those in valley 1. Thus there will be a net transfer of electrons by intervalley scattering from the hotter valley to the cooler valley, and the electron population becomes larger in valley 1 than in valley 2. The decrease in the population of the higher-mobility valley 2 then leads to a negative contribution to the current in the y-direction, which may result in a transverse negative differential conductivity (NNDC).

A more recent example of a drift instability is the *real-space transfer* of electrons in a modulation-doped heterostructure [21]. The conduction band edge of a GaAs/Al$_x$Ga$_{1-x}$As heterolayer is shown schematically in Fig. 5a. The band edge discontinuity ΔE_{c} is due to the larger band gap of Al$_x$Ga$_{1-x}$As as compared to GaAs, e.g., $\Delta E_{\mathrm{c}} = 250$ meV for $x = 0.3$. The fine structure associated with band-bending and with the potential barrier ϕ is caused by the space charge of the carriers. The AlGaAs layer is heavily doped, while the GaAs layer is undoped. Therefore, ionized impurity scattering leads to a much smaller mobility (μ_2) in AlGaAs than in GaAs (μ_1). The ionized electrons fall into the GaAs quantum well. Thus the mobile carriers (density n_1) are separated in real space from their parent donors in the AlGaAs layer, which are responsible for scattering. This results in extremely high current densities $j = en_1\mu_1\mathcal{E}$ in the GaAs layer if an electric field $\mathcal{E}$ is applied parallel to the layers (Fig. 5b). As $\mathcal{E}$ is increased, the carriers in the GaAs are heated up strongly due to their large current density. Thermionic emission of electrons from GaAs into AlGaAs becomes effective, which leads to a transverse current density $J_{1\rightarrow 2} \sim \exp\left(-\Delta E_{\mathrm{c}}/kT_1\right)$, where T_1 is the electron temperature in GaAs. This real-space transfer increases the electron density n_2 in the AlGaAs layer and reduces the averaged mobility $\mu_{\mathrm{eff}} = (\mu_1 n_1 + \mu_2 n_2)/(n_1 + n_2)$ and thus may lead to NNDC. The result is similar to the Gunn effect, but the underlying physical mechanism is *real-space* transfer rather than intervalley transfer in *k-space*. The extremely high mobility ($\mu_1 \geq 5000$ cm^2/Vs at 300 K) singles out this effect for applications in fast electronic switches and oscillators.

A quantitative dynamic theory [22] has been based upon time-dependent balance equations taking into account thermionic emission currents $J_{1\to2}$, $J_{2\to1}$, energy exchange through convection, heat flow and pressure-induced energy flow, phonon scattering and Joule heating, and the nonlinear dynamics of the longitudinal electric field $\mathcal{E}$ and the space-charge potential ϕ. Self-generated GHz oscillations in the regime of NNDC (at several kV/cm) and a period-doubling route to chaos have been found numerically from the dynamic equations [22].

3.2 Generation–Recombination Instabilities

The large class of g–r instabilities is distinguished by a nonlinear dependence of the steady-state carrier concentration n upon the field $\mathcal{E}$, which yields a nonmonotonic current density vs field relation $j = e\mu n(\mathcal{E})\mathcal{E}$ of either NNDC or SNDC type. This dependence is due to a redistribution of electrons between the conduction band and bound states (and possibly the valence band, in the case of ambipolar currents) during the heating of the electron gas. The microscopic transition probabilities of the carriers between different states, and hence the g–r coefficients, generally depend upon the electric field. A particularly strong dependence is expected for the rate constants of the following g–r processes: *field-enhanced trapping* and *Poole–Frenkel emission*, which often leads to NNDC, and *impact ionization*, which is the key process for SNDC. Field enhanced trapping occurs e.g. in gold-doped n-Ge [7]. The gold atoms form deep impurity levels, corresponding to singly or doubly charged negative ions. Trapping of electrons into these levels requires the penetration of a Coulomb potential barrier. Therefore the trapping coefficient increases with field $\mathcal{E}$ while the (thermal) emission of electrons is practically field-independent, as long as $\mathcal{E}$ is well below the threshold for field ionization. Thus the free-carrier density decreases with rising field, $dn/d\mathcal{E} < 0$, and the differential conductivity $dj/d\mathcal{E}$ may become negative. In larger fields the ionization coefficient increases and causes the carrier density to rise again with the field, leading to a positive differential conductivity branch. Thus, an N-shaped j–$\mathcal{E}$ characteristic is produced. Self-generated low-frequency oscillations (LFO) in semiinsulating GaAs at room temperature [23,24] have been attributed to traveling field domains induced by field-enhanced trapping, but field-enhanced emission from deep trap levels due to the lowering of the ionization energy by a field (Poole–Frenkel effect [13]) has also been proposed as a possible explanation [11], accounting in particular for the observed very low frequencies of 50 – 100 Hz.

Impact ionization of carriers from impurity levels (shallow donors, acceptors, or deep traps) or across the bandgap (avalanching) is another process which may lead to NDC. If a free carrier has gained enough kinetic energy in the electric field, it may transfer this energy in a collision to a bound carrier, which is then released to the conduction band (or valence band, in the case of a bound hole); thereby an additional free carrier is generated which may, in turn, impact-ionize other carriers. Such a positive feedback ("autocatalysis") leads to a rapid increase of the free-carrier density. The impact-ionization coefficient increases strongly with $\mathcal{E}$ beyond a threshold field required to heat up the carriers, so that they have enough energy for ionization. Models of this type are relevant in a variety of materials and in different temperature ranges, for instance in low-temperature impurity breakdown or in threshold switching [1]. The view of impurity breakdown as a nonequilibrium phase transition was advanced first in [25]. In a systematic survey [26], impact ionization of electrons or holes across the bandgap or from localized levels was recognized as the main "autocatalytic" process necessary for g–r-induced nonequilibrium phase transitions. SNDC can be produced if the excited level of the impurities is also considered [1]. This simple g–r mechanism can explain the self-organized formation of current filaments [27], as well as self-generated oscillations displaying a period-doubling route to chaos when coupled with dielectric relaxation of the field [28].

Two important devices are also based upon g–r-induced bulk negative differential conductance, but the coupling with junction effects is essential in these cases: pin diodes [18,29] and IMPATT diodes. The *pin diode* involves double injection of electrons and holes, and field-enhanced trapping. It consists of an intrinsic (undoped) layer adjacent to a p-doped and an n-doped region on either side. If the n-layer is connected to a negative and the p-layer to a positive voltage, electrons and holes are injected from the n-side and the p-side, respectively, into the central intrinsic (i) layer. We assume that the i-region contains deep acceptor-like recombination centers with a large cross section (or capture coefficient) for hole capture and a much smaller cross section for electron capture, and that these are completely occupied by electrons in thermal equilibrium. At low injection currents almost all of the injected holes will be captured by the recombination centers near the injecting pi junction, while the injected electrons freely traverse most of the i-layer. Thus, there is a "recombination barrier" to the passage of holes. The resulting electron current is limited by the space charge that the injected electrons build up. At high injection currents, where the injected electron and hole concentrations exceed that of the recombination centers, all centers have trapped a hole, and the excess holes as well as the electrons traverse the i-layer. They are approximately equal in concentration, if the injection level is sufficiently high. The current is then carried by a quasi-neutral semiconductor plasma. Thus, at a given value of the applied voltage, there are two stable steady states: a low-current state, where the recombination centers (except for a narrow region near the pi junction) are occupied by electrons and the current is a single-carrier space-charge-limited (SCL) current, and a high-current state, where the recombination centers are filled with holes, and the current is carried by an injected plasma. In between, there is an NDC state in which both the single-carrier SCL current and the semiconductor plasma extend some way into the i-region from both sides (from the in and the pi junction, respectively), and they are connected by regions of more complex carrier and field distributions [29].

3.3 Electron Overheating Instabilities

Negative differential conductivity can also arise from changes in the nature of the dissipation of energy and momentum of the carriers as they are heated up by the electric field [5,7]. In such an electron overheating instability the energy and momentum relaxation times (τ_e and τ_m, respectively) depend in a nonlinear way upon the average energy per carrier E, which can be related to an effective electron temperature T_e by $E = \frac{3}{2} k_B T_e$. The mobility (in the simplest approximation, $\mu = e\tau_m/m^*$) is thus a nonlinear function of T_e, which is specified by the particular scattering mechanisms, e.g. by acoustic phonons, optical phonons, or impurities. The electron temperature T_e as a function of $\mathcal{E}$ follows in the steady state from a balance of the electrical power density $j\mathcal{E} = en\mu(T_e)\mathcal{E}^2$ with the power density $n(E - E_0)/\tau_e(T_e)$ dissipated by the electrons to the lattice. From this the differential conductivity $dj/d\mathcal{E}$ with $j = en\mu(T_e(\mathcal{E}))\mathcal{E}$ can be calculated; both SNDC and NNDC are possible, but inverted S- and loop-type characteristics are also possible [46]. Regular [30] and spatially chaotic [31] stationary current filaments have been studied on the basis of this mechanism. Self-generated oscillations are also found if the optical-phonon-induced nonlinearity of $\tau_e(T_e)$ and $\tau_m(T_e)$ is coupled with impurity impact ionization [32].

There are many other NDC mechanisms, like the acousto-electric effect [7], or plasma instabilities involving a magnetic field, e.g., helicon waves [9], but these will not be treated here.

4. Theoretical Description

A theoretical understanding of current instabilities should aim at developing simple physical models which reproduce the dynamic behavior observed in the experiments. A full understanding would require the derivation of a macroscopic nonlinear dynamic system exhibiting such behavior starting from the microscopic level of elementary transport processes. While this is not yet available, there exist various attempts to identify the relevant physical mechanisms at different levels of hot-electron transport theory [1,10,12].

Generally, the modeling of a spatially extended nonlinear dynamic system, such as a semiconductor, may be cast into one of the following common forms, depending on whether space, time, and the dynamic variables are chosen as continuous or discrete:

(i) *A set of partial differential equations*
   (space, time, and dynamic variables all continuous).
(ii) *A set of ordinary differential equations*
   (space discrete, time and dynamic variables continuous).
(iii) *Iterated maps*
   (space and time discrete, dynamic variables continuous).
(iv) *Cellular automata*
   (space, time, and dynamic variables all discrete).

Semiconductor transport theory is usually formulated in terms of (i) or (ii), e.g., balance equations or coupled rate equations, and the existing models for current instabilities are mostly of these types. However, a model for the onset of chaos in semiconductors was recently derived in the form of an iterated map (iii), starting from a stochastic Chapman–Kolmogoroff equation [33]. Work geared towards modeling current filaments in terms of a cellular automaton (iv) has recently been initiated [34]. In the following, we shall focus on approach (i).

4.1 Hydrodynamic Balance Equations

A frequently used approach to hot-carrier transport is based upon the Boltzmann equation for the semiclassical carrier distribution function $f(r, k, t)$, where r is the spatial coordinate and $\hbar k$ is the (crystal) momentum [13]:

$$\frac{\partial f}{\partial t} + v_g \cdot \nabla_r f + q\hbar^{-1}\vec{\mathcal{E}} \cdot \nabla_k f = (\frac{\partial f}{\partial t})_{\text{coll}} \ . \tag{4.1}$$

Here v_g is the group velocity, $\vec{\mathcal{E}}$ is the electric field, and the subscript coll denotes the collision integral, which includes all dissipative processes like phonon and impurity scattering, carrier–carrier scattering, and impact ionization. The carrier charge is $q = \pm e$ for holes or electrons, respectively.

The Boltzmann equation relies on the assumption that (i) the distribution function varies little over the de Broglie wavelength, (ii) the carrier density is sufficiently low such that only binary collisions occur, (iii) the time between successive collisions is much longer than the duration of a collision, (iv) the density gradients are small over the range of the interparticle potential. Failing this, quantum transport equations shoud be used [35].

Since a space- and time-dependent solution of the Boltzmann equation is difficult to obtain, a moment expansion is often employed to derive a set of hydrodynamic balance equations for the slow macroscopic observables like the carrier density $n(r, t) \equiv \int f(r, k, t)z\mathrm{d}^3 k$, the mean momentum per carrier $p(r, t) \equiv < \hbar k >$, and the mean energy per carrier $E(r, t) \equiv < E(k) >$, with the ensemble average

$$< A > \equiv n^{-1}\int Af(r, k, t)z\mathrm{d}^3 k$$

and the density of states $z = 2(2\pi)^{-3}$ in k-space. In the following a nondegenerate, isotropic parabolic band structure $E(k) = \hbar^2 k^2/(2m^*)$ with effective mass m^* will be assumed. In this case the mean momentum and energy are related to the mean group velocity $v(r, t) \equiv < v_g >$ and the electron temperature $T_e \equiv (m^*/3k_B) < (v_g - < v_g >)^2 >$ by

$$p(r, t) = m^* v \ , \tag{$\dots$}$$

$$E(r, t) = \frac{m^*}{2} v^2 + \frac{3}{2}k_B T_e \ . \tag{4.3}$$

The mean energy is thus composed of a convective and a thermal contribution. Multiplying (4.1) with appropriate powers of k and integrating over the first Brillouin zone, we obtain a coupled hierarchy of moment equations [12]:

$$\dot{n} + \nabla \cdot (nv) = \phi(n, E) \ , \tag{4.4}$$

$$\dot{p} + (v \cdot \nabla)p + \frac{1}{n}\nabla(nk_B T_e) - q\vec{\mathcal{E}} = -p\phi/n - p/\tau_m(E) \ , \tag{4.5}$$

$$\dot{E} + (v \cdot \nabla)E + \frac{1}{n}\nabla(nk_B T_e v) - \frac{\kappa}{n}\Delta T_e - qv \cdot \vec{\mathcal{E}} = -E\phi/n$$
$$- (E - E_0)/\tau_e(E) \tag{4.6}$$

if the following assumptions [36] are used:

(i) The electron temperature is a scalar:

$$< (v_{gi} - <v_{gi}>)(v_{gj} - <v_{gj}>)> = \frac{3}{m^\star} k_B T_e \delta_{ij} , \tag{4.7}$$

i.e., the momentum-flux tensor reduces to the scalar electron pressure $n k_B T_e$.

(ii) In the energy-flow density

$$\frac{m^\star}{2} n <v_g^2 v_g> = n v E + n v k_B T_e + j_Q , \tag{4.8}$$

where $j_Q = -\kappa \nabla T_e$

with thermal conductivity κ.

(iii) The collision integrals are evaluated approximately [37] yielding the generation–recombination $(g-r)$ rate, including impact ionization

$$\phi(n, E) = \int \left(\frac{\partial f}{\partial t} \right)_{\text{coll}} z d^3 k , \tag{4.9}$$

the momentum relaxation rate

$$-n \frac{p}{\tau_m(E)} = \int \hbar k \left(\frac{\partial f}{\partial t} \right)_{\text{coll}} z d^3 k , \tag{4.10}$$

and the energy relaxation rate

$$-n \frac{E - E_o}{\tau_e(E)} = \int \frac{\hbar^2 k^2}{2m^\star} \left(\frac{\partial f}{\partial t} \right)_{\text{coll}} z d^3 k \tag{4.11}$$

with mean energy-dependent momentum and energy relaxation times τ_m, τ_e, respectively, and $E_0 \equiv (3/2) k_B T_L$, T_L = lattice temperature. This is equivalent to a self-consistent parametrization of the distribution function by n, p, and E. Equations (4.6–6) with (4.2,3) constitute a closed set of nonlinear partial differential equations for n, p, E. All information about the microscopic physical processes is contained in the functions $\phi(n, E)$, $\tau_m(E)$, $\tau_e(E)$, and κ. In particular, impact ionization contributes to the $g-r$ rate *and* to the energy relaxation rate. The latter may be approximated by

$$n \frac{E - E_0}{\tau_e} = n \frac{E - E_0}{\tau_e\prime} + E_{th} \phi_{ii}(n, E) , \tag{4.12}$$

where $\tau_e\prime$ includes phonon scattering, and ϕ_{ii} and E_{th} are the impact ionization rate and threshold energy, respectively [37]. The impact ionization rate of impurities is given by

$$\phi_{ii}(n, E) = X(E) n n_t , \tag{4.13}$$

where n_t is the density of trapped carriers, and $X(E)$ is the impact ionization coefficient which depends on the mean energy in a threshold-like way. The impact-ionization and capture coefficients involving shallow impurity levels can be calculated from analytical approximations of the distribution function or by Monte Carlo techniques [13,37].

4.2 Time-scale Separation

The transport equations (4.4–6) are coupled to the local electric field $\vec{\mathcal{E}}$ by Maxwell's equations

$$\varepsilon_0 \varepsilon_S \nabla \cdot \vec{\mathcal{E}} = \varrho \qquad (4.14)$$

with local charge density $\varrho = e(N_D - N_A) + q(n + n_t)$ (N_D and N_A are donor and acceptor densities, and ε_0 and ε_S are absolute and relative permittivity, respectively), and

$$\nabla \times H = \varepsilon_0 \varepsilon_S \dot{\vec{\mathcal{E}}} + j \equiv j_0 \qquad (4.15)$$

with conduction-current density $j = qnv$ and magnetic field H. From (4.15) it follows by applying ∇ that the total external current density j_0, which may be regarded as a control parameter, is composed of displacement and conduction currents:

$$\varepsilon_0 \varepsilon_0 \dot{\vec{\mathcal{E}}} = j_0 - j \ . \qquad (4.16)$$

The time-scale on which the variables $n, p, E, \vec{\mathcal{E}}$ occurring in (4.4–6,16) change is determined by the g–r lifetime, the momentum relaxation time, the energy relaxation time, and Maxwell's dielectric relaxation time, resprectively. Since the macroscopic dynamics is governed by the slow variables, which of the above variables must be considered as relevant dynamic transport variables depends on the relations between these time-scales. The fast variables may be eliminated adiabatically from the dynamics [38].

Denoting the set of relevant, i.e. slow hydrodynamic, variables by $x = (x_1, \ldots, x_N)$, we can write (4.4–6) in the general form

$$\dot{x} = \Phi(x, \vec{\mathcal{E}}) \ , \qquad (4.17)$$

where $\Phi \equiv (\Phi_1, \ldots, \Phi_N)$ is a set of nonlinear functions. Equations (4.16,17) constitute a macroscopic nonlinear dynamic system with variables $x, \vec{\mathcal{E}}$. General results on instabilities of this system, which may occur either in negative or in positive differential conductivity regimes, will be briefly discussed below.

A number of special cases of impact-ionization-induced instabilities at low temperatures have been analyzed with respect to chaotic behavior, self-generated oscillations, and spatio-temporal pattern formation. These can be classified according to the different time-scales involved as follows.

Momentum relaxation generally occurs faster than the other processes, so that p can be eliminated adiabatically from (4.5) by setting $dp/dt \equiv \dot{p} + (v \cdot \nabla)p = 0$. Introducing the mobility

$$\mu(E) \equiv \frac{e}{m^\star} \tau_m'(E), \qquad \frac{1}{\tau_m'} \equiv \frac{1}{\tau_m} + \frac{\phi}{n} \ , \qquad (4.18)$$

the drift-diffusion equation

$$j = qn\boldsymbol{v} = en\mu(E)\vec{\mathcal{E}} - qD(E)\nabla n - \frac{q}{e}\mu(E)nk_B\nabla T_e \qquad (4.19)$$

is recovered if the diffusion coefficient D is defined via the Einstein relation $eD = \mu k_B T_e$. Further simplifications arise if additional fast variables can be eliminated, or if spatial inhomogeneities are neglected. Oscillation mechanisms which involve the following slow variables have been studied:

(i) **Carrier Density and Electric Field.** This refers to the case of a high-purity "relaxation semiconductor", i.e. the dielectric relaxation of the field occurs on a slow time-scale. The fast variable E can be expressed as a function of the field by the energy balance (4.6) in the steady-state condition in case of r-independent T_e. In the regime of impurity impact ionization, dielectric relaxation oscillations displaying a period-doubling route to chaos have been found in spatially homogeneous models, either driven by an ac-field [39,40] or self-generated [28,41,62]. Spontaneous spatio-temporal oscillations based on such a mechanism will be discussed in Sect. 5.

(ii) **Carrier Density, Mean Energy, and Electric Field.** Self-generated oscillations involving these dynamic variables may be induced by a combination of dielectric relaxation, impact ionization, and optical-phonon emission in the post-breakdown regime of p-Ge [32], or by real-space transfer of electrons, energy transfer, and space-charge dynamics in modulation-doped GaAs/AlGaAs heterostructures [22]. A spatially homogeneous model for the first kind of oscillation [32] uses mean-energy-dependent momentum and energy relaxation times $\tau_m(E)$, $\tau_e(E)$ obtained from a Monte Carlo simulation rather than phenomenological functions, and is thus founded at a microscopic level. The dramatic decrease of the mobility and the energy relaxation time when the threshold energy for emission of optical phonons is exceeded leads to strong nonlinearities and to NNDC in the post-breakdown regime. In the preceding positive differential conductivity regime, a subcritical Hopf bifurcation generates an unstable limit cycle which annihilates with a stable limit cycle in a global bifurcation. The latter undergoes a second global bifurcation with critical slowing-down of the oscillation frequency. Hysteresis between oscillatory and stationary states is also found. Both phenomena, hysteresis and global bifurcations, have been observed in p-Ge at low-temperature impurity breakdown by Peinke and coworkers [17].

(iii) **Carrier Density and Mean Energy.** If the electric field relaxes fast, (4.16) with (4.19) reduces to the static current density vs field relation. Self-generated energy relaxation oscillations described by (4.4,6) may arise as a result of nonlinear energy loss through impact ionization [42,43]. If two or more such oscillating cells are coupled by energy exchange (via phonons), this may serve as a simple model for the excitation and interaction of different spatio-temporal degrees of freedom. In a two-cell model, resonance transitions between linearly correlated oscillations [44], mode-locking structures obeying the Farey tree ordering, and quasi-periodicity [45] have been found, in accordance with experiments [46,47]. The bit-number cumulants of the invariant density of the underlying dynamic system have been shown to represent a characteristic measure for the build-up of internal correlations between different localized oscillation centres [48]. In particular, the bit-number variance may be viewed as a generalized specific heat in the framework of the thermodynamic formalism of dynamic systems [49].

4.3 General Results on Oscillatory Instabilities

The different oscillation mechanisms presented in Sect. 4.2 (i–iii) may be discussed from a unified viewpoint [11] by analyzing the bifurcations of the general dynamic system (4.16,17). To this end we restrict ourselves to the simple, spatially homogeneous form, although extensions to include current filaments are possible [11]. If the semiconductor is operated in a resistive external circuit as given by (2.5), and connected in parallel to an external capacitance C [1], (4.16) with (4.19) can be cast into the form

$$\dot{\mathcal{E}} = [J_0 - (\sigma_L + \sigma(\boldsymbol{x}, \mathcal{E}))\mathcal{E}]/[\varepsilon_0 \varepsilon_s (1 + C/C_i]) \, , \tag{4.20}$$

where $\sigma \equiv ne\mu$ is the conductivity which depends on the field $\mathcal{E}$ and the additional transport variables $\boldsymbol{x}$, $C_i \equiv \varepsilon_0 \varepsilon_s A/L$ is the intrinsic capacitance, and

$$J_0 \equiv U_0/(R_L A) \text{ and } \sigma_L \equiv L/(R_L A)$$

are control parameters.

The static differential conductivity is given by

$$\sigma_{\text{diff}} = \frac{d}{d\mathcal{E}^{\star}}[\sigma(\boldsymbol{x}(\mathcal{E}^{\star}), \mathcal{E}^{\star})\mathcal{E}^{\star}] \, , \tag{4.21}$$

where $\star$ denotes the steady-state values obtained from $\dot{\boldsymbol{x}} = 0$, $\dot{\mathcal{E}} = 0$ in (4.17,20). The stability of the steady state with respect to small fluctuations $(\delta\boldsymbol{x}, \delta\mathcal{E}) \sim \exp(\lambda t)$ follows from linearizing (4.17,20) around $(\boldsymbol{x}^{\star}, \mathcal{E}^{\star})$, which yields a secular equation of degree $N+1$ for the eigenvalues $\lambda_1, \ldots, \lambda_{N+1}$. One can prove the general relation [11,50]

$$\prod_{i=1}^{N+1} \lambda_i = -\det \tilde{A} \cdot \lambda_M^{\text{diff}} \, , \tag{4.22}$$

where $\tilde{A}_{ij} \equiv (\partial \Phi_i/\partial x_j)_{\star}$ $(i, j = 1, \ldots, N)$ is determined by the transport equations (4.17), and

$$\lambda_M^{\text{diff}} \equiv (\sigma_L + \sigma_{\text{diff}})/[\varepsilon_0 \varepsilon_s (1 + C/C_i)] \tag{4.23}$$

is an inverse differential dielectric relaxation time.

Saddle-node bifurcations of fixed points are given by a vanishing real eigenvalue λ_i, hence by (4.22) $\det \tilde{A} = 0$ or $\sigma_L + \sigma_{\text{diff}} = 0$ must hold. Hopf bifurcations of limit cycles occur when a pair of complex conjugate eigenvalues crosses the imaginary axis. Here we consider only the simplest case $N = 1$. The conditions for a pair of purely imaginary eigenvalues can then be stated with the use of (4.22):

43

$$\tilde{\lambda} = \tilde{\nu} \tag{4.24}$$

$$\tilde{\lambda}(\sigma_L + \sigma_{\text{diff}}) < 0 \ ,$$

where $\tilde{\lambda} \equiv \det \tilde{A}$ and $\tilde{\nu} \equiv (\sigma_L + \frac{\partial}{\partial \mathcal{E}}[\sigma(\boldsymbol{x},\mathcal{E})\mathcal{E}])/[\varepsilon_0\varepsilon_S(1 + C/C_i)]$. For $\sigma_L = 0$, $\tilde{\nu}$ is proportional to the differential mobility $\mathrm{d}v/\mathrm{d}\mathcal{E}$. Equation (4.24) can be satisfied by either

(i) $\quad \sigma_L + \sigma_{\text{diff}} < 0 \quad$ and $\tilde{\lambda} = \tilde{\nu} > 0$

or

(ii) $\quad \sigma_L + \sigma_{\text{diff}} > 0 \quad$ and $\tilde{\lambda} = \tilde{\nu} < 0$

This demonstrates that oscillatory instabilities are possible for either *negative* differential conductivity (and *positive* differential mobility, in the case of $\sigma_L = 0$, or *positive* differential conductivity (and *negative* differential mobility, for $\sigma_L = 0$). With $N > 1$, more complex situations are possible, but $\sigma_{\text{diff}} < 0$ and $\sigma_{\text{diff}} > 0$ may still occur [11,50]. Note that a Hopf bifurcation is not the only way to create self-generated oscillations: global bifurcations are another possibility [17,32]. Various models for oscillations falling in this general scheme have been worked out [11].

5. Breathing Current Filaments as Spatio-temporal Structures

In this section we shall apply the theoretical framework outlined in Sect. 4 to discuss a particular example of self-organized spatio-temporal structure formation: *breathing current filaments*. Current filaments have recently attracted much interest in connection with low-temperature impurity breakdown in n-GaAs [27,51-53] and p-Ge [17], and with pin-diodes at room temperature [18,54,55]. "Breathing" current filaments were suggested as a possible mechanism for self-generated current oscillations [1,28]. Recent space- and time-resolved experimental investigations have indeed demonstrated that in a certain regime of the S-shaped current–voltage characteristic, spontaneous nonlinear oscillations arise which are localized at the filamentary boundaries, and which represent a breathing motion of the filament walls [52,53,56].

The theoretical description of current filaments based on the semiconductor transport equations of Sect. 4 may proceed in three steps [1]:

(i) A linear mode analysis around the spatially uniform steady state for small space- and time-dependent fluctuations, yielding a dynamic differential-conductivity tensor. Unstable modes can be shown to lead to the bifurcation of stationary current filaments or field domains [27,57].

(ii) A nonlinear analysis of the transport equations in the steady state, using generalized equal-areas rules [27,58,59]. Thus, fully developed stationary filaments can be modeled.

(iii) An analysis of the nonlinear time-dependent transport equations, using a projection technique to derive simple dynamic equations for the filament radius and other relevant dynamic variables, whose solutions describe breathing filaments [60].

In the present article we shall sketch approach (iii).

We shall adopt dielectric relaxation oscillations as the essential oscillation mechanism, such that the slow dynamic variables are given by the electric field and the concentration of free carriers (we confine attention to electrons), in accordance with (i) in Sect. 4.2. Current filaments will be modeled within the following approximations:

(i) There is cylindrical symmetry, and $\vec{\mathcal{E}} = \vec{\mathcal{E}}_{||} + \vec{\mathcal{E}}_{\perp}$, where $||$, $\perp$ denote components parallel and perpendicular to the applied voltage.

(ii) Longitudinal inhomogeneities are neglected. Assuming planar Ohmic contacts gives $n = n(R,t)$, $\mathcal{E}_{||} = \mathcal{E}_{||}(t)$, $\mathcal{E}_{\perp} = \mathcal{E}_{\perp}(R,t)$, where R is the radial coordinate.

(iii) The occupation of the different impurity levels relaxes on a faster time-scale than $\mathcal{E}$ and n, such that the concentration of trapped carriers distributed over the ground state and the excited states of the impurities can be expressd as $n_t(n, \mathcal{E})$ in (4.14) through the steady-state g–r rate equations [1].

Equations (4.4,14,16,19) then reduce to

$$\dot{n} = D(n'' + \frac{n'}{R}) + \mu(\mathcal{E}_{||})[\mathcal{E}_{\perp}n' + n\varrho(n,\mathcal{E}_{||})/(\varepsilon_0\varepsilon_S)] \ , \tag{5.1}$$

$$\varepsilon_0\varepsilon_S\dot{\mathcal{E}}_{\perp} = -en\mu(\mathcal{E}_{||})\mathcal{E}_{\perp} - eDn' \ , \tag{5.2}$$

$$\varepsilon_0\varepsilon_S\dot{\mathcal{E}}_{||} = j_0 - en\mu(\mathcal{E}_{||})\mathcal{E}_{||} \ , \tag{5.3}$$

$$\varepsilon_0\varepsilon_S(\mathcal{E}'_{\perp} + \frac{\mathcal{E}_{\perp}}{R}) = \varrho(n,\mathcal{E}_{||}) \ . \tag{5.4}$$

The prime denotes the derivative with respect to the radial coordinate. The spatially homogeneous steady states are given by

$$\varrho(n, \mathcal{E}_{||}) = 0, \ \mathcal{E}_{\perp} = 0 \ . \tag{5.5}$$

The essential nonlinearity is contained in $\varrho(n, \mathcal{E}_{||})$ due to the g–r processes implicit in $n_t(n, \mathcal{E}_{||})$. Impact ionization from at least two impurity levels may induce a nonmonotonic dependence of ϱ on n in some range of applied fields $\mathcal{E}_0$, resulting in bistability of the uniform steady states (low conductivity n_1, high conductivity n_3), and in S-shaped negative differential conductivity [1].

The time-independent, spatially nonuniform solutions $n^*(R)$, $\mathcal{E}^*_{\perp}(R)$ satisfy the equations

$$\mathcal{E}^{*\prime}_{\perp} + \frac{\mathcal{E}^*_{\perp}}{R} = \varrho(n^*, \mathcal{E}^*_{||})/(\varepsilon_0\varepsilon_S) \ , \tag{5.4'}$$

$$\mu n^* \mathcal{E}^*_{\perp} + Dn^{*\prime} = 0 \ . \tag{5.6}$$

They include filamentary solutions fulfilling the boundary conditions $n^*(0) \simeq n_3, n^*(\infty) = n_1$, $\mathcal{E}^*_{\perp}(0) = 0$ (Fig. 6a). The filament has a radius $R^*_0(\mathcal{E}_0)$, $\mathcal{E}_0 := \mathcal{E}^*_{||}$, and a thin wall (transition layer) of thickness $\Delta R \ll R^*_0$. The filamentary solution leads to an additional, monotonically decreasing branch in the stationary current–voltage characteristic, corresponding to negative differential conductance (Fig. 6.b).

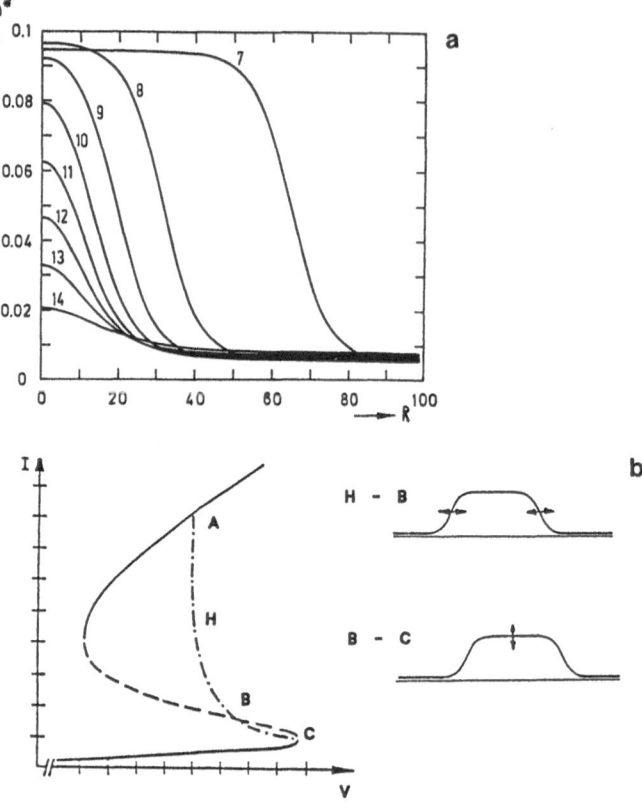

Fig. 6. (a) Stationary filamentary electron density profiles $n^*(R)$ calculated for n-GaAs for various control parameters $\mathcal{E}_0$ (denoted by 7 to 14). (b) Stationary current–voltage characteristic including uniform states (*full lines*: stable, *broken lines*: unstable) and filamentary states (*dot-dashed*), calculated with the parameters of (a). The inset shows the breathing mode (H–B) and a bulk-dominated oscillation mode (B–C).

Breathing filaments are characterized by a rigid shift of the filament wall with negligible changes in the profile (Fig. 7). In order to model such dynamics, we identify as the most relevant dynamic variables the radius of the current filament $R_0(t)$, the position of the peak of the transverse field profile $R_\varepsilon(t)$, and the longitudinal field $\mathcal{E}_{||}(t)$. For stationary filaments, $R_0 = R_\varepsilon = R_0^*$ holds by (5.6), but, for time-dependent filaments, $R_0(t)$ and $R_\varepsilon(t)$ must in general be allowed to vary independently because of the inertia of dielectric relaxation.

We use the ansatz

$$n(R, t) = n^*(R - R_0(t) + R_0^*) \,, \tag{5.7}$$

$$\mathcal{E}_\perp(R, t) = \mathcal{E}_\perp^*(R - R_\varepsilon(t) + R_0^*) \,, \tag{5.8}$$

where $\mathcal{E}_\perp^*$ is given by (5.6) with the approximation $D/\mu \simeq k_B T_L/e$. The spatial degrees of freedom may be projected out in (5.1–2) by multiplying with a suitable weight function G and integrating over $\int_0^\infty dR$. The choice of G is guided by

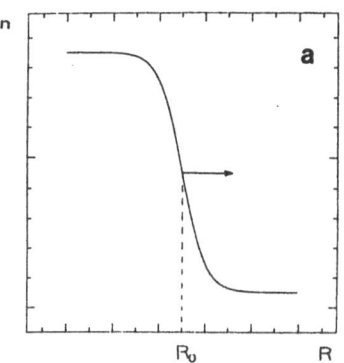
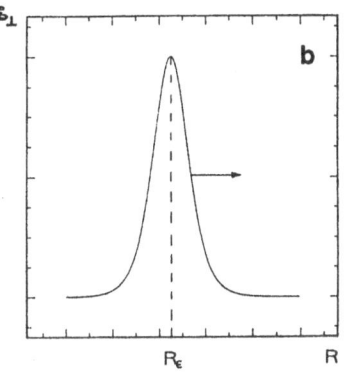

Fig. 7a,b. Breathing current filaments: Radial profile of (a) carrier density, (b) transverse electric field (schematic)

the requirement that it should be sharply peaked at the filament wall in order to project out the *relevant* contribution of the profile, and moreover G should allow for analytical evaluation of some of the integrals occurring in (5.1) without explicit use of $n^*(R)$. A detailed analysis with realistic parameters [60] shows that for Ge the transverse field profile is adiabatically enslaved by $R_0(t)$, and therefore we confine attention to $R_e = R_0$ here. With $G = n^{*\prime}(R - R_0 + R_0^*)$ we obtain from (5.1) approximately [60]

$$\dot{R}_0(t) = \frac{k_B T_L}{e} \mu(\mathcal{E}_{||}) \left[\frac{B_q(\mathcal{E}_{||})}{R_0^*(\mathcal{E}_{||})} - \frac{B_0(R_0)}{R_0} \right] \ , \tag{5.9}$$

where $B_0(R_0)$ and $B_q(R_0^*(\mathcal{E}_{||}))$ are functions which incorporate the vanishing boundary conditions at $R_0 = \triangle R$ and $R_0^* = 0$, respectively, and are unity in the bulk.

Integrating (5.3) over the cross-section A of the current flow and connecting a load R_L and an external capacitance C as in (4.20) gives

$$\dot{\mathcal{E}}_{||} = [J_0 - (\sigma_L + e\bar{n}(R_o)\mu(\mathcal{E}_{||}))\mathcal{E}_{||}]/[\varepsilon_0\varepsilon_S(1 + C/C_i)] \tag{5.10}$$

with

$$\bar{n}(R_0) = \frac{1}{A} \int_A n(R,t)\mathrm{d}f \simeq n_1 + (n_3 - n_1)\pi R_0^2/A \ . \tag{5.11}$$

Equations (5.9,10) represent a nonlinear dynamic system, which may be regarded as an expansion of the partial differential equations (5.1,3) in terms of the "breathing" mode.

Self-generated oscillations induced by periodically breathing current filaments correspond to limit cycles in the $(R_0, \mathcal{E}_{||})$ phase space. The condition for a Hopf bifurcation of such a limit cycle can be obtained analytically by linearizing (5.9,10) around the state $R_0 = R_0^*$, $\mathcal{E}_{||} = \mathcal{E}_0$. This is a special case of the general conditions discussed in Sect. 4.3. A Hopf bifurcation occurs for sufficiently large load resistance R_L and negative differential conductance with numerical parameters for p-Ge and n-GaAs [60]. The obtained limit cycle for p-Ge is shown in Fig. 8.

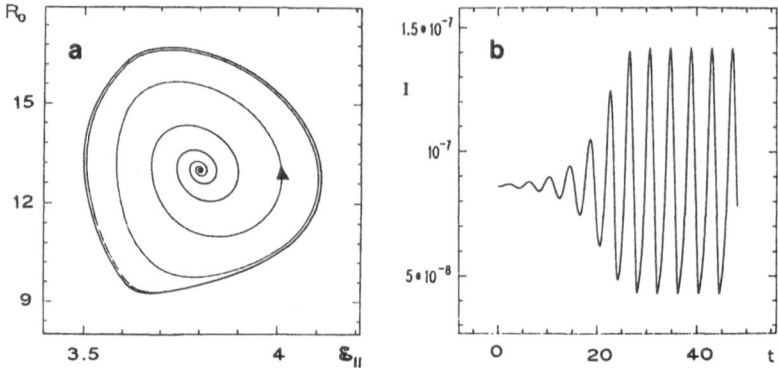

Fig. 8a,b. Phase portrait and time series of a limit cycle corresponding to a breathing current filament for p-Ge at 4K. (a) Filament radius R_0 (in μm) versus longitudinal field $\mathcal{E}_{||}$ (in V/cm). (b) Current (a.u.) versus time (in μs) for $C/C_i = 1000$

The oscillatory mechanism is based upon two features:
(i) An instability of the filament radius $R_0(t)$, such that R_0 tends to increase further if R_0 is above a critical radius $R_0^\star(\mathcal{E}_{||})$, and to decrease for $R_0 < R_0^\star(\mathcal{E}_{||})$. Thus $R_0^\star$ is analogous to the critical droplet radius (*critical nucleus*) in equilibrium or nonequilibrium phase transitions [1]. The microscopic mechanism in our case is impact ionization of impurities, but this enters only implicitly into the dynamics through the function $R_0^\star(\mathcal{E}_{||})$. Any other autocatalytic mechanism which yields a *decreasing* function $R_0^\star(\mathcal{E}_{||})$ will furnish similar results.
(ii) A restoring force, which is provided by dielectric relaxation of the longitudinal electric field $\mathcal{E}_{||}$. It is essentially controlled by the average carrier density $\bar{n}(R_0)$ which forces $\mathcal{E}_{||}$ to decrease with increasing R_0.

The physical mechanism underlying the breathing oscillations is thus similar to the spatially homogeneous model for dielectric relaxation oscillations [28]. The two models are simple approximations of two different modes of dielectric relaxation oscillations [1]: breathing or bulk-dominated, respectively, as sketched in the inset of Fig. 6b. Both types of oscillation are predicted to occur on the falling branch of the (filamentary or uniform, respectively) current–voltage characteristic, generated by Hopf bifurcations with decreasing applied current. This agrees well with the experiments by RAU et al. [61] who found – with increasing current – two different types of oscillation in different regimes of the falling current–voltage characteristic, associated with large-amplitude circuit-limited (CLO = bulk-dominated) modes and small-amplitude structure-limited (SLO = breathing) modes with lower frequency, followed by a regime of stable stationary current filaments. It appears that this behavior can be consistently explained by a crossover between the filamentary and the uniform NDC branches within the same model I–V characteristic, as shown in Fig. 6(b). H denotes the Hopf bifurcation of breathing filaments. If for larger applied currents I_0 the behavior is governed by the filamentary branch A–H–B, while at B a crossover occurs to the homogeneous unstable branch B–C, all four regimes of the static and dynamic behavior can be understood within a single consistent model: With increasing J_0, a stable homogeneous low-conductivity state up to the onset of breakdown (O–C), bulk-dominated limit-cycle oscillations (C–B), a breathing current filament (B–H), and a stable stationary current filament (H–A) subsequently occur.

The oscillation frequency can be scaled by the effective dielectric relaxation time

$$\tau_M = \varepsilon_0 \varepsilon_S (1 + C/C_i)/[e\mu_0(N_A - N_D)] , \tag{5.12}$$

which is proportional to the external circuit capacitance C for $C \gg C_i$. Thus the very slow observed frequencies might be explained by including a sufficiently large C.

6. Conclusions

The present survey has shown that current instabilities in semiconductors represent a fascinating example of a nonlinear dynamic system. Some understanding of the observed dynamic behavior can already be gained from simple physical models which are derived from hot-electron transport theory. A comprehensive quantitative derivation from a microscopic level, however, is not yet available. But the examples discussed are intended to demonstrate that the modern tools of nonlinear dynamics can be usefully applied to the intrinsically nonlinear semiconductor transport equations. Open problems to be investigated in the future include the role of a magnetic field [62], realistic boundary conditions, contact effects, nonlinear phenomena in low-dimensional geometries, and fully developed solid-state turbulence.

Acknowledgements

I would like to thank K. Aoki, D. Drasdo, G. Hüpper, D. Jäger, J. Parisi, J. Peinke, and U. Rau for discussions.

1. E. Schöll: *Nonequilibrium Phase Transitions in Semiconductors* (Springer, Berlin 1987)
2. H. Haken: *Synergetics, An Introduction*, 3rd ed. (Springer, Berlin 1983)
3. G. Nicolis, I. Prigogine: *Self-Organization in Non-Equilibrium Systems* (Wiley, New York 1977)
4. W. Ebeling, R. Feistel: *Physik der Selbstorganisation und Evolution* (Akademie-Verlag, Berlin 1982)
5. A.F. Volkov, Sh.M Kogan: Sov.Phys.Usp. 11, 881 (1969)
6. H. Thomas: In *Synergetics*, ed. by H. Haken (Teubner, Stuttgart 1973) p. 87
7. V.L. Bonch-Bruevich, I.P. Zvyagin, A.G. Mironov: *Domain Electrical Instabilities in Semiconductors* (Consultant Bureau, New York 1975)
8. M.P. Shaw, H.L. Grubin, P. Solomon: *The Gunn-Hilsum Effect* (Academic Press, New York 1979)
9. J. Pozhela: *Plasma and Current Instabilities in Semiconductors* (Pergamon, Oxford 1981)
10. E. Schöll: Appl. Phys. **A48**, 95 (1989)
11. E. Schöll: Physica Scripta **T29**, 152 (1989)
12. E. Schöll: Solid-State Electron. **32**, 1129 (1989)
13. L. Reggiani, and V.V. Mitin: Riv. Nuovo Cimento 12, N.11,1 (1989)
14. J. Shah, G.J. Iafrate (eds.): Proc. 5th Int. Conf. on Hot Carriers in Semiconductors, Solid-State Electron. **31**, 319–820 (1988)
15. Y. Abe (ed.): "Special Issue on Nonlinear and Chaotic Transport in Semiconductors". Appl. Phys. A, Vol. **48**, pp. 93–192 (1989)
16. L. Akers (ed.): Proc. 6th Int. Conf. on Hot Carriers in Semiconductors, Solid-State Electron. **32**, 1051–1923 (1989)

17. J. Peinke: see this volume
18. D. Jäger: see this volume
19. M.P. Shaw, V.V. Mitin, E. Schöll and H.L. Grubin: *The Physics of Instabilities in Solid State Electron Devices.* (Plenum Press, New York 1992)
20. M. Asche, V.M. Ivashenko, V.V. Mitin: Z. Phys. B **59**, 265 (1985)
21. Z.S. Gribnikov: Sov. Phys.-Semicond. **6**, 1204 (1973)
    K. Hess, H. Morkoç, H. Shichijo, B.G. Streetman: Appl. Phys. Lett. **35**, 469 (1979)
22. K. Aoki, K. Yamamoto, N. Mugibayashi and E. Schöll: Solid-State Electron. **32**, 1149 (1989);
    E. Schöll, K. Aoki: Appl. Phys. Lett. **58**, 1277 (1991)
23. G. Maracas et al.: Solid-State Electron. **32**, 1887 (1989)
24. W. Knap, M. Jeżewski, J. Lusakowski, W. Kuszko: Solid-State Electron. **31**, 813 (1988)
25. E. Schöll, P.T. Landsberg: Proc. 14th Int. Conf. Physics of Semiconductors, Edinburgh 1978, ed. by B.L.H. Wilson (Institute of Physics, Bristol 1979) p. 461
26. E. Schöll: Proc. Roy. Soc. **A365**, 511 (1979)
    E. Schöll, P.T. Landsberg: Proc. Roy Soc. **A365**, 495 (1979)
27. E. Schöll: Z. Phys. **B48**, 153 (1982)
28. E. Schöll: Physica **134B**, 271 (1985); Phys. Rev. **B34**, 1395 (1986)
29. M.A. Lampert, P. Mark: *Current Injection in Solids* (Academic, New York 1970)
30. B.S. Kerner, V.V. Osipov: Sov. Phys. JETP **52**, 112 (1980)
31. A.L. Dubitskij, B.S. Kerner, V.V. Osipov: Sov. Phys. Semicond. **20**, 755 (1986)
32. G. Hüpper, E. Schöll, L. Reggiani: Solid-State Electron. **32**, 1787 (1989)
33. P.T. Landsberg, E. Schöll, P. Shukla: Physica D **30**, 235 (1988)
34. M. Rieger, P. Vogl: Solid-State Electron. **32**, 1399 (1989)
35. J.R. Barker: In *Physics of Non-linear Transport in Semiconductors*, ed. by D.K. Ferry, J.R. Barker, C. Jacoboni (Plenum, New York 1980)
36. G. Baccarani, M. Rudan, R. Guerrieri, and P. Ciampolini: In *Advances in CAD for VLSI*, Vol. 1, ed. by W. Engl (North-Holland, Amsterdam 1986) pp. 107–158
37. E. Schöll, W. Quade: J. Phys. C**20**, L861 (1987)
38. N.G. van Kampen: Phys. Reports **124**, 69 (1985)
39. S.W. Teitsworth, R.M. Westervelt: Phys. Rev. Lett. **53**, 2587 (1984)
40. K. Aoki, N. Mugibayashi, K. Yamamoto: Phys. Scripta **T14**, 7 (1986)
41. K. Piragas, Yu. Pozhela, A. Tamshyavichyus, and Yu. Ulbikas: Sov. Phys. Semicond. **21**, 335 (1987)
42. E. Schöll: Proc. 18th Int. Conf. Physics of Semiconductors Stockholm 1986, ed. by O. Engström (World Scientific, Singapore 1987), pp. 1555–1558
43. E. Schöll: Solid-State Electron. **31**, 539 (1988)
44. E. Schöll, H. Naber, J. Parisi, B. Röhricht, J. Peinke, S. Uba: Z. Naturforsch. **44a**, 1139 (1989)
45. H. Naber, E. Schöll: Z. Phys. B **78**, 305 (1990)
46. E. Schöll, J. Parisi, B. Röhricht, J. Peinke, R.P. Huebener: Phys. Lett. **A119**, 419 (1987)
47. U. Rau, J. Peinke, J. Parisi, R.P. Huebener, and E. Schöll: Phys. Lett. **124**, 335 (1987)
48. H. Naber, E. Schöll: Z. Phys. B **78**, 301 (1990)
49. F. Schlögl, E. Schöll: Z. Phys. B **71**, 231 (1988)
50. G. Hüpper: Master Thesis, RWTH Aachen (1989)
51. K. Aoki, K. Yamamoto: Phys. Lett. **98A**, 72 (1983)
52. K.M. Mayer, J. Parisi, R.P. Huebener: Z. Phys. B **71**, 171 (1988)
53. A. Brandl, M. Völcker, W. Prettl: Appl. Phys. Lett. **55**, 238 (1989)
54. D. Jäger, H. Baumann, R. Symanczyk: Phys. Lett. **A117**, 141 (1986)
    H. Baumann, R. Symanczyk, C. Radehaus, H.G. Purwins, D. Jäger: Phys. Lett. **123**, 421 (1987); R. Symanczyk, D. Jäger, E. Schöll: Appl. Phys. Lett. **59**, 105 (1991)
55. H.G. Purwins, G. Klempt, J. Berkemeier: Festkörperprobleme **27**, 27 (1987)
56. K.M. Mayer, J. Parisi, U. Rau, R.P. Huebener: Proc. 19th Int. Conf. Phys. of Semicond. Warsaw 1988, ed. by M. Grynberg (Pol. Acad. Sciences, Warsaw 1988) p. 1411
57. E. Schöll: "Instabilities in Semiconductors: Domains, Filaments, Chaos", in *Festkörperprobleme* (Advances in Solid-State Physics), ed. by P. Grosse (Vieweg, Braunschweig 1986) Vol. 26, 309–333
58. E. Schöll: Solid-State Electron. **29**, 687 (1986)
59. E. Schöll, P.T. Landsberg: Z. Phys. **B72**, 515 (1988)
60. E. Schöll: Proc. Int. Conf. Phys. of Semicond. Warsaw 1988, ed. by M. Grynberg (Pol. Acad. Sciences, Warsaw 1988) p. 1407
    E. Schöll, D. Drasdo: Z. Phys. **B81**, 183 (1990)
61. U. Rau, K.M. Mayer, J. Parisi, J. Peinke, W. Clauss, and R.P. Huebner: Solid-State Electron. **32**, 1365 (1989)
62. G. Hüpper, E. Schöll: Phys. Rev. Lett. **66**, 2372 (1991)

Current Instabilites in the Interplay Between Chaos and Semiconductor Physics

J. Peinke

C.N.R.S.-Centre de Recherches sur les Très Basses Températures,
25, avenue des Martyrs, 166X-Centre de Tri,
F-38042 Grenoble Cedex, France

1. Introduction

To get a theoretical understanding of experimentally observed current instabilities in a semiconductor system, the experimentalist will first of all be guided by the desire to explain the whole system by a single appropriate model. In such an approach, one would like to start with elementary semiconductor physics and end up with a comprehensive understanding, without requiring the experimentalist to exclude any experimental facts from his mind. This article intends to show how far it is possible to explain the instabilities on the basis of semiconductor physics, and to point out where this becomes impossible with present knowledge. In contrast to the ansatz based on semiconductor physics, different models based on nonlinear dynamics are presented that explain and predict nonlinear features of current instabilities.

Experimental systems requiring for their explanation two different models are common in science. This may remind the reader of a dualistic principle. I would not dare to proclaim a new type of dualism for semiconductor experiments, but I am convinced that only with such an approach can the true character of a whole set of nonlinear phenomena in semiconductors, and more generally in solid-state physics, be obtained. This problem may have its origin in the different qualities of a microscopic world, modeled by semiconductor or solid-state physics, and a macroscopic world, composed of an infinite number of microscopic elements requiring nonlinear models. This transition from a microscopic to a macroscopic level has to be seen in close relation to the concepts of self-organization and synergetics [1].

Using these ideas as the main theme, this article is organized as follows. In Sect. 2, an introduction and a short historic survey of research done in the field of instabilities in semiconductors are given. One intention of this survey is to show that there are many different semiconductor systems which can be seen in the light of semiconductor physics and nonlinear dynamics. This implies the generality of the problem mentioned above. Another intention is to provide the interested reader with references giving access to this research field. In Sect. 3, one specific exemplary system is presented, and the observed instabilities of this system are discussed on the basis of semiconductor physics. Finally, in Sect. 4 the behavior of the same system is taken as a "black box" providing signals which can be used as a testing ground for nonlinear dynamics. It is important to note that even as a "black box" this semiconductor is a piece of nature. But at the same time it may obey scaling behavior predicted by nonlinear dynamics, a product of a mathematical artificial world. In this sense, universalities of the theory of nonlinear dynamics can be verified by the experiment.

2. Survey of Instabilities in Semiconductors

Before specific semiconductor instabilities are presented, I want to start with some considerations of principle. The nonlinearities and instabilities discussed here are based on electric charge transport. This transport is described by the local "Ohm's law":

$$j(\mathcal{E}) = \sigma(\mathcal{E})\mathcal{E} \ , \tag{1}$$

where j is the current density, σ the electric conductivity, and $\mathcal{E}$ the electric field. Nonlinearities are due to the field dependence of the conductivity. In the simples case, the conductivity σ can be written as the product of the elementary charge q, the density n, and the mobility μ of the charge carriers (for simplicity, only one type of charge carrier is considered),

$$\sigma = qn\mu \ . \tag{2}$$

Nonlinearities may be based on the dependence of n and μ on the electric field.

So far, the description of charge transport holds for any material. Next, I shall discuss why nonlinearities occur with a greater likelihood in semiconductors than in other materials. Insulators are characterized by a very small number of mobile charge carriers so that hardly any charge transport takes place. Applying high electric fields, charge transport may start, but this usually leads to a destruction of the material owing to a breakdown. The main reason why insulators are not really suitable for investigation can be seen in the high binding energies of the charge carriers. This is different in semiconductors, where the binding energies come down to the range of thermal energies of practicable temperatures, i.e., between meV and eV. Charge carriers in semiconductors arise from the thermal excitation of bound carriers, and their density is typically in the range of 10^{12}–10^{15} carriers per cm^3, depending on the concentration and the kinds of impurity. In contrast, metals typically have 10^{22} mobile charge carriers per cm^3, and charge transport is based on this high density of charge carriers moving relatively slowly. In semiconductors, relatively few charge carriers are very mobile (with mobilities up to 10^6 cm^2/Vs at low temperatures). This high mobility and the resulting high velocity of charge carriers in semiconductors leads easily to nonequilibrium states suitable for exhibiting nonlinearities. (Here it is common to speak of hot charge carriers.)

On the basis of these simplified arguments, it becomes clear that nonlinearities are likely to occur in semiconductors. These nonlinearities may be reflected in the integral current–voltage (I–V) characteristics. Incidentally, this effect is used to construct various semiconductor devices. A further consequence of these nonlinearities may be the occurrence of instabilities leading to self-organized electric structures in a semiconductor. For example, an inhomogeneous current distribution in a homogeneous material, like a current filament, is a spatial structure, whereas a spontaneous oscillation is a temporal structure.

Investigations in this field began at the end of the 1950s with the work of IVANOV and RYVKIN [2]. During the 1960s, both experimental and theoretical research activities on semiconductor instabilities reached a peak in their intensity as can be concluded from the huge number of papers written in that decade. Access to these works can be found in the Proceedings of the International Conferences on Semiconductor Physics (since the middle of the 1960s called the International Conferences on the Physics of Semiconductors), and in [3–8]. The main interest was attracted by the occurrence of the so-called current instabilities. Semiconductors exhibit, under certain conditions which are kept temporarily constant, spontaneously alternating electric signals. Since the frequencies of these current instabilities range even to some GHz, they became important for developing semiconductor devices, e.g. the Gunn oscillator. But it became more and more difficult to classify all current instabilities and even the following classification was suggested. (1) observed but incomprehensible, (2) comprehensible but not observed, (3) comprehensible and observed. Major progress in our understanding was made with the introduction of the plasma approach, which describes the instabilities as a collective response of the charge carriers to a perturbation. Thus not only was a better understanding of the instabilities of semiconductors achieved, but simultaneously a better understanding of plasma physics was obtained. It appeared that semiconductors are very suitable for studying the plasma effect and are at the same time very cheap compared to thermonuclear experiments [7].

But even with the plasma ansatz it was not possible to reach a comprehensive understanding of these instabilities. Thus, the activities declined at the beginning of the 1970s. In my opinion, effects which are nowadays attributed to nonlinear dynamics confused researchers in those days. If one looks carefully through the publications of the 1960s, one finds many indications of typical nonlinear phenomena. In Fig. 1 one example is shown to underline this statement. This measurement presents an observed spontaneous oscillation where, owing to the chosen time scale, only the variation of the amplitude can be seen.

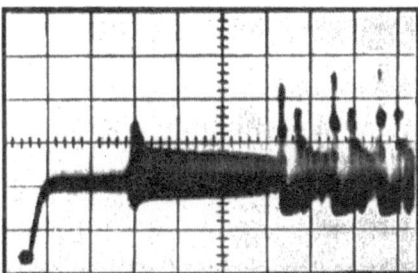

Fig. 1. Oscillations in p-Ge at room temperature during application of an electric field pulse. Horizontal scale 2 μs/div. Vertical scale: 4 mA/div. (After [9]).

The amplitude does not stay constant as expected for a periodic oscillation, but varies as is nowadays known from intermittency. Thus it is no surprise that the advances in the theory of nonlinear dynamical systems gave a new impetus to the research of instabilities in semiconductors in the 1980s. During the last five years, it has become clear that current instabilities in semiconductors represent an interesting and highly productive area for investigating nonlinear dynamics experimentally.

A recent survey of these activities is given in a special issue of Journal of Applied Physics [10]. A list of published articles can be found in [11]. It is important to note that the majority of semiconductor systems investigated from the point of view of nonlinear dynamics displays an avalanche breakdown of shallow impurities. This avalanche breakdown can be regarded as a nonequilibrium phase transition [8] from a low conducting state to a high conducting state. It is well known that spontaneous structure formation is expected in such situations [1]. A central point of this avalanche breakdown is its autocatalytic nature, (see the next section): free charge carriers produce further free charge carriers. This process, which may be thought of as positive feedback, is a good candidate for causing an instability.

There is another important aspect under which these instabilities in semiconductors can be classified. Here it can be asked to what degree these instabilities are self-generated, that is, to what degree these instabilities are dependent on external working conditions. It is evident that certain external parameters must be set to drive the system into an unstable state. Thus, a power supply is always required because of this dissipative nature. But sometimes other, special external parameters are required, such as a strong longitudinal magnetic field. In some cases even periodically changing control parameters are essential to find special nonlinear dynamics. In such case, one has a periodically driven system, and it may be the experimentalist who creates the nonlinear dynamics. In the extreme case of an analog computer, nonlinear phenomena are trivially expected if these phenomena are known from the simulated equations and if the analog computer works. In contrast to this are those experiments where only constant external parameters are applied, and where all structures are generated by the semiconductor alone. The external capacitors and resistors have no fundamental influence. These remarks conclude the more general discussion of current instabilities.

3. Spatial and Temporal Self-generated Structures in an Exemplary System

In the following, the electric avalanche breakdown of a homogeneous p-germanium sample at low temperatures serves as an exemplary system which will be discussed more extensively. As material, p-germanium with an acceptor concentration of about 10^{14} cm^{-3} and with a compensation ratio smaller than 10^{-2} was used. From this material, samples with typical dimensions of some

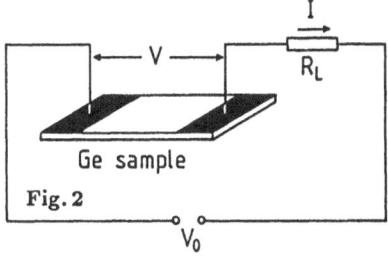

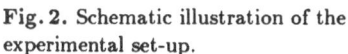

Fig. 2

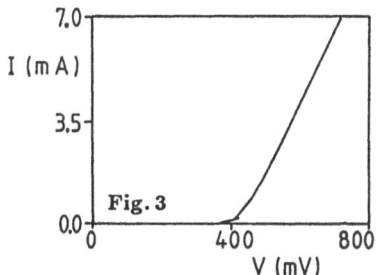

Fig. 3

Fig. 2. Schematic illustration of the experimental set-up.

Fig. 3. Current–voltage characteristic of a 0.9 mm long sample at $T = 4.2$ K, $B = 6.4$ G, and $R_L = 1.0$ k Ω.

mm were made. To investigate the electric bulk features, ohmic contacts (p^+ regions) were placed on the surface. A typical arrangement is shown in Fig. 2. The sample had a thickness of 0.2 mm while the contact distance was some mm. The black areas correspond to the contact regions. As contact material, either aluminium was evaporated and subsequently alloyed, or boron was ion-implanted. Further details of the characterization of the prepared samples are given in [12].

Such a sample was connected over a load resistor to a dc voltage source and cooled down to liquid-helium temperatures by placing the sample directly into the liquid helium. By means of vapor–pressure regulation, the temperature could be varied between 1.7 and 4.2 K. Since at these temperatures the samples showed a very high sensitivity to external irradiation (visible, far infrared), a metal shield, kept at liquid-helium temperature, was used for shielding, to make sure that apart from the electrical energy due to the bias voltage no further energy contribution was imposed on the sample.

As a further variable external parameter, a magnetic field was used. It was oriented perpendicular to the broad side of the sample, i.e. perpendicular to the electric field inside the sample (see Fig. 2). Thus, for the experimental set-up, the following external parameters were available: the temperature T ranging from 1.7 to 4.2 K, the bias voltage V_0 ranging from 0 to 20 V, the load resistor R_L ranging from 1 to $10^7 \Omega$, and the magnetic field B ranging from 0 to 10^4 G. All parameters were kept temporarily as constant as possible – typically in the relative range 10^{-4}–10^{-5}.

With such an experimental system, a current–voltage (I-V) characteristic could easily be measured. A typical form is shown in Fig. 3. Up to nearly 400mV, no current could be detected with this resolution; the actual current value is in the nA range. At a threshold value of about 400 mV, the current increases abruptly by several orders of magnitude to values in the mA range. This is the onset of electric breakdown. Note the low power dissipation of less than 1 mW required to achieve a breakdown in this system. The threshold value corresponds to an electric field value of about 4.5 V/cm. It is important to point out that this field strength remains unchanged if the sample length is larger than 0.1 mm [13]. Thus, this electric breakdown is a field effect which can be explained by local bulk features of the material.

For the explanation of this breakdown let us remember the basic transport equations (1) and (2). The breakdown may be either due to an increase of the density of charge carriers n (here labeled as n although holes are present), or due to an increase of the mobility μ. From Hall measurements [14] it becomes clear that the breakdown is mainly due to an increase of n, while μ only changes within one order of magnitude. Thus, a first approach to explain the breakdown just takes the dependence of n on temperature T and electric field $\mathcal{E}$ into account.

Firstly, with decreasing temperature (thermal energy) the probability of free charge carriers decreases. In particular, if the thermal energy becomes smaller than their binding energy to the impurities, n decreases rapidly. This process is called a freeze-out of the charge carriers. In the above-mentioned p-germanium, the binding energy to the acceptors is 10 meV, and the freeze-out sets in at about 10 K. In the picture of generation and recombination processes described by rate equations [8], the freeze out can be understood as a decreasing generation coefficient of trapped charge carriers, while the recombination rate of the free charge carriers is assumed to remain constant.

Secondly, an electric field may invert the temperature effect in such a way that the last few free charge carriers are accelerated and thus gain kinetic energy. The energy gain increases with increasing electric field if the mean life time of the charge carriers remains constant (corresponding to constant mobility μ). When the kinetic energy gain surpasses the ionization energy of trapped charge carriers, an avalanche-like multiplication (of the charge carriers) takes place. This effect is called an avalanche breakdown due to impurity impact ionization. In terms of rate equations [8], this corresponds to an increase of the impact ionization coefficient such that the total generation rate becomes larger than the recombination rate of the free charge carriers. In linear approximation, the generation rate $\dot{n}$ is proportional to n (now with a positive coefficient) leading to an exponential growth. So far, the electric breakdown can be explained as a pure field effect.

Investigating the I-V characteristic in the breakdown region more carefully, some fine structures appear; see Fig. 4. If a sufficiently high load resistor is used, even negative differential parts in the I-V characteristic are found, which require a modification of the breakdown model. On the basis of generation and recombination processes, the occurrence of a negative differential conductivity may be explained by the presence of an excited trapped state of the charge carriers [8]. That impurities in p-Ge trap the free charge carriers not only in the ground state but also in excited states is shown by photoconductivity measurements in [12].

In Fig. 5, a calculated current-density versus electric-field (j-$\mathcal{E}$) characteristic is shown. The region of negative differential conductivity ($\mathrm{d}j/\mathrm{d}\mathcal{E} < 0$ is indicated by a broken curve. From stability arguments, it follows directly that the negative differential part of the characteristic is unstable against fluctuations. Comparing this j-$\mathcal{E}$ characteristic with the measured I-V characteristic,

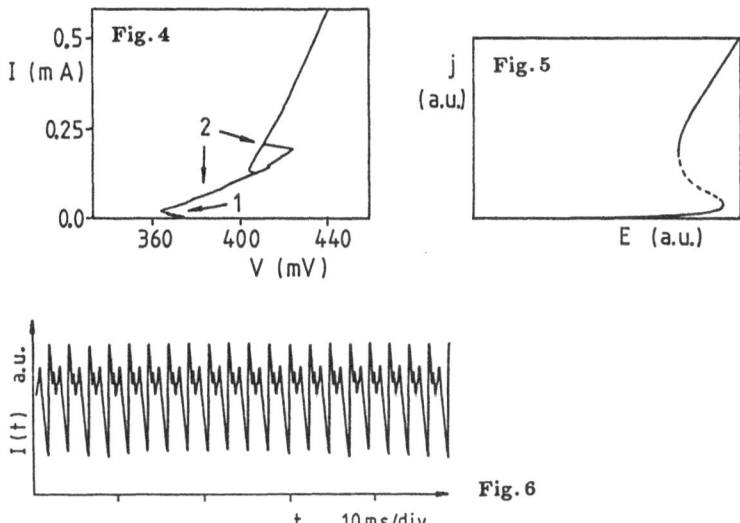

Fig. 4. Blow-up of the I–V characteristic of Fig. 3.

Fig. 5. Calculated current-density–electric-field characteristic on the basis of a generation-recombination process involving excited trapped states. The unstable negative differential part is shown by a broken line. (After [8]).

Fig. 6. Temporal structure of a spontaneous current oscillation in the post-breakdown region (sample-length was 5 mm, $I = 3.2004$ mA, $T_C = 1.98$ K, $B = 1.5$ G).

apart from the appearance of negative differential parts, no real accordance can be found. One reason for the difference between these two characteristics lies in the fact that the measured I–V characteristic is obtained by time and space averages of the sample current I and the sample voltage V.

A time-resolved measurement shows that in the whole breakdown region spontaneous current and voltage oscillations are present, although the bias voltage V_0 was kept constant in time. In the first negative differential part of the characteristic marked "1" in Fig. 4, spontaneous oscillations of quite large amplitude occur (up to 100 mV along a 1 MΩ load resistor and consequently along the sample, too). The nature of these oscillations is a firing and quenching of the breakdown. The time constant of this oscillation is mainly determined by the capacity of the electric set-up, leading to the name circuit-limited oscillation [15]. At the end of this first negative differential part in the characteristic these switching oscillations disappear, and a new type of oscillation occurs with different frequency and amplitude. In Fig. 6, a periodic form of this oscillation is shown. The frequency is typically in the kHz region and independent of the external electric set-up. The amplitude typically ranges in the μA region. While for the circuit-limited oscillations the current signal corresponds to switching the breakdown on and off, now the current oscillation is just an offset in the permille region of the flowing current. (Here the time-averaged current is a very good control parameter.) This second type of oscillation is linked to a spatial

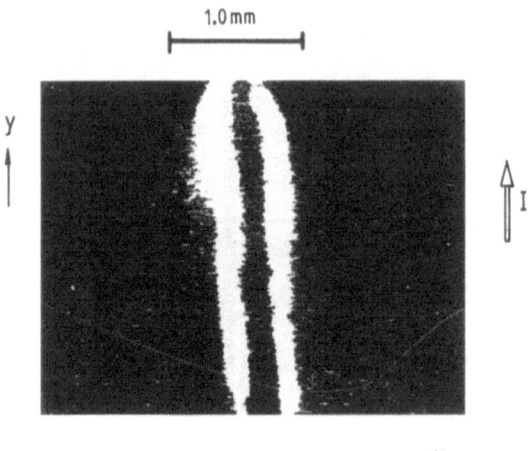

Fig. 7. Brightness-modulated image of a current filament in a p-Ge sample ($I = 50\mu$A, $T = 4.2$ K, $B = 0$ G). The dark area between the two bright regions corresponds to a filament.

inhomogeneity of the flowing current, namely, the occurrence of a current filament. Thus, this inhomogeneity is known as structure-limited oscillation [15]. The region where this phenomenon occurs is marked by "2" in the characteristic in Fig. 4.

With the help of a low-temperature scanning electron microscope [16], it is possible to detect the current distribution within a semiconductor. From this we know that after the first negative differential part in the I–V characteristic (see Fig. 4) a stable current filament is present. This transition from a circuit-limited oscillation to the appearance of a stable current filament with structure-limited oscillations seems to be closely related to the principle of a minimal radius for a stable current filament [17]. In Fig. 7, such a current filament is shown. The bright regions reflect the boundary of the current filament. For further details see [18].

With the electron scanning microscope it was also possible to detect the location of oscillation centers [18,19]. These observations show that the structure-limited oscillations are located at the boundary of a current filament, leading to the picture of a breathing current filament. This is in line with the relation between the amplitude of the oscillation and the value of the time-averaged current.

These experimental findings of self-generated structures can be explained by rate equations of generation and recombination processes including an excited trapped state. If a semiconductor is driven into the unstable part of the j–$\mathcal{E}$ characteristic (see Fig. 5), then current oscillations and structure formation of current filaments become possible [8]. Furthermore, SCHÖLL [20] showed that the current filaments exhibit temporal instabilities like breathing. So far, it seems that the experimental findings are in good agreement with the con-

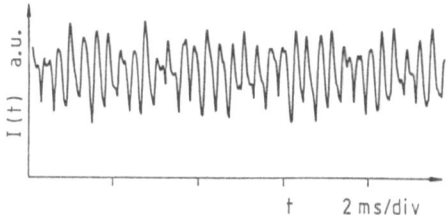

Fig. 8. Temporal structure of an irregular spontaneous current oscillation in the post-breakdown region (sample-length was 5 mm, $I = 8.3248$ mA, $T = 4.2$ K, $B = 20.1$ G).

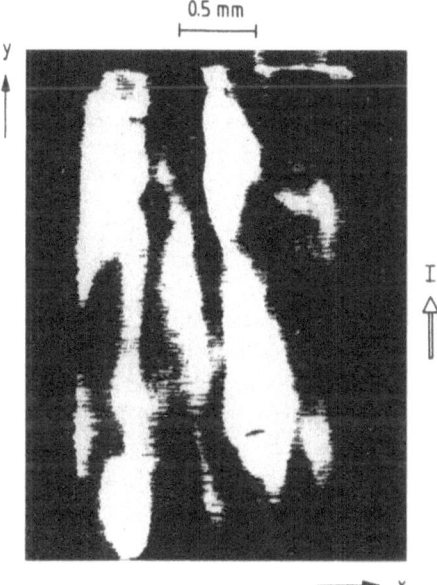

Fig. 9. Image of current filaments in analogy to Fig. 7. $R_L = 1\Omega$, $V_0 = 2.00$ V, $T = 4.2$ K, $B = 0$ G. (After [12]).

sequences of the generation–recombination model developed by SCHÖLL [8]. Well, there are some discrepancies between experiment and theory. The frequencies of the structure-limited oscillation are too small compared with the theoretically expected values. The signals of these oscillations are not exclusively periodic but show a wide variety of different forms; see for example Fig. 8. Minute changes in the control parameters may cause large changes in the form of the oscillation. Along a single characteristic, hundreds of different oscillation modes could be found. Changes in the external magnetic field in the range of only 0.1 Gauss may also affect the oscillation perceptibly. Noisy and very periodic oscillations alternate with each other. Also the form of the current filament changes drastically with the control parameter; see Fig. 9.

For the current filaments, it was possible to adjust the sample shapes and the control parameters in such a way that nice and simple filamentary structures were detected. The same was not possible for the structure-limited oscillations. The form of these oscillations changed from sample to sample. Up to now, no elementary oscillating cell showing only a simple form of oscillation was found. All experiments showed that these oscillations always display many nonlinear effects. Thus, today it appears to be hopeless to try to get a comprehensive understanding of this type of oscillation from semiconductor theory. To conclude this section, I should point out that it has been shown that semiconductor theory explains in principle the existence of an instability in electric breakdown leading to spontaneous current oscillations and to the formation of filaments. But the complex forms especially of the current oscillations cannot be explained in detail. In the next section, a discussion on the basis of nonlinear dynamics is presented, showing that these different oscillation forms are not totally accidental but obey the rules on nonlinear dynamics.

4. Nonlinear Dynamics of the Structure-Limited Oscillations

In this section, the structure-limited oscillations are discussed from the viewpoint of nonlinear dynamics. It is a remarkable feature of this experimental system that by very small changes of the control parameters, a whole zoo of different dynamical states can be found [21]. Here, two exemplary states are picked out. First, a transition from a nonoscillatory state (fixed-point attractor) to a periodic oscillatory state (limit-cycle attractor) is discussed. Second, the characterization of a chaotic state by various numerical methods is given. Further details of these two nonlinear phenomena may be found respectively in [22,23]. The emergence of an oscillatory state is picked out to illustrate that nonlinear dynamics is not merely chaos. There also occur very interesting nonlinear phenomena in the nonchaotic states. It is even easier to investigate the nonchaotic dynamics, because here well-established methods (linear analysis), like the measurement of power spectra, are sufficient to characterize a phenomenon. On the other hand, a chaotic signal is presented to demonstrate some methods, giving new insight into the mechanism causing chaos. Before the development of modern nonlinear dynamics, chaotic states were just classified as noisy states.

To begin with, the onset of a periodic oscillation is shown in Fig. 10 by a series of four different dynamical states obtained for a slight increase in the time-averaged current. In the upper trace the time signals are shown. In Fig. 10a no oscillation is seen, while in Fig. 10b a spontaneous oscillation with some irregularities (interruptions) is presented. Figures 10c and d show periodic oscillations with decreasing period. The corresponding phase portraits (middle trace) have been reconstructed by the time-delay method [24]. The power spectra of the lower trace reflect the frequencies of these oscillations. One remarkable feature of this bifurcation (transition) is the unchanging form of the phase portrait and the unchanging amplitude of the oscillation, whereas the frequency changes a lot.

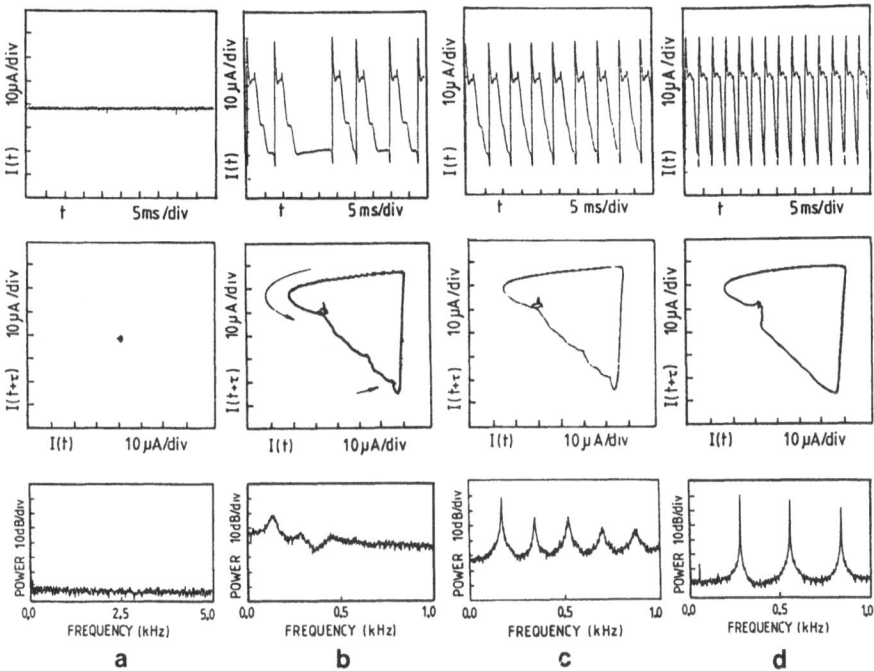

Fig. 10a-d. Time signals (*upper trace*), phase portraits (*middle trace*), and power spectra (*lower trace*) of the alternating current $I(t)$ obtained for the control parameters (a) $I = 3.000$ mA; (b) $I = 2.8638$ mA, (c) $I = 2.8581$ mA; (d) $I = 2.8080$ mA, while $B = -0.33$ G and $T = 1.97$ K were kept constant.

The best-known bifurcation from a fixed point to a limit cycle is the Hopf bifurcation. Here the amplitude grows from zero to a finite value, while the frequency stays constant in the lowest-order approach [25]. Thus, the experimental findings described above are in contrast to the Hopf bifurcation. It is the saddle-node bifurcation on a limit cycle [26] which explains this experiment. This bifurcation can be pictured as a bowl running in a valley around a hill (like a Mexican hat). The oscillation is hindered by a growing saddle in the valley. When the energy of the bowl becomes too small to pass the saddle, the bowl comes to rest. From this picture it becomes clear that this bifurcation predicts a constant amplitude, while the frequency slows down to zero. The idea of such a bifurcation can also be expressed by a differential equation [26]:

$$\dot{r} = r - r^3$$
$$\dot{\theta} = \mu - \cos\theta \ , \tag{3}$$

where r denotes the radius and θ the angle, and μ is a control parameter. The equation for r just gives a constant solution $r = 1$. The whole bifurcation is expressed by the angle variable. Owing to the polar coordinates, the simplest nonlinear term in θ is $\cos\theta$. For $\mu < 1$, a constant θ value solves (3). When μ becomes larger than 1, a time-constant solution of the angle equation no longer

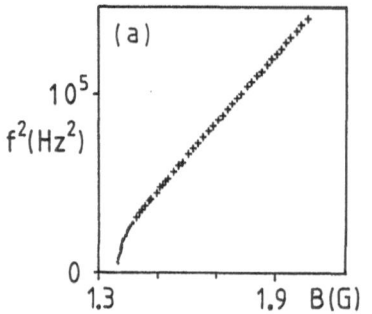
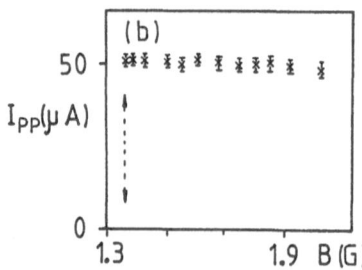

Fig. 11a,b. Square of the frequency (a) and peak-to-peak amplitude I_{PP} (b) as a function of the magnetic field B. Constant control parameters $I = 3.1010$ mA, $T = 1.97$ K. The arrow marks the bifurcation point. (After [22]).

exists. Thus, θ starts growing, leading to an oscillation. The integration of the angle equation over 2π yields the period and the frequency. As a characteristic feature of this bifurcation, a square-root scaling of the frequency with the control parameter is obtained [26]. In Fig. 11, the scaling of the frequency and of the amplitude is shown as a function of the magnetic field. (In contrast to Fig. 10, now the magnetic field is changed, causing the same behavior.)

In Fig. 11a, the square of the frequency is plotted as a function of the control parameter. Except for the lowest 15 experimental points marked by dots, all other frequency data accurately follow the square-root law. For this presentation the statistical examination gave a correlation coefficient of 0.9998. Figure 11b shows clearly that the amplitude remains unchanged under the variation of the control parameter.

Very close to the bifurcation point, a departure from the predicted behavior was observed, although the best accordance is predicted here [26]. Simultaneously with the departure from the square-root scaling, irregularities in the time signals were detected. Instead of a periodic oscillation, a stochastic spiking of single oscillation events is seen, as shown in Fig. 10b. The spiking rate becomes smaller and smaller as the control parameter approaches the bifurcation point. The frequency obtained from these oscillations corresponds to the average spiking rate. As a consequence of stochasticity, the error bars of the frequency increased from less than 5 Hz in the nonstochastic case up to at most 50 Hz close to the bifurcation point. There are two suggestive mechanisms which might lead to the observed behavior. First, near the bifurcation point the system might be sensitively influenced by noise. For clarity, we looked at the influence of control-parameter noise on the generic saddle-node bifurcation by computer simulation. In this way, however, the observed departure of the frequency scaling near the bifurcation point could not be remodeled. Second, hitherto suppressed variables may now have a growing influence. Thus, the observed dynamics might be the result of some underlying deterministic chaotic process. It should be emphasized that similar spiking behavior of single oscillatory events has been reported recently for different experimetal systems.

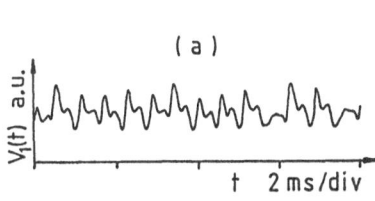
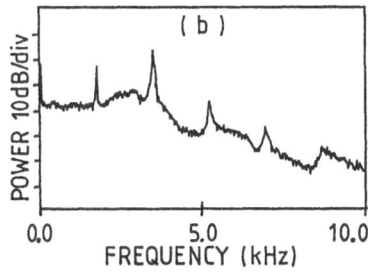

Fig. 12a,b. Temporal structure (a) and corresponding power spectrum (b) of a spontaneous voltage oscillation over a part of a semiconductor sample for the control parameter. $V_0 = 2.145\,\text{V}$, $R_L = 100\,\Omega$, $B = 31.5$ G, $T = 4.2$ K. (After [23]).

In the case of a laser experiment, these oscillation forms could be ascribed to Shilnikov-type chaos [27]. However, there is definitely no evidence of Shilnikov-type chaotic behavior in the present semiconductor experiment [22]. Note that the departure of the predicted features of this bifurcation was only found in the frequency behavior, while the shape of the phase portrait and the amplitude of the oscillation did not show any discrepancy. Thus, it can be concluded that the experimental system exhibits a saddle-node bifurcation on a limit cycle.

So far, one bifurcation from a fixed point to a limit cycle was characterized by one universal bifurcation. As a further case, a chaotic dynamics is discussed. In the investigation of bifurcations leading to chaos, also known as routes to chaos, similar methods can be used like those characterizing the bifurcation from a fixed point to a limit cycle. But for chaotic dynamics, new methods of characterization are required. To get a quantitative characterization of chaos, the signals have to be digitalized and CPU-time-consuming numerical calculations have to be done. In the following, the result of such a characterization is presented for the signal of Fig. 12a. The power spectrum of this signal, shown in Fig. 12b, shows a somewhat noisy state, but no other information can be detected. Thus a question arises about the origin of this noise. The answer is that this noise is due to a nonlinear interaction of a few degrees of freedom causing deterministic chaos. This result is based on numerical analysis [23].

To begin with, the form of the attractor has to be reconstructed from the time signal. For this purpose, new time series can be created by using a time shift by an arbitrary time constant τ. Thus, signals $x(t)$, $x(t+\tau)$, $x(t+2\tau) \ldots$ are obtained. The attractor is now reconstructed by plotting these signals versus each other [24]. This attractor can be analyzed with respect to its geometric and dynamic features [28]. To evaluate the geometry of the attractor the trajectories of the attractor are transformed to points by a stroboscopic presentation. Thus, a point distribution in space is given. As a next step, a quality Q of these points is defined, for instance the number of neighboring points. The next question is how this quality scales with an increasing diameter l of a neighborhood at any point on the attractor. This question is evaluated under the following scaling assumption: $Q(l) \sim l^{d_f}$. The scaling index d_f is called the fractal dimension;

note that d_f does not have to be an integer. Obviously for different qualities different fractal dimensions may be obtained. The evaluation of the capacity dimension and the information dimension, labeled $D(0)$ and $D(1)$, gave for the signal of Fig. 12 [23]:

$$D(0) = 2.6 \pm 0.1 \ ,$$
$$D(1) = 2.5 \pm 0.1 \ .$$

Here the algorithm proposed by BADII and POLITI [29] was used. These dimensions are noninteger and are larger than two. This shows that the attractor is fractal and that the underlying dynamics must be due to at least three independent variables.

Besides these geometrical features, the dynamical ones are of special interest in characterizing chaos. Here, instead of the distribution of the points on an attractor, the evolution of nearby trajectories on the attractor is considered. One remarkable feature is the exponential separation of neighboring initial conditions under their evolution on the attractor (running along trajectories), which causes the sensitive dependence of the system on initial conditions, a main feature of chaos leading to disorder and noise. A quantitative measure for this separation is the Lyapunov exponent λ. These exponents are based on the following scaling hypothesis: there are distinct directions on the attractor where the distance of neighboring points Δ scales like $\Delta \sim e^{\lambda \bullet t}$ in the time average. Thus, the sensitive dependence on initial conditions corresponds to the existence of finite and non-negative Lyapunov exponents. Depending on the order of chaos, there are one or more positive Lyapunov exponents [30]. Using an algorithm of STOOP, descibed in [30], the following Lyapunov exponents (per arbitrary time units) were evaluated for the experimental signal corresponding to Fig. 12 [23]:

$$\lambda_1 = 0.095 \pm 0.005 \ ,$$
$$\lambda_2 = 0.003 \pm 0.005 \ ,$$
$$\lambda_3 = -0.72 \pm 0.02 \ .$$

These three different Lyapunov exponents reflect the following structure of the dynamics on the investigated attractor. There is one eigendirection which corresponds to $\lambda_1 > 0$ and displays the exponential divergence of neighboring trajectories. The direction corresponding to the zero-valued Lyapunov exponent is nothing but the direction of the flow on the attractor. Neighboring points on one trajectory cannot separate in the time average if the system is autonomous. The negative Lyapunov exponent marks a direction where the trajectories come closer. This is due to a folding back, leading to an attractor of finite size.

Up to now, two independent methods have been presented, both showing that the underlying dynamics of the experimental signal is chaotic. For the numerical investigation of time signals it is important that a consistent picture is obtained by different methods. This will be discussed next.

So far, we have distinguished between geometrical and dynamical aspects. But because of the construction of the attractor by the time-delay method, some dynamical aspects have got into the "geometric" dimension evaluation. A point taken at time t_i in phase space has the components $(x(t_i), x(t_i + \tau), x(t_i + 2\tau) \ldots)$. Thus, this point contains some information of its own temporal development in the time steps τ, 2τ ... From this it can be shown that simultaneously with the evaluation of the fractal dimensions the entropies $K(0)$, $K(1)$ can be obtained [32].

Roughly speaking, these entropies correspond to the sum over all positive Lyapunov exponents. Consequently, a positive entropy indicates chaos. Furthermore it is shown that $K(1)$ should be smaller or equal to the sum over the positive Lyapunov exponents [28]. The evaluation for our signal gave for $K(0)$ and $K(1)$ 0.09 ± 0.01. This is in very good agreement with the evaluation of the Lyapunov exponents.

As a last point, it should be mentioned that it is possible to calculate from the Lyapunov exponents a dimension [33]. If the Lyapunov exponents are ordered in such a way that $\lambda_j > \lambda_{j+1}$ (for $j = 1, 2, \ldots$), a dimension can be defined as follows:

$$D = j + \left(\sum_{i=1}^{j} \lambda_i \right) \Big/ \lambda_{j+1} \ .$$

Here the integer j is defined by the requirement that the sum over λ_i from λ_1 to λ_j is positive while the sum from λ_1 to λ_{j+1} is negative. For the evaluated Lyapunov exponents, a dimension of 2.1 ± 0.1 is obtained, again fitting well with the other results.

To conclude, the numerical analysis of an experimental signal obtained in the breakdown regime yielded several quantities proving an underlying chaotic dynamics. In addition to this, a consistent picture of these quantities was obtained. Thus, these results have become reliable. But is should be pointed out that such an evaluation is very time consuming. A lot of problems arise when one comes to the details; see [34]. There is still no algorithm available which can be applied in a straightforward manner to experimental signals. In this respect the "old" methods like FFT have big advantages. Because of this, a characterization of chaos by bifurcations is quite often much easier [25].

5. Conclusion

In the present work two aspects of spontaneously arising structure formation have been presented. On the one hand, it was shown to what degree the formation of the current filament and the spontaneous oscillation can be explained. On the other hand, it was shown that a comprehensive understanding of the oscillations cannot be achieved without the nonlinear dynamics. So far, it has become obvious that these two methods complement each other. But there was also a fruitful interaction between the field of nonlinear dynamics and semiconductor physics. Thus, it was essential that, right from the beginning, we looked at this system from these two perspectives. For example, as one result of nonlinear dynamics, the question arose how the system changes its numbers of independent variables by minute variations of the control parameter. Quite soon it became obvious that a more and more complicated generation-recombination process can never be the truth. Moreover, the picture of turbulence in fluids leads us to the idea that a spatial interaction between different regions of the semiconductor, organized by spontaneous symmetry breaking, is a good candidate explanation of this system [1,35]. This idea could be verified experimentally [36] and initiated new theoretical models on the basis of semiconductor physics [37].

Well, I want to end this lecture with some speculations for the future. It may become possible to achieve such an understanding of the self-generated nonlinear dynamics in semiconductor systems that it becomes possible to prepare samples in such a way that they show predicted nonlinear phenomena. Here, the self-generated dynamics becomes important to guarantee that the phenomena are caused by the semiconductor itself. Then it will be possible to make measurements of the highest precision thanks to the extreme sensitivity of nonlinear phenomena. This may lead to the best measurements of the properties of the semiconductor, like the effective mass and scattering cross-section.

Acknowledgements

The main progress in understanding the semiconductor system was achieved by a fruitful and intensive cooperation of the "Semiconductor-Chaos Group" in Tübingen, of which I was a member. Thus, I am very grateful for the cooperation of W. Clauß, R.P. Hübener, K.M. Mayer, J. Parisi, B. Röhricht, O.E. Rössler, R. Richter, U. Rau, E. Schöll, and R. Stoop. Furthermore, I want to thank the Stiftung Volkswagenwerk for financial support.

References

1. H. Haken: *Advanced Synergetics*, in Springer Ser. Syn., Vol. 20 (Springer, Berlin 1983); I. Prigogine, R. Lefever:"Theory of Dissipative Structures", in *Synergetics*, ed. by H. Haken (Teubner, Stuttgart 1973) pp. 124–135
2. I.L. Ivanov, S.M. Ryvkin: Sov. Phys. Techn. Phys. **3**, 722 (1958)
3. B. Ancker-Johnson: "Plasmas in Semiconductors and Semimetals"., in *Semicond. and Semimetals*, Vol. 1, ed. by R.K. Willardson and A.C. Beer (Academic Press, New York 1966) pp. 379–481
4. H. Hartnagel: *Semiconductor Plasma Instabilities* (Heinemann Educational Books, London 1969)
5. M. Glicksman: "Plasma in Solids", in *Solid State Phys.*, Vol. 26, ed. by H. Ehrenreich, F. Seitz, and D. Turnbell (Academic Press, New York 1971) pp. 275–427
6. V.L. Bonch-Bruevich, I.P. Zvyagin, A.G. Mironov: *Domain Electrical Instabilities in Semiconductors* (Consultants Bureau, New York 1975)
7. J. Pozhela: *Plasma and Current Instabilities in Semiconductors* (Pergamon, Oxford 1981)
8. E. Schöll: *Nonequilibrium Phase Transition in Semiconductors*, Springer Ser. Syn., Vol. 35 (Springer, Berlin, Heidelberg 1987)

9. M. Cardona, W. Ruppel: Jour. Appl. Phys. **31**, 1826 (1967)
10. Y. Abe (ed.): *Nonlinear and Chaotic Transport Phenomena in Semiconductors*, Applied Physics, Vol. A48, No. 2 (Springer, Heidelberg 1989)
11. R.P. Huebener, J. Peinke, J. Parisi: Appl. Phys. **A48**, 107 (1989)
12. J. Peinke, J. Parisi, B. Röhricht, K.M. Mayer, U. Rau, W. Clauß, R.P. Huebener, G. Jungwirt, W. Prettl: Appl. Phys. **A48**, 155 (1989)
13. W. Clauß, U. Rau, J. Parisi, J. Peinke, R.P. Huebener, H. Leier, A. Forchel: J. Appl. Phys. **67**, 2980 (1990)
14. J. Parisi, U. Rau, J. Peinke, K.M. Mayer: Z. Phys. **B72**, 225 (1988)
15. U. Rau, K.M. Mayer, J. Parisi, J. Peinke, W. Clauß, R.P. Huebener: Proc. 6th Int. Conf. on Hot Carriers in Semicond., Scottsdale 1989; Solid State Electron. **32**, 1365 (1989)
16. R.P. Huebener: Rep. Prog. Phys. **47**, 175 (1984)
17. K.M. Mayer, R.P. Huebener, U. Rau: J. Appl. Phys. **67**, 1412 (1990)
18. K.M. Mayer, R. Gross, J. Parisi, J. Peinke, R.P. Huebener: Solid State Commun. **63**, 55 (1987)
19. K.M. Mayer, J. Parisi, J. Peinke, R.P. Huebener: Physica D**32**, 306 (1988)
20. E. Schöll: Proc. 19th Int. Conf. Phys. of Semicond., Warsaw 1988, ed. by W. Zawadzki; Inst. of Physics, Polish Academy of Sciences (1988)
21. J. Peinke, J. Parisi, U. Rau, W. Clauß, B. Röhricht, R.P. Huebener, R. Stoop: "Experimental Progress on the Ladder Towards Higher Chaos", in *A Chaotic Hierarchy*, ed. by M. Klein and G. Baier (World Scientific, Singapore 1991) pp. 317–340
22. J. Peinke, U. Rau, W. Clauß, R. Richter, J. Parisi: Europhys. Lett. **9**, 743 (1989)
23. R. Stoop, J. Peinke, J. Parisi, B. Röhricht, R.P. Huebener: Physica D**35**, 425 (1989)
24. F. Takens: "Detecting Strange Attractors in Turbulence", in *Dynamical Systems and Turbulence*, ed. by D.A. Rand and L.-S. Young, Springer Lect. Notes in Math., Vol. 898 (Springer, New York 1981) pp. 366–381; N.H. Packard, J.P. Crutchfield, J.D. Farmer, and R.S. Shaw: Phys. Rev. Lett. **45**, 712 (1980)
25. H.G. Schuster: *Deterministic Chaos* (Physik Verlag, Weinheim, 1984)
26. H.B. Stewart: Z. Naturforsch. **41a**, 1412 (1986)
27. F.T. Arecchi, A. Lapucci, R. Meucci, J.A. Roversi, P.H. Coullet: Europhys. Lett. **6**, 677 (1988)
28. J.-P. Eckmann, D. Ruelle: Rev. Mod. Phys. **57**, 617 (1985)
29. R. Badii, A. Politi: J. Stat. Phys. **40**, 725 (1985)
30. O.E. Rössler: Z. Naturforsch. **38a**, 788 (1983)
31. R. Stoop, P.F. Meier: J. Opt. Soc. Am. **B5**, 1037 (1988); R. Stoop, J. Parisi, J. Peinke: "Lyapunov Exponent Calculations in High Dimensional Embedding Space, a Singular Value Approach", in *A Chaotic Hierarchy*, ed. by M. Klein and G. Baier (World Scientific, Singapore 1991) pp. 341-352
32. P. Grassberger: in *Chaos*, ed. by A.V. Holden (Manchester Univ. Press, Manchester 1986) p. 291ff;
    K. Pawelzik, H.G. Schuster: Phys. Rev. **A35**, 2207 (1987)
33. J.L. Kaplan, J.A. Yorke: in *Lecture Notes in Mathematics*, Vol. 730, ed. by H.O. Peitgen and H.O. Walther (Springer, Berlin 1979) p. 204
34. G. Mayer-Kress (ed.): *Dimensions and Entropies in Chaotic Systems*, in Springer Ser. Syn., Vol. 32 (Springer, Berlin 1986)
35 J. Peinke, B. Röhricht, A. Mühlbach, J. Parisi, Ch. Nöldeke, R.P. Huebener, O.E. Rössler: Z. Naturforsch. **40a**, 562 (1985)
36 J. Peinke, A. Mühlbach, B. Röhricht, B. Wessley, J. Mannhart, J. Parisi, R.P. Huebener: Physica **23D**, 176 (1986);
    J. Parisi, J. Peinke, B. Röhricht, K.M. Mayer: Z. Naturforsch. **42a**, 329 (1987);
    J. Peinke, J. Parisi, B. Röhricht, B. Wessley, K.M. Mayer: Z. Naturforsch. **42a**, 841 (1987);
    U. Rau, J. Peinke, J. Parisi, R.P. Huebener: Z. Phys. **B71**, 305 (1988)
37. E. Schöll, J. Parisi, B. Röhricht, J. Peinke, R.P. Huebener: Phys. Lett. **119A**, 419 (1987);
    E. Schöll, H. Naber, J. Parisi, B. Röhricht, J. Peinke, S. Uba: Z. Naturforsch. **44a**, 1139 (1989)

Current Density Filaments in Semiconductor Devices

D. Jäger and R. Symanczyk

Institut für Angewandte Physik, Universität Münster,
Corrensstrasse 2-4, W-4400 Münster, Fed. Rep. of Germany
Present Address:
Fachgebiet Optoelektronik, Universität -GH- Duisburg,
Kommandantenstr. 60, W-4100 Duisburg, Fed. Rep. of Germany

1. Introduction

In recent years the study of nonlinear dynamic systems has become a major topic in different areas of physics, chemistry and biology [1]. Undoubtedly, the interest was further stimulated by an increasing number of experimental observations of characteristic instabilities, such as pattern formation, self-oscillation phenomena, bistability and chaos (see other parts of this book).

Today, semiconductor devices play a keyrole as qualified physical systems because they are readily available or can easily be prepared, and because the theoretical modelling is greatly advanced. This is due to the fact that special device structures are already of utmost technological importance or are very promising components for future information technology.

The origin of electrical instabilities in semiconductor materials is commonly traced back to negative differential conductivity (NDC) where different physical mechanisms may be responsible, see for example [2]-[4]. From a phenomenological point of view one can distinguish between two basic types of current density (j) vs. electrical field (E) characteristics, see Fig. 1. In the case of N-type j–E characteristics (Fig. 1a) one easily sees that for a given current density j_0, different electrical fields can coexist inside the device. In regions where the electrical field is E_0, instabilities will occur because the NDC will lead to a negative dielectric relaxation. As a result, spatial fluctuations of the charge carrier density are amplified, and electrical field domains are generated where the contacts and the external circuit determine the boundary conditions. In case of an impressed voltage, for example, the sum of the potential drops over the low-field regions, E_l, and the high-field regions, E_h, must be a constant.

The dual case of these field domains are the current density filaments for an S-type characteristic as shown in Fig. 1b. Similar arguments lead to the conclusion that for a given electrical field E_0, different regions of low current density, j_l, and high current density, j_h, can coexist, and that the state with j_0 is unstable. Two further basic properties of devices as shown in Fig. 1 should be noted at this point. First, the spatial structures generated by the NDC mechanism may change in time or even move through the device so that electrical oscillations can occur. Second, from the bistable behaviour of the

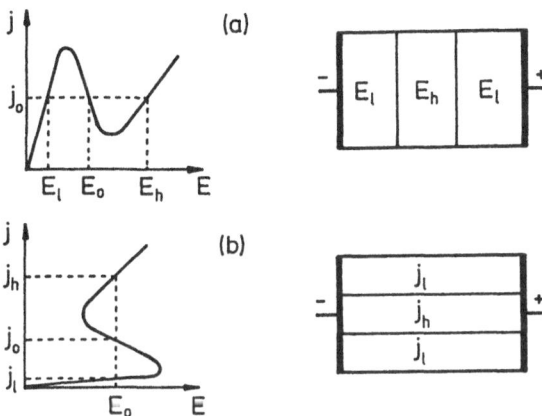

Fig. 1: Generation of electrical field domains (a) and current density filaments (b) in semiconductors with N-type or S-type j–E characteristics, respectively.

characteristics it is obvious that the device can exhibit switching properties. It should also be pointed out at this place that due to pattern formation the current - voltage relations of the devices in Fig. 1 will differ significantly from the j–E characteristics.

In the past, the physics of electrical field domains (Fig. 1a) was investigated thoroughly because the generation of microwaves in the famous Gunn-Diode [5] can be traced back to an N-type j–E characteristic based upon a negative differential mobility, for example in GaAs. By way of comparison, however, there is only little knowledge about the properties of current density filaments. From the theoretical viewpoint one reason for this situation can shurely be found in the at least two-dimensional nature of this kind of pattern formation. Additionally, the characteristic time constants of available devices with S-type NDC behaviour are in the range of 10 ns at best, which is of less technical significance. Breakdown phenomena on the other hand, are usually attributed to contact effects instead of material properties, so that the ideas of Fig. 1b may not be applicable.

At present, several open questions concerning the S-type NDC materials and the corresponding devices are discussed. For example, there are still uncertainties about the physical mechanism and the microscopic picture leading to a j–E characteristic as in Fig. 1b. Furtheron, the generation of electrical structures as well as their stability or instability have yet to be analysed, and the exact definition of what a filament really is has to be given. Detailed experimental results are also needed, together with a theoretical description of the spatio-temperal structures and the global current - voltage characteristic, for example. It is also an open question whether filaments of the kind as shown in Fig. 1b exist at all and how big the influences of contact effects are. Moreover, it is expected that the investigations on current density filaments in semiconductors can lead to novel device concepts for technical applications.

In the following, we review typical experimental results on filamentary current flow in semiconductor devices. Special emphasis is then laid upon silicon pin diodes, where double injection of electrons and holes lead to the formation of filaments at room temperature and where different experimental techniques have been employed. Finally, we will discuss very recent preliminary steps in the direction of a theoretical description of pattern formation in pin diodes due to current density filaments.

Other devices showing filamentation are discussed by PEINKE in another part of this book. The investigated material is p-Ge and n-GaAs, and the NDC behaviour is due to impact ionisation of shallow impurities at low temperatures. Here, the current flow is based on one type of charge carriers only.

2. Review of filamentary device studies

The experimental observation of filamentary phenomena in semiconductor devices dates back more than 30 years. In 1955, for example, silicon pn contacts were operated in the reverse direction to breakdown, and small spots of light about 10 μm or less in diameter were detected in the junction region [6]. Clearly, the intensity of the emitted light is a measure for the recombination rate and therefore also for the density of free charge carriers, provided that the material properties are sufficiently uniform. Similar experiments were performed in 1956 [7] where the recombination radiation was observed directly through the contact area. A photograph of the light emission shows a two-dimensional arrangement of more than 1000 spots, which is exactly reproducible with respect to the position and time sequence when the current is changed.

In the following years, a large amount of experimental work was published aiming towards the detection of electrical inhomogeneities perpendicular to the direction of current flow. In Table 1 a variety of results are summarized with emphasis on characteristic experimental methods and device structures. For example, pn diodes on GaAs, InP and GaP [10,11] were also analysed by observation of the recombination radiation due to the formation of a microplasma by avalanche breakdown. In [9] S-type current (I) - voltage (U) characteristics are also shown. In 1966 filamentary current flow in semi-insulating silicon diodes was studied [12]. The underlying mechanism was theoretically attributed to double injection phenomena in pin diodes, and the analytical results were compared with measurements. GaAs double injection devices were also investigated in detail [14]. Hysteretic I–U characteristics were recorded, and the results were compared with theoretical predictions. Again, light spots were observed through the contacts, but for the first time also from the side, i.e. perpendicular to the contact plane. This was an important step because the light emission is now spatially resolved between the contacts, which can provide at least a qualitative insight into the structure of a filament, cf. Fig 1b. Further similar investigations on pnpn structures [17,18] have pointed out that switching in these four-layer devices may be accompanied by a nonuniform distribution

Table 1: Review on experimental filament observations in semiconductor devices. *) This column shows the orientation of the observation: = means parallel and ⊥ perpendicular to the contact plane.

Exp. Method	Device	Material	*)	Result	Ref.
recombination radiation	pn diode	Si	=	multiple spot-like light emission from the contact number or Ø increasing with current	6-9
		GaAs	=		10
		InP	=		10
		GaP	=		11
	pin diode	Si:In	=		12
		GaAs:O	=,⊥		13,14
		GaAs:Cr	=,⊥		14
	pnpn	Si	=,⊥		17,18
	thin layer	a-Si	=		19
		SOS	⊥		20
		n-GaAs	⊥		21
optical beam induced current (OBIC)	pin diode	Si:Au	⊥	filament (-wall) high photosensitivity, influence of magn. field	15
		Ge:Au	⊥		15
	thin layer	n-GaAs	⊥		24
optical transmission	thin layer	SOS	⊥	hose-like filaments with temperature increase	22
		VO$_2$	⊥		26,27
electron beam induced current (EBIC)	thin layer	n-GaAs	⊥	filament (-wall) high EBIC-sensitivity, hose-like filaments	23
		VO$_2$	⊥		28
	bulk	p-Ge	⊥		25
potential probe	pin diode	Si:Au	=	influence of magn. field	16

of the current density. This is on the other side a clear drawback for the turn-off properties of technical power thyristors [18]. Formation of filaments was also detected in evaporated silicon films [19], in silicon-on-sapphire (SOS) diodes [20] and in thin epitaxial layers of GaAs [21] by measuring recombination radiation. Another technique which was used to study nonuniform properties inside a device is called the OBIC (optical beam induced current) method. In this case an optical beam is scanned across the side of the sample between the contacts [15,24]. The photoinduced current is now a measure of the local photosensitivity, and the observed inhomogeneity is said to be a result of a filament. Very interesting results on the spatial structures have recently been found in GaAs epitaxial layers where also the influence of a magnetic field has been analysed [24]. Filaments in thin VO$_2$ films can directly be discovered as dark areas where the optical transmission seems to be largely reduced [26,27]. A clear strip-shaped pattern is found between the contacts indicating a long and narrow region of increased temperature.

The EBIC method (electron beam induced current) is similar to the OBIC technique except that a focussed electron beam is used to generate free carriers [23,25,28]. Undoubtly, this method was proven to be an excellent tool for the study of inhomogeneities in different materials. Hose-like and strip-shaped patterns were observed in p-Ge [20] similar to those in VO_2 [26]-[28]. A final technique which is included in Table 1 uses a metallic probe to measure the surface potential of a semiconductor sample with an inhomogeneous current distribution [16]. By the action of a magnetic field the filament has been deformed, and oscillation phenomena were generated.

From the examples listed in Table 1 it is obvious that very different methods have been applied up to now to study nonuniform properties due to current instabilities in semiconductor devices. There are experiments showing pattern formation in the contact plane or between the contacts. Clearly, there is no direct measurement of the current density. On the other side, the EBIC technique offers a very sensitive method with high spatial resolution, but it is not clear which physical parameter is ultimately detected. Large internal electrical fields in the filament walls, for example, could lead to a high local sensitivity indicating that the edges of a filament may be formed by space-charge layers. As can be seen, further systematic measurements are needed in order to get a clear view of the properties of a filament and to give an answer to the above questions. Nevertheless, in the special case of pin diodes, a series of different techniques has been applied to the same samples in order to get information about various physical parameters. The experimental results are given in the following chapter.

3. Investigations with Si pin diodes

To study current density filaments, silicon pin diodes have been choosen because the microscopic theory of such devices is available (see below). Moreover, they are easily fabricated by common semiconductor technology. The preparation steps from the Si wafer up to the finished device are explained briefly in the first part of this Section, followed by a description of the measured I–U characteristics, which are determined by the formation of filaments. In the last part of this Section we present some experimental methods which permit a spatially and temporally resolved observation, and discuss the obtained results concerning the structure and the properties of filaments.

3.1 Technology

The pin diodes are fabricated from commercial phosporus-doped Si wafers with resistivities $\rho = 10...100\,\Omega cm$. The steps up to the finished device are particular standard processes as described in more detail for example in [29]. First, the wafer is lapped, polished and cleaned with organic solvents to get a proper surface. To remove non-organic pollution, a thin layer is then oxidized with hot HNO_3 and subsequently etched with HF. The high resistivity 'i'-layer is now made by compensation with gold, that means by producing deep trap levels in the forbidden gap. For that purpose Au is evaporated on the wafer in a high vacuum ($p \approx 10^{-3}\,Pa$) system and a conventional diffusion is carried out

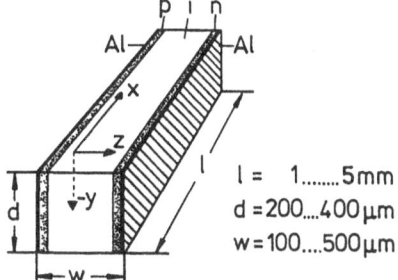

Fig. 2: Structure of a pin diode with coordinate-system and typical dimensions. Thickness of the n^+- and p^+-layer $\leq 1\,\mu m$.

$l = 1........5\,mm$
$d = 200....400\,\mu m$
$w = 100....500\,\mu m$

($T \approx 1000^\circ C$) afterwards. After removing the gold on the surface with aqua regia, a phosporus containing liquid dopand solution is spun on one side of the wafer and the n^+-layer is formed by a repeated diffusion process. The p^+-layer is formed either by using a boron containing liquid dopand solution in the same manner on the other side of the wafer, or by evaporating Al and alloying the layer at $600^\circ C$. Finally, Al is evaporated to provide metallic contacts. A sandwich diode is cut out and the four surfaces are polished. The schematic structure of such a pin diode and the typical dimensions are sketched in Fig. 2.

3.2 Current - voltage characteristics

First hints to filamentation can be obtained from the course of the I–U characteristic. A dc-voltage supply is applied to the device via a load resistor ($R \approx 1...10\,k\Omega$) in a way that the diode is forward biased. A typical I–U characteristic is shown in Fig. 3 where U and I denote the dc-parts of the voltage across the diode and the current, respectively, even in the case of temporal instabilities (see below). As can be seen, a pronounced multistability is observed, where different values of the current are measured at the same voltage level, here at about $15\,V$. The different states are clearly separated by jumps when particular threshold values are reached.

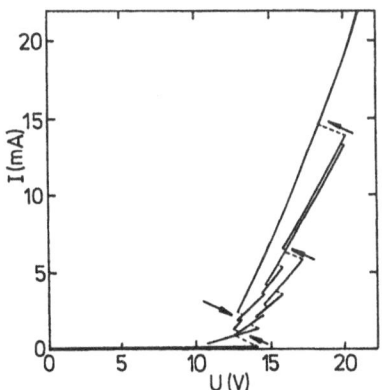

Fig. 3: Typical I–U characteristic of a Si:Au pin diode.

73

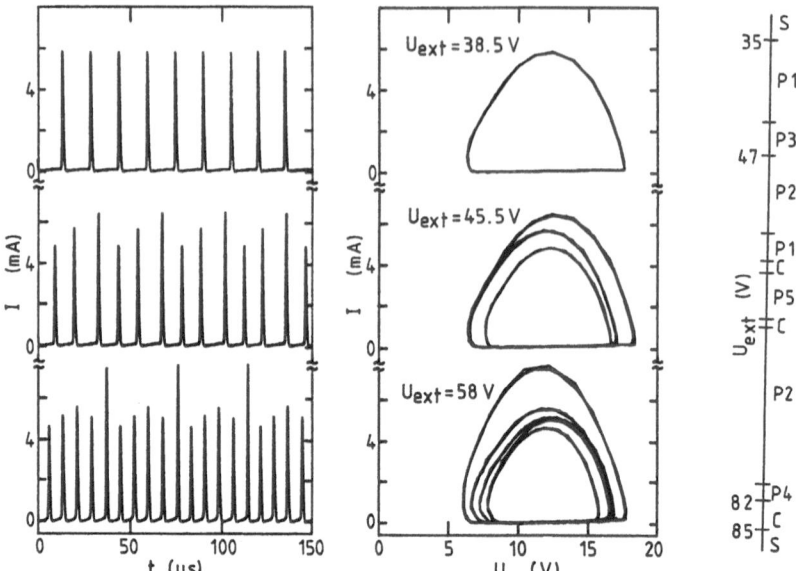

Fig. 4: Measured selfgenerated oscillations of the current and the voltage. The graphs on the left hand side show the current vs. time at different values of the external voltage. The curves on the right hand side show the corresponding phase portraits.

Moreover, in the vicinity of these jumps self-generated oscillations of the current and the voltage can often be observed. Although an experimental proof is still lacking, these oscillations are supposed to be connected with instabilities of the filaments [4,30]. As an example, in Fig. 4 periodic oscillations are plotted as have been measured close to the first jump in the I–U characteristic of a Si:Au pin diode. As a further experimental result, the temporal period of the oscillations is varied if the external voltage U_{ext}, i.e. the voltage across the diode plus the load resistor, is changed. The rightmost graph shows the different modes as a function of U_{ext}, where C means chaotic oscillations, S a stationary characteristic and P1,...,P5 a periodic behaviour. For a further study of the chaotic behaviour, we have computed the Fourier spectra of the current vs. time signals at the transition from periodic to non-periodic oscillations. They show either an increase in noise and a spreading of the frequency peaks, or the generation of additional peaks with a non-integral frequency ratio. Plots of the (n+1) th maximum vs. the n th maximum of the current in these cases form a return map which shows a clear correlation of successive maxima, which reveals a deterministic chaotic behaviour.

In the following we will explain some experimental methods which permit the investigation of stable filaments as well as effects in connection with oscillations of current and voltage.

3.3 Experimental observation of filament formation

Most of the experimental methods allowing spatially resolved detection of electrical properties in semiconductor devices are limited to surface measurements. We have found that the lateral extension of our Si pin diodes in y-direction is small (see Fig. 2) and comparable to the diameter of a filament as will be shown below. Therefore any variation of the material properties due to filament formation is detectable even on the surface of the sample.

The first method is based on the idea that the higher current density inside a filament will produce excess Joule's heat. Thus the measured temperature distribution on the surface of a pin diode images the current density distribution [31]. Fig. 5 represents the results using a commercial liquid-crystal film and a black coating on top of the sample. These liquid crystals selectively reflect incident white light to show bright iridescent colours depending on the temperature. The photograph (Fig. 5a) qualitatively shows localized regions with an excess temperature of $1 - 2° C$ at three different operating points in the corresponding I–U characteristic (Fig. 5c). The sketch (Fig. 5b) displays the observable colour spots when increasing the current through the diode. It should be pointed out that the formation of colour spots occurs abruptly when passing a jump in the I–U curve, and that each jump is accompanied by the

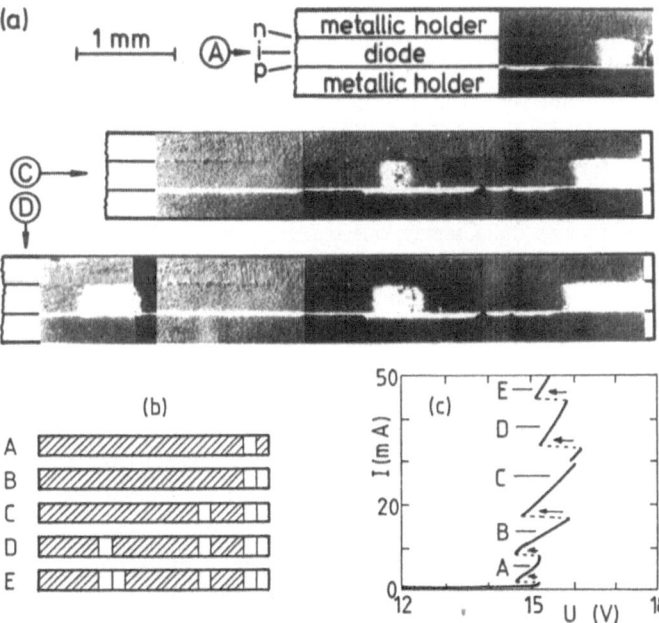

Fig. 5: Qualitative temperature distribution on the surface of a Si pin diode using a liquid-crystal film (a), sketch of the observable colour spots (b) and corresponding I–U characterictic (c). The photograph was taken from colour prints.

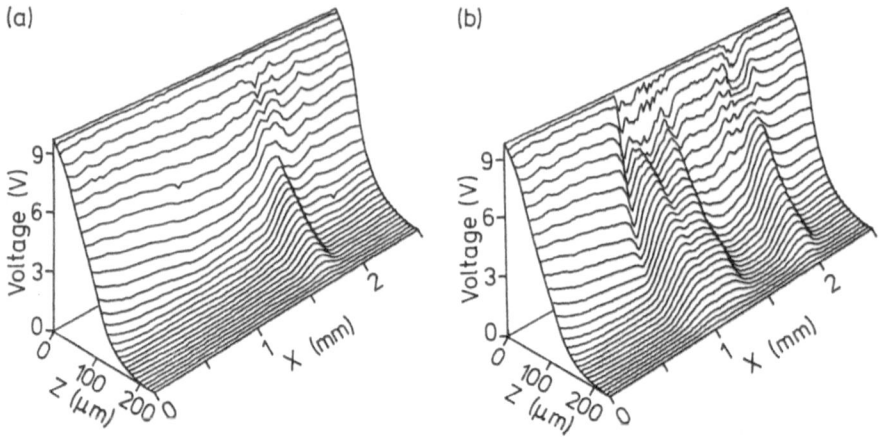

(a)   (b)

Fig. 6: Measured potential distributions on the top of a Si pin diode at two different operating points: Beyond the first jump (a) and beyond the third jump (b) in the I–U curve.

generation of one spot with similar diameter. So, the rightmost region in the sample at operating points B-E is connected with the existence of two adjacent inhomogeneities. It is concluded that these temperature inhomogeneities point out in a direct way the pattern formation due to nonuniform current flow in form of localized regions wich high current density.

To get additional experimental indications we present the results of spatially resolved potential measurements on the surface of the diodes [31]. The measurements are carried out by using a tungsten probe with a tip diameter of about $10\,\mu m$, operated by a computer-controlled micromanipulator, and a high-impedance voltmeter. The obtained potential distribution between the contacts of a Si pin diode at an operating point beyond the first jump in the I–U curve is illustrated in Fig. 6a. Clearly, one can see a well defined inhomogeneity perpendicular to the current flow at $x \approx 1700\,\mu m$. This potential hump disappears abruptly when the external current is decreased to the lowest state. In the other part of the sample, the longitudinal potential distribution is qualitatively that of a usual pin diode. Fig. 6b shows the results of the same diode at an operating point beyond the third jump in the I–U curve. One can observe the generation of additional potential humps with each jump. This kind of pattern formation is reproducable when decreasing and increasing the current. But it should be noted that the detailed course of the I–U characteristic and the locations of filament generation in the sample may vary if experimental conditions such as the temperature are changed. Similar results concerning the potential inhomogeneities in Si pin diodes may be obtained when using the potential contrast method in a scanning electron microscope instead of a mechanical probe [33,34].

For a more quantitative analysis we have determined the half-value width (FWHM) of the potential inhomogeneities. As a first result, we found that

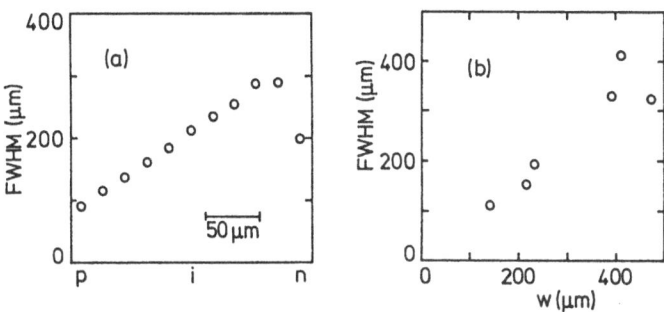

Fig. 7: FWHM of a potential inhomogenity vs. the z-coordinate (a) and vs. the contact distance w at a fixed distance from the n-contact for different samples (b).

each hump exhibits similar structure and the FWHM is almost independent of the current through the device [32]. Therefore it is expedient to specify the FWHM vs. the z-coordinate and vs. the contact distance w of different samples regardless of the operating point as executed in Fig. 7. One can see an increase of the diameter from the p- to the n-contact as well as an increase with the contact distance. In a first-order approximation a proportionality FWHM $\sim$ w is observed.

It has to be regarded as a disadvantage that the presented methods can detect only signals on the surface of the samples, and give only a qualitative and indirect image of the electrical properties of the semiconductor material. Now we introduce two optical methods yielding signals from the bulk material inside the device. These techniques even allow quantitative calculations of the free carrier density [35]. For that purpose, the optical absorption coefficient α , which is a function of the free carrier density, is determined by measuring the transmitted light when the diode is irradiated and scanned by a focussed laser beam. Experimentally, we chose a wavelength of $3,39\,\mu m$ from a HeNe-laser to prevent transitions of carriers between the bands and the trap level. To improve the signal to noise ratio we have applied conventional lock-in technique where the current has been modulated with an additional ac-voltage signal (for details of the method see [35]). Accordingly, we get the spatially resolved variation of the absorption coefficient $\Delta\alpha$ between the contacts as shown in Fig. 8. Clearly, an absorption inhomogeneity at $x = 800\,\mu m$ is detected with an enhanced absorption near the contacts. The properties of these inhomogeneities are comparable to the results mentioned above: Inhomogeneities disappear and arise abruptly when passing a jump in the I–U curve. Furtheron, each inhomogeneity has qualitatively the same structure, and the diameter is larger near the n-contact. From the optical signal and further experimental data (for example potential-probe measurements to determine the electric field) we can calculate the variation ΔN of the free carrier density. In particular, for given z coordinate we determine ΔN in the maximum - with respect to x - of

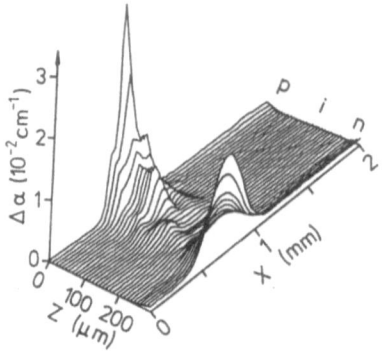

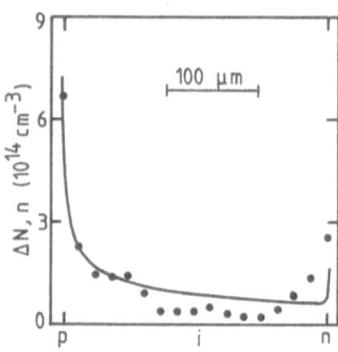

Fig. 8: Measured variation of the absorption coefficient $\Delta\alpha$ between the contacts of a Si pin diode. Operating point on the first branch of the I–U curve.

Fig. 9: Calculated free carrier density n and ΔN vs. z-coordinate. The solid line is a result of a numerical calculation for a homogeneous pin diode with a current density comparable to the estimated value inside the filament. The dots are experimental results for the inhomogeneity shown in Fig. 8.

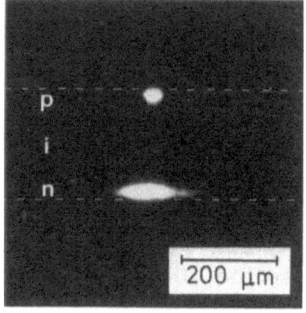

Fig. 10: Spatially resolved distribution of the recombination radiation emitted from a Si pin diode.

the observable structure as plotted in Fig. 9. We compare these values with numerical calculations of the free-electron density n in homogeneous pin diodes under the condition of high injection (see Section 4.1), and use a current density estimated from the I–U data and the width of the structure. As can be seen from Fig. 9, we find satisfactory agreement. Consequently, the experimental results show that inside the sample a region exists where double injection occurs and a high current density is present.

The second optical setup provides the possibility to detect temporally and spatially resolved recombination radiation emitted from the sample [35]. Fig. 10 shows the intensity of the emitted radiation in the x-z-plane when operating the diode on a stable branch beyond the first jump in the I–U curve. One sees localized regions of high irradiance near the contacts and a wider structure near the n-contact. These results are likewise in agreement with the other experiments because the recombination rate scales with the product of free electron and hole concentrations.

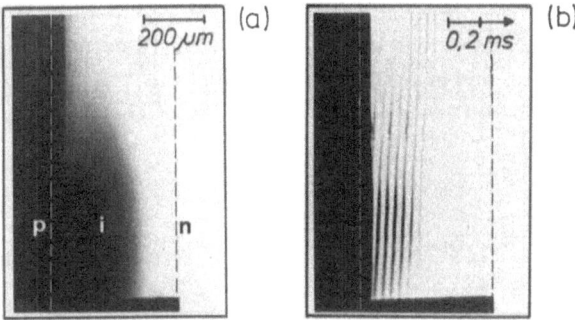

Fig. 11: Measured potential contrast of a Si:Au pin diode using a SEM. The diode is biased to a stable operating point (a) and to a region with self-generated oscillations (b). The scanning direction is indicated by the arrow.

Most of the presented experimental methods, after some minor modifications, are suitable for spatially and temporally resolved investigations in connection with self-generated oscillations of the pin diode. Here we discuss exemplarily some of the typical results. In the case of the second optical setup, we observe a spatially limited region of radiation emission which turns off and on periodically. The location of the unstable inhomogeneity is the same as the observable emission region when operating the diode on a stable branch following the oscillations. The image of the spatially resolved variation of the absorption coefficient is qualitatively identical to the plot in Fig. 8 if the external current modulation is turned off and the sample is biased into a region of periodic self-generated oscillations of the current. Finally, Fig. 11 presents the measured potential contrast on the surface of a Si:Au pin diode with the help of a scanning electron microscope (SEM) [34]. A positive potential is indicated by a low detector signal, thus leading to a dimmed scan. Therefore the p-contact is dark and the n-contact bright. The filamentary inhomogeneity is visible by a dark bump comparable to the potential hump in Fig. 6b. This case is demonstrated in the left photograph (Fig. 11a) where a stable operating point is chosen beyond the first jump in the I–U curve. Fig. 11b shows the situation where the sample produces self-generated oscillations. One observes that the potential bump turns off and on periodically with progressing beam and therefore likewise progressing time. The position of the oscillating inhomogeneity is that of Fig. 11a. Therefore, these results, like the above mentioned measurements, indicate a switching of a filament in connection with self-generated oscillations.

In summary, we have found up to now the following features of filaments in Si:Au pin diodes. The pronounced relation between the multistability in the I–U curve and filament formation could be distinctly confirmed. All presented measurements reveal comparable diameters of the inhomogeneities with an increasing width from the p- to the n-contact. The filaments show a solitary behaviour, that means all inhomogeneities in one sample have qualitatively the same structure, and the diameter is nearly independent of the current through the device. Additionally, we have seen that the experiments easily permit the investigation of dynamic processes due to self-generated oscillations. At operating points with periodic oscillations, the results can be interpreted as switching processes of a filament.

In the following chapter we will present some ideas which allow an understanding of the course of the I–U curve and the S-shaped j–E characteristic in Si:Au pin diodes. Furtheron, electrical circuits are discussed to explain the filament formation in these materials.

4. Theory of filament formation

An exact theoretical description of a three-dimensional filamentary semiconductor device is not available until now. Some basic concepts which exist today are described elsewhere in this book and in Ref. [4]. In the following a preliminary model is given for a pin diode because of three reasons. First, the microscopic analysis of a pin diode is well advanced, and experimental I–U characteristics can be compared with numerical results. Second, various measuring techniques have been applied to yield detailed results on spatio-temporal inhomogeneities of numerous physical parameters. Third, it is expected that the study of pin diodes will ultimately lead to a general understanding of current density filaments. This is because nonlinear dynamical systems are known to exhibit very general phenomena which can be described by only a few basic mechanisms, (see for example the role of dispersion, dissipation and nonlinearity in wave propagation). Accordingly, the model to be discussed consists of three steps, the homogenous pin diode, a pin diode with transverse inhomogeneity and the general activator - inhibitor principle.

4.1 Double injection and j–E characteristic

The basic principle of current flow in a homogeneous pin diode is double injection, i.e. the injection of electrons and holes by suitable contacts into an intrinsic semiconductor region. The excess charge carriers recombine and the current is determined by at least three mechanisms [36,37]: (i) At low injection levels a space-charge limited current occurs due to the injected mobile charge near the contacts. (ii) At high injection levels electrons and holes form an almost uncharged plasma, and the current is determined by the recombination process. (iii) The recombination rate depends critically on the density of charge carriers where instabilities can arise. The detailed theory on the basis of the Poisson equation, the continuity equation and the expressions for the current density is outlined in [36,38]. In the following we only give an intuitive analysis of the fundamental physical principle of S-type NDC where the recombination behaviour plays a keyrole. In order to be specific, we take Si:Au pin diodes for any quantitative results.

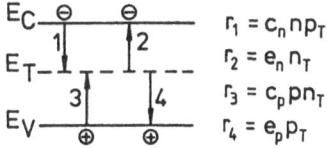

$r_1 = c_n n p_T$

$r_2 = e_n n_T$

$r_3 = c_p p n_T$

$r_4 = e_p p_T$

Fig. 12: Shockley-Read-Hall- (SRH-) model, c_n, c_p, e_n, e_p are capture and emission coefficients for electrons and holes, respectively. n and p are the densities of free electrones and holes, respectively.

In pin diodes the intrinsic region is, for example, an n-type semiconductor material compensated by deep trapping centers - acceptors - of density N_T (see Section 3.1). In other words, the excess electrons supplied by the shallow donors are trapped by the compensating atoms to produce a high-resistivity material, n_T is the density of those negatively charged impurities, and $p_T = N_T - n_T$ is the density of empty states. The basic interaction processes between mobile electrons and holes in the conduction and valence bands, respectively, and the deep centers are sketched in Fig. 12. According to the famous Shockley-Read-Hall- (SRH-) model, the rates for capture and emission of electrons, r_1 and r_2, and capture and emission of holes, r_3 and r_4, are proportional to the densities of states concerned in the process. In Fig. 13 a numerical computation of the I–U curve using these rates is compared with experimental results of a pin diode showing excellent agreement. It is therefore concluded that the microscopic picture mentioned above is very well suited to describe the properties of an experimental device. It should, however, be mentioned that the area of the measured pin diode was sufficiently small, so that the current flow remained homogeneous, see the corresponding arguments in Section 3. Clearly, the pin diode of Fig. 13 reveals an S-type I–U characteristic, but with respect to Section 1, the basic problem of the structure of the j–E characteristic is still unsolved. For that purpose, the electrical-field distributions have been calculated for various values of the current density. The results are presented in Fig. 14. As the most important results, j vs. E depends on z and is S-shaped in the major part of the device, whereas j(E) is a unique function only in the vicinity of the cathode. Hence, the j–E characteristic seems to be a combination of some bulk properties and additional contact effects, where the latter ones are needed in any case as boundary conditions.

The S-shaped j–E relationship may qualitatively be understood from the following two arguments. (i) From the SRH mechanism the stationary recombination rate R can be obtained by setting $\partial n/\partial t = \partial p/\partial t$ to yield

$$R = r_1 - r_2 = r_3 - r_4 \quad . \tag{1}$$

The stationary density of negatively charged trapping centers is then given by

$$n_T = N_T \frac{c_n n + e_p}{c_n n + c_p p + e_n + e_p} \quad . \tag{2}$$

81

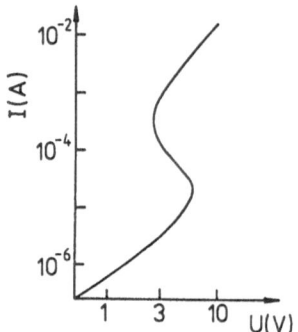

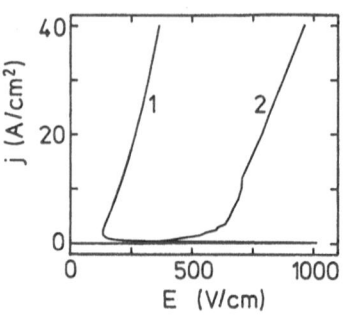

Fig. 13: Theoretical and experimental I–U characteristic of a Si:Au diode [36]. Both curves are plotted showing only negligible deviation.

Fig. 14: Numerical j–E characteristics of a Si:Au pin diode at different z-coordinates: curve 1 near the p-, curve 2 near the n-contact. The following parameters were used: $N_D = 1.8 \cdot 10^{14}\,cm^{-3}$, $N_T = 2.1 \cdot 10^{14}\,cm^{-3}$, $w = 200\,\mu m$, $T = 300\,K$, $c_p = 1 \cdot 10^{-7}\,cm^3 s^{-1}$, $c_n = 3.4 \cdot 10^{-8}\,cm^3 s^{-1}$.

Now R can be expressed by

$$R = N_T c_n c_p \frac{pn - n_i^2}{c_n n + c_p p + e_n + e_p} = \frac{\Delta n}{\tau(n)} \tag{3}$$

where $n = n_0 + \Delta n$ and $p = p_0 + \Delta p$ are the sums of the equilibrium values, n_0 and p_0, and the injected densities, Δn and Δp. We have further assumed $\Delta n = \Delta p$ because the current is assumed to be limited by recombination and not by space-charge effects. From Eq. (3) we get as a first result the dependence of the lifetime τ on the density n. (ii) Using the arguments of ref. [39] we obtain roughly the following expression for the current density

$$j = \frac{1}{2z}\sigma\mu\tau E^2 \tag{4}$$

where σ is the conductivity of the i-layer and μ the mobility. Note, however, that $\tau = \tau(n)$ is not a constant parameter. Eq. (4) can be reformulated to give

$$j = \frac{1}{2}\sigma\frac{\tau}{T}E \sim n(\tau, E)E \quad . \tag{5}$$

Here $T = z/(\mu E)$ is the transit time and Eq. (5) is formally equivalent to a photoconductor equation, where the photoconductivity is increased by the factor τ/T under the condition of stationary illumination. Hence the effective carrier density is enhanced by this factor, i.e. n may be regarded as a linear function of τ with E as a parameter. To solve simultaneously for $\tau(n)$ from Eqs. (3) and (5), a graphical method as depicted in Fig. 15 is used. For a given field E the intersection points between the two curves give the solution. As can be seen, at low values of E, there is first a unique solution and n is small. With increasing

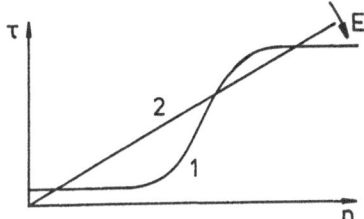

Fig. 15: Graphical solution of $\tau(n)$ from Eq. (3), curve 1, and $n(\tau, E)$ from Eq. (5), curve 2, with E as a parameter for the slope.

E, there are three values of n until at higher fields n is large since τ reaches the high injection limit. As a result, for a given z, n vs. E is S-shaped and similarly j vs. E. The associated instability can be seen from the following flow chart:

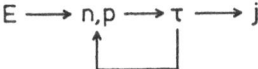

The electric field determines the carrier densities which in turn control the lifetime and the current density. As a key point, the carrier densities depend on the lifetime in such a way that a positive feedback occurs. This leads to bistability and switching, as outlined by KLINGSHIRN et. al. in another part of this book. Note, however, that the j–E characteristic depends on z as found numerically in Fig. 14.

4.2 Equivalent circuit and filament equation

The results in Fig. 14 have shown that the pin diode can electrically be seperated into two regions, the first with an S-shaped j–E characteristic which covers the major part of the device, and the second with a unique j(E) relation near the cathode. Without loss of generality we characterize these two areas as the nonlinear active and the ohmic regions, respectively. Consequently, we replace the pin diode by a series representation of two resistances and introduce a transverse resistance to account for inhomogeneous current distribution.

The resulting equivalent circuit of a strip-shaped, one-dimensional pin diode is shown in Fig. 16. Here G' is the conductance per unit length of the ohmic region near the cathode. U denotes the voltage drop across the nonlinear active part with S-shaped I'–U characteristic. C' is the capacitance per unit length which introduces a finite relaxation time of the active material. The resistance per unit length R' introduces a diffusive coupling between adjacent cells. From the circuit in Fig. 16 one easily derives the following equations

$$\frac{\partial U}{\partial x} = IR' \quad , \tag{6}$$

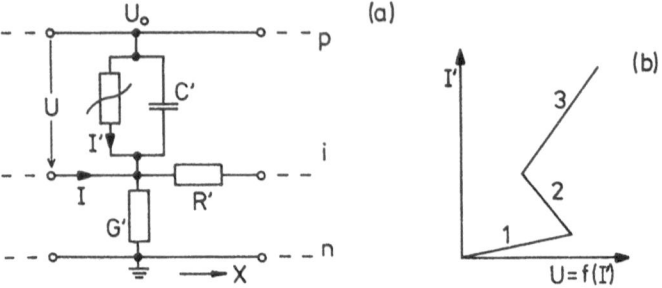

Fig. 16: (a) Electrical unit cell of a pin diode with one spatial dimension and (b) piecewise linear representation of $f(I')$ in Eq. (8), i=1,2,3.

$$\frac{\partial I}{\partial x} = I' - (U_0 - U)G' + C'\frac{\partial U}{\partial t} \tag{7}$$

$$\text{with} \qquad U = U(I') = f(I') \quad . \tag{8}$$

Eliminating the current I from Eqs. (6) and (7) we obtain

$$\frac{\partial^2 U}{\partial x^2} = R'I' - (U_0 - U)G'R' + C'R'\frac{\partial U}{\partial t} \quad . \tag{9}$$

It should be noted that this equation has no unique solution because I' cannot be written as a function of U (see Eq. (8)). Therefore we introduce $U(I') = f(I')$ into Eq. (9) which yields

$$f''(I')\left(\frac{\partial I'}{\partial x}\right)^2 + f'(I')\frac{\partial^2 I'}{\partial x^2} = R'I' - (U_0 - f(I'))\,G'R' + C'R'f'(I')\frac{\partial I'}{\partial t} \tag{10}$$

where $f' = \partial f/\partial I'$ and $f'' = \partial^2 f/\partial I'^2$. Eq. (10) is called filament equation because the solution $I'(x,t)$ denotes the spatial and temporal behaviour of the current per unit length for any boundary conditions. For $\partial I'/\partial t = 0$ a stationary filament is obtained. The corresponding potential distribution $U(x,t)$ can directly be judged from the given $U(I')$ relation of Eq. (8). As can be seen, in the special case of a pulse-like $I'(x)$, the potential $U(x)$ in the nonlinear active regions is W-shaped, i.e. $U(x)$ exhibits two minima in the vicinity of the filament walls.

Eq. (10) is a nonlinear active diffusion equation. It can be solved by a piecewise linear relation $f(I') = a_i + b_iI'$ (see Fig. 16b) and by matching the solutions of the different regions $i = 1, 2, 3$. Obviously, in this case $f' = const$ and $f'' = 0$, so that Eq. (10) turns out to be a diffusion equation, well-known from the propagation of electrical impulses on a nerve axon [42].

4.3 Activator - inhibitor principle

The activator - inhibitor principle has been applied to a large number of biological and chemical systems to describe spatially inhomogeneous stable structures (see for example [40]). The model is based upon two coupled nonlinear differential equations where one dependent variable acts as an activator and the other as an inhibitor. An attempt was made to transfer this principle to the current flow in semiconductor diodes in order to describe the observed pattern formation due to filamentation [41]. For that purpose the device is divided into two layers with different electrical properties, one layer with an ohmic relation while the second layer is nonlinear and obeys an S-shaped relation. From the continuity and current equations two differential equations are derived to calculate the potential and current density distributions at the interface of the two layers [41]:

$$\frac{\partial j^*}{\partial t} = \delta_j \Delta j^* + f(j^*) - u^* \quad , \tag{11a}$$

$$\delta_u \frac{\partial u^*}{\partial t} = \Delta u^* + j^* - u^* - T \quad . \tag{11b}$$

Here u^* and j^* denote the normalized voltage and current at the interface, $f(j^*)$ expresses the S-shaped j–E characteristic of the nonlinear layer, and δ_u, δ_j and T are parameters. Eqs. 11a and 11b are reaction - diffusion equations of the activator - inhibitor type where j^* can be regarded as an activator because of the NDC properties. On the other hand, the voltage drop across the linear layer will limit the increase of current and, consequently, u^* may be interpreted as an inhibitor.

If the parameters are selected properly, numerical calculations reveal a multistable I–U characteristic of the device and the formation of stable pulse-like current density distributions at the interface. The diameter of the inhomogeneities increases with the thickness of the linear layer. This result is comparable to the experimental plot in Fig. 7b if the thickness of the nonlinear layer is small compared to the linear layer. At modified values of the parameters, one can observe self-generated periodic or chaotic oscillations in the I–U characteristic. But a combined behaviour as measured experimentally in diodes with alternating regions of stable branches and oscillations has not been found up to now. A further lack until now is the arbitrary choice of the parameters which cannot be fitted to experimental semiconductor device data.

5. Conclusions

It has been shown in the past that semiconductor devices are excellent samples to study nonlinear dynamics in solid state physics both theoretically and experimentally. Especially, pattern formation in form of current density filaments has been observed by various methods. Various physical parameters have been measured yielding characteristic inhomogeneities in the plane transversal to the current flow.

Detailed experimental results have been obtained from pin diodes where spatially resolved measurements of surface temperature, potential distribution, recombination radiation and optical absorption due to free carriers have been carried out. As a clear result, the jumps in the I–U characteristic are associated with the generation or disappearance of a filament. These filaments have a solitary structure and may be stable or unstable. The observed oscillations of the current or the voltage are traced back to switching properties of the filaments where adjacent spatial regions can exhibit different temporal behaviour. It is found that periodic doubling, for example, is probably due to out-of-phase oscillations of two filaments. The electrical properties of the pin diodes are described by double injection phenomena which can explain the S-type j–E characteristic and which gives excellent agreement between theoretical and experimental results. On the basis of an equivalent circuit, a filament equation is finally deduced which can be solved for the spatial current density distribution.

It is obvious that a large number of problems are still unsolved such as the mechanisms of stability or instability of filaments. Until now, the solitary behaviour, as well as the three-dimensional structure of current inhomogeneities cannnot be described theoretically. And it is still an open question whether material effects or the influence of contacts are responsible for the generation of filaments. Further detailed experimental observations are also needed where for example a direct detection of the current density would be advantageous.

Since semiconductor devices with breakdown mechanisms and active properties are important for many applications, it is foreseen that filaments may play a keyrole in future technical components. In case of thyristors the negative influence of filaments on the switching behaviour has been discussed in the literature [18]. Similar effects hold for light-emitting or laser diodes and in transistors where impact ionization can occur via nonuniform current distributions. On the other hand, a filament might serve as a controllable inductance and instabilities may be useful for oscillators. The high carrier densities may lead to luminescence devices with high local radiation intensities. Finally, these devices exhibit simultaneously the features of switching components with memory. It is thus expected that the study of nonlinear dynamics and pattern formation phenomena in semiconductors might lead to novel device concepts of technical significance.

References

[1] H. Haken: "Synergetics. An Introduction" (Springer-Verlag, Berlin 1978)

[2] F. Stöckmann: in Festkörperprobleme 9, ed. O. Madelung (Vieweg, Braunschweig 1969)

[3] V.L. Bonch-Bruevic, I.P. Zvyagin, A.G. Mironov: "Domain electrical instabilities in semiconductors" (Consultant Bureau, New York 1975)

[4] E. Schöll: "Nonequilibrium phase transitions in semiconductors" (Springer-Verlag, Berlin 1987)

[5] S.M. Sze: "Physics of semiconductor devices" (John Wiley & Sons, New York 1981)

[6] R. Newman: Phys. Rev. **100** 700 (1955)

[7] A.G. Chynoweth, K.G. McKay: Phys. Rev. **102** 369 (1956)

[8] A. Goetzberger, C. Stephens: J. Appl. Phys. **32** 2646 (1961)

[9] R.H. Haitz, A. Goetzberger, R.M. Schultz, W. Shockley: J. Appl. Phys. **34** 1581 (1963)

[10] A.E. Michel, M.I. Nathan, J.C. Marinace: J. Appl. Phys. **35** 3543 (1964)

[11] M. Gershenzon, A. Ashkin: J. Appl. Phys. **37** 246 (1966)

[12] A.M. Barnett, A.G. Milnes: J. Appl. Phys. **37** 4215 (1966)

[13] A.M. Barnett, H.A. Jensen: Appl. Phys. Lett **12** 341 (1968)

[14] A.P. Ferro, S.K. Ghandhi: J. Appl. Phys. **42** 4015 (1971)

[15] M.E. Alekseev, I.V. Varlamov, E.A. Poltoratskii, V.P. Sondaevskii: Sov. Phys.–Semicond. **3** 1514 (1970)

[16] K. Homma, Y. Kobayashi, T. Fukami: Appl. Phys. Lett. **21** 154 (1972)

[17] M.E. Alekseev, L.F. Rodionov, V.P. Sondaevskii: Sov. Phys.–Semicond. **8** 970 (1975)

[18] M. Stoisiek, R. Sittig: in Festkörperprobleme **26**, ed. P. Grosse (Vieweg, Braunschweig 1986)

[19] M. Braunstein, A.I. Braunstein, R. Zuleeg: Appl. Phys. Lett. **10** 313 (1967)

[20] D.J. Dumin: IEEE–ED **16** 479 (1969)

[21] B.S. Kerner, V.F. Sinkevich: JETP Lett. **36** 436 (1982)

[22] D.H. Pontius, W.B. Smith, P.P. Budenstein: J. Appl. Phys. **44** 331 (1973)

[23] K.M. Mayer, J. Parisi, R.P. Huebener: Z. Phys. B.–Cond. Matter **71** 171 (1988)

[24] A. Brandl, M. Völcker, W. Prettl: Appl. Phys. Lett. **55** 238 (1989)

[25] K.M. Mayer, R. Gross, J. Parisi, J. Peinke, R.P. Huebener: Solid State Commun. **63** 55 (1987)

[26] C.N. Berglund: IEEE–ED **16** 432 (1969)

[27] J. Duchene, M. Terraillon, P. Pailly, G. Adam: Appl. Phys. Lett. **19** 115 (1971)

[28] R.P. Beaulieu, D.V. Sulway, C.D. Cox: Solid-State Electr. **16** 428 (1973)

[29] I. Ruge: "Halbleitertechnologie" (Springer-Verlag, Berlin 1984)

[30] K. Aoki, O. Ikezawa, K. Yamamoto: Phys. Lett. A **106** 343 (1984)

[31] D. Jäger, H. Baumann, R. Symanczyk: Phys. Lett. A **117** 141 (1986)

[32] H. Baumann, R. Symanczyk, C. Radehaus, H.-G. Purwins, D. Jäger: Phys. Lett. A **123** 421 (1987)

[33] T. Pioch, H. Baumann, D. Jäger: BEDO **18** 133 (1985)

[34] H. Baumann, T. Pioch, H. Dahmen, D. Jäger: SEM **II/1986** 441 (1986)

[35] R. Symanczyk, E. Pieper, D. Jäger: Phys. Lett. A **143** 337 (1990)

[36] I. Dudeck, R. Kassing: J. Appl. Phys. **48** 4786 (1977)

[37] K.L. Ashley, A.G. Milnes: J. Appl. Phys. **35** 369 (1964)

[38] W.H. Weber, G.W. Ford: Solid-State Electr. **13** 1333 (1970)

[39] P. Migliorato, G. Margaritondo, P. Perfetti: J. Appl. Phys. **47** 656 (1976)

[40] F. Rothe: Math. Biol. **7** 375 (1979)

[41] C. Radehaus, K. Kardell, H. Baumann, D. Jäger, H.-G. Purwins: Z. Phys. B - Cond. Matter **65** 515 (1987)

[42] A.C. Scott: Rev. of Modern Physics **47** 487 (1975)

Optical Instabilities in Passive Semiconductors

C. Klingshirn, J. Grohs, and M. Wegener*

Fachbereich Physik, Universität Kaiserslautern,
W-6750 Kaiserslautern
*Present Address: Fachbereich Physik, Universität Dortmund,
  W-4600 Dortmund 50

1. Introduction

1.1. General Remarks

The nonlinear optical properties of matter and the associated phenomena of optical instabilities and of optical bistability gained increasing interest in the last years [1-12]. This evolution is based on the fact that aspects of applied research e.g. for optical data handling as well as aspects of fundamental research in the area of nonlinear dynamics can be investigated with these systems. We will concentrate in this contribution on the second point, and, to keep things easy, we will treat only a single material, CdS, and a single group of optical nonlinearities, namely the photo-thermal ones.

The term "optical nonlinearity" means all reversible changes of the optical properties like the spectra of transmission, reflection or luminescence induced by sufficiently strong illumination. In general, the optical nonlinearities can be divided into two groups: The so-called photo-electronic optical nonlinearities which are created by dense real or virtual excitations in the electronic system, leading to many-particle effects like band renormalization or the transition to an electron-hole plasma (see e.g. [9,13-16]), and the photo-thermal optical non-linearities which are due to the temperature rise in the sample caused by the absorbed optical power. To understand the basic mechanisms of this second group of effects, we will treat in the following subsection the linear optical properties of semiconductors and especially their temperature dependence. Section 2 is devoted to various aspects of optical bistability, section 3 to regular and irregular self-oscillations in (hybrid) ring resonators, and section 4 finally gives an outlook on further investigations.

1.2. The Temperature Dependence of the Absorption

The basic process during the absorption or emission of a photon in a semiconductor in the spectral region between the frequencies of optical phonons and of valence-band plasmons is the creation or annihilation of an electron-hole pair. This means that all optical transitions are two-particle transitions which involve an occupied and an empty (hole) state. This is in contrast to electronic transport properties described e.g. in the contributions to this book by Schöll,

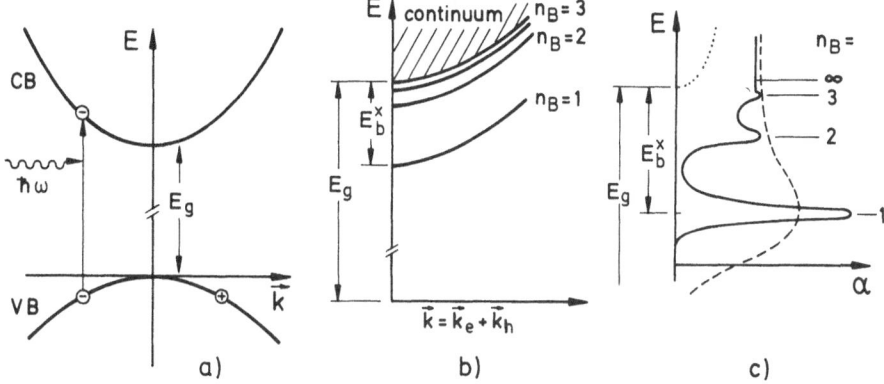

Fig. 1: Idealized band-structure of a direct-gap semiconductor (a), corresponding dispersion relation of excitons (b) and resulting absorption spectra for
low (full line) and high (dashed line) temperatures (c). The dotted line gives
the square-root dependence of α expected for simple band-to-band transitions.
From [18].

Peinke and Jäger. There is usually one type of carriers (e.g. electrons) predo-
minant, and the particles are created by doping, injection or impact ionization.

Due to the opposite charges, the two particles involved in an optical transition interact and form a new quasiparticle, the exciton. These processes are
illustrated in an idealized band-structure of Fig. 1.

The wave functions and eigen-energies of such an exciton can be calculated
in first approximation in analogy to those of a hydrogen atom by inserting the
Coulomb interaction into the Schrödinger equation. Nevertheless, the different
masses and the dielectric function in the semiconductor modify the well-known
formulas. It turns out that the excitons form roughly a hydrogen-like series of
resonances in the spectrum that begins energetically some tens of meV below
the band gap (Fig. 1b+c). These resonances and the continuum states determine the optical properties of a semiconductor in the vicinity of the bandgap.
A closer examination would have to take into account the dispersion relation
of the excitons and the mixed states between photon and exciton, the so-called
polaritons. Details on this topic can be found e.g. in [17,18].

Essential for our purposes is the interaction of the excitons with thermally excited lattice vibrations, which are described by another type of quasi-
particles, the phonons. This interaction leads to a broadening of the exciton
resonances, and thus to an increase of the absorption in the spectral region below the first resonance with increasing temperature. A formula that describes
the connection between the absorption coefficient α, the lattice temperature T_L
and photon energy $\hbar\omega$, respectively, is the Urbach-Martienssen rule [19,20,21]

$$\alpha(\omega, T_L) = \alpha_0 \, exp\left(\sigma \, \frac{\hbar\omega - \hbar\omega_0}{k_B T_L}\right) \qquad (1)$$

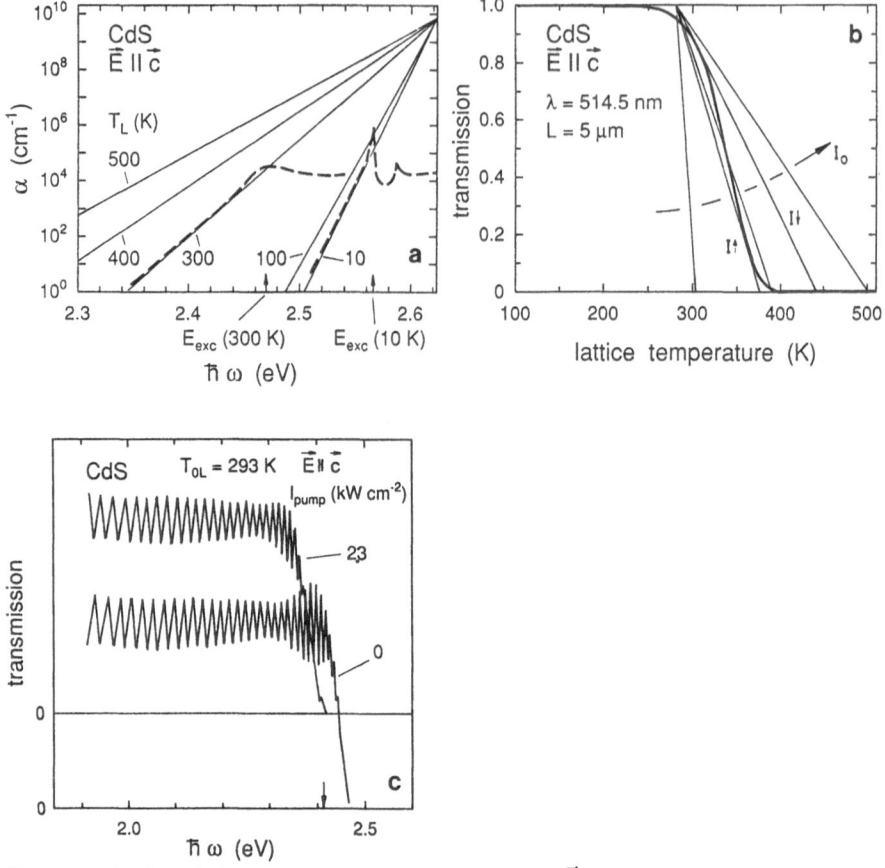

Fig. 2: Calculated absorption spectrum of CdS for $\vec{E} \parallel \vec{c}$ for various lattice temperatures T_L (full lines), and schematic drawing of the experimentally observed ones (dashed lines) (a), the transmission as a function of T_L at $\lambda = 514.5$ nm and various straight lines to construct the static hysteresis loop (b), experimental transmission spectra of CdS at room temperature with and without Ar$^+$ laser excitation (c). From [12].

where σ, $\hbar\omega_0$ and α_0 are material parameters. Deviations from this form are often described using a temperature dependence of σ. Examples are presented in Fig. 2, where the calculated spectrum of $\alpha(\omega, T_L)$ for a CdS crystal (thickness 5 μm) at different temperatures (a) and the transmission at a wavelength of 514.5 nm as a function of temperature (b) are shown. Deviations of the experiment are indicated schematically by dashed lines. One can see that the transmission $T(T_L)$ decreases from a value near 1 (without reflection!) at $T_L \approx 300$ K very rapidly to a value near 0 at $T_L \approx 400$ K.

The detailed derivation of the Urbach-Martienssen rule is still a topic of modern science and requires many instruments of theoretical physics. We mention two models which are considered as the most relevant ones:

According to Sumi and Toyozawa [22], the exponential dependence of α is caused by a momentary localization of excitons in the randomly fluctuating potential introduced by thermally excited phonons. Another model by Dow and Redfield [23] attributes the Urbach-Martienssen rule to the electric field that is mainly produced by optical phonons. It tilts the bands, and this tilting leads to a tunneling of electron and hole wave functions from the exciton state through the forbidden gap into the continuum states. So, absorption at lower photon energies becomes possible, exhibiting an exponential tail in the spectrum.

More recent experimental and theoretical investigations indicate that both effects contribute with a weight determined by the material parameters of the semiconductor or insulator under consideration [24].

1.3. Nonlinear Optics

As we have already mentioned, nonlinear optics describes the change of the optical properties of matter under high illumination. An example for a thermal nonlinearity is presented in Fig. 2c, where the transmission spectrum of a CdS crystal is shown with and without additional laser excitation with an intensity of $I_0 = 2.3 \; kW/cm^2$ at $\lambda = 514.5 \; nm$ ($\hbar\omega = 2.408 \; eV$). This wavelength is marked with an arrow in Fig. 2c. One can see that the transmission spectrum is shifted under excitation to lower energies, or with other words, the absorption at the laser wavelength is strongly increased (compare with Fig. 2b). Besides, one observes a slight red shift of the Fabry-Perot modes (FPM) in the transparent region of the spectrum. The FPM are caused by the as-grown, parallel surfaces of the platelet-type samples with a thickness of some μm. The red shift of the FPM indicates an increase of the refractive index connected with the red shift of the absorption edge due to the Kramers-Kronig relations.

To understand this behaviour we again have to stress the microscopic processes that are connected with the absorption of light in a semiconductor. The first interaction between photons and matter takes place in the electronic system. We excite real or virtual excitons, or more generally speaking electron-hole pairs. Their direct interaction can lead to electronic nonlinearities if their density is sufficiently large. Nevertheless, more important for our purposes are the relaxation processes after a real excitation. It has been found that the last step of this relaxation, the interband recombination of electron and hole, is in most semiconductors predominantly nonradiative. The absorbed incident optical energy is consequently transformed into other forms of energy, usually heat. The increase of lattice temperature, which is a consequence of this heat-production, depends on the material via the specific heat and heat conduction and on the excitation conditions like pulse duration, size of illuminated volume, heat sink etc. If the rise in sample temperature $\Delta T_L = T_L - T_{L0}$ caused by the absorbed light is sufficiently large, then ΔT_L results in noticable changes of the optical properties, as can be seen in Fig. 2c.

This means we have two coupled effects for the coefficients of absorption α and diffraction n: as we have seen they depend on the sample temperature, but on the other hand the sample temperature depends via nonradiative recombination on the excitation conditions like the excitation intensity I_{exc}. This can be summarized schematically in the following equation

$$n = n(T_L), \quad \alpha = \alpha(T_L) \tag{2}$$

with

$$T_L = T_L(I_{exc}). \tag{3}$$

2. Optical Bistability

In the previous section we have seen that photothermal optical nonlinearities generally lead to an increase of the absorption and of the refractive index under strong illumination in the spectral region below the exciton resonance. In the present section we describe how these features can be used to establish optical bistability.

2.1. The Basic Mechanism

An optically bistable device has two different, reversible states of transmission and/or reflection for the same input parameters, so that in a plot of transmitted intensity I_t versus incident intensity I_0 one gets a stationary hysteresis loop. A necessary but not sufficient condition for optical bistability (OB) is an optical nonlinearity connected with a suitable feedback.

The optical nonlinearity can be an excitation-induced increase of absorption (induced absorptive OB) or a change of the refractive index (dispersive OB), as described above. A third possible mechanism is the bleaching of absorption which, however, generally plays no role in systems based on photo-thermal effects, but often in photo-electronic optical nonlinearities. The feedback can either be an intrinsic one as in the case of induced absorptive OB, or can be produced by a Fabry-Perot resonator that is necessary in the two other cases. The as-grown surfaces of our crystals often form such a Fabry-Perot, but because of its small finesse we will neglect its effect as long as the sample is not coated with additional dielectric reflection layers (see however [25]). Indeed, in the experiment we observe a hysteresis loop typical for induced absorptive OB, which is shown in Fig. 3a.

The experimental observation of absorptive OB is rather simple. By some type of modulator a smooth pulse of duration τ_P is cut out of the cw-Ar$^+$ laser beam. With two photo-diodes one measures the temporal evolution of the incident and transmitted pulse $I_0(t)$ and $I_t(t)$, respectively, and deduces the time-free plot $I_t(I_0)$.

If one starts at low values of the incident intensity I_0 the sample is in a highly transmitting state and switches at a certain value $I_{\downarrow}$ to a state of low transmission. If one reduces I_0, the system remains over a certain range in this low transmitting state and switches back only at a lower value $I_{\uparrow}$. The

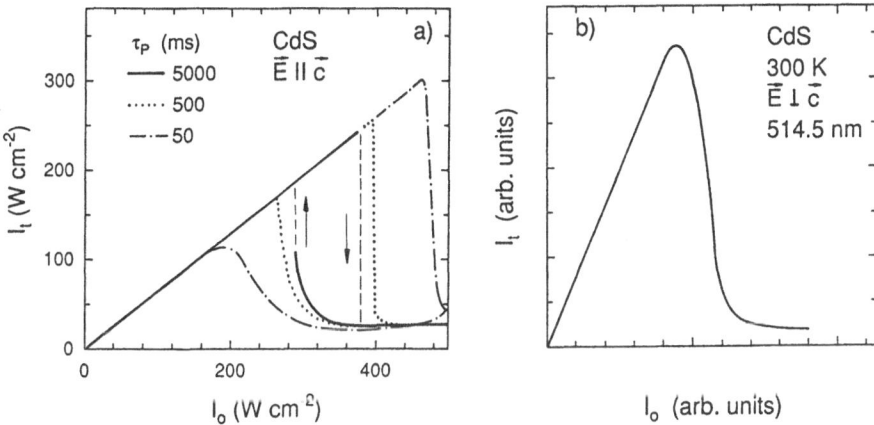

Fig. 3: Measured hysteresis loops for various pulse lengths τ_P of the incident intensity in the case of photo-thermally induced absorptive OB with $\vec{E} \parallel \vec{c}$ (a), nonlinear input-output characteristics for $\vec{E} \perp \vec{c}$ (b). From [30]

difference between $I_\downarrow$ and $I_\uparrow$ defines the bistable range. The hysteresis-loop is traced clockwise. For long input pulses $\tau_P \gg \tau$ (thermal relaxation time) one follows the stationary hysteresis loop. For short input pulses one observes a dynamical overshooting of the transmitted intensity.

The origin of the OB is as follows: Only a small fraction of I_0 is absorbed for small values of I_0. If, with increasing I_0, this small fraction is sufficient to increase α (in our case by an increase of temperature), a positive feedback sets in since an increasing α leads to an increase of T_L, and so on. If the characteristics $A(T_L)$ is sufficiently steep, the system may switch into a state of high absorption A (i.e. low transmission T). Now almost all the incident light is absorbed and consequently I_0 can be lowered, and the system remains on the lower branch until the switch-back occurs for $I_0 = I_\uparrow < I_\downarrow$, resulting in a loop as shown in Fig. 3a. This type of induced absorptive OB has been found independently by various groups for photo-electronic, photo-thermal and electrooptic systems [16,26-33]. In the following we present a quantitative approach according to [28].

Neglecting reflection losses on the surfaces, we have to fulfill two conditions for the transmission $T(T_L)$. First, the equation

$$T(T_L) = exp(-\alpha(T_L)L) \qquad (4)$$

holds. This relation can be measured by putting the sample in an oven. For the other relation, we take the heat conduction equation. It reads in our case:

$$\frac{dT_L}{dt} = \frac{I_0 A(T_L)}{C} - \frac{T_L - T_{L0}}{\tau} \qquad (5)$$

with $A(T_L) = 1 - T(T_L)$. The content of this equation is that the sample temperature changes under the influence of a generation term which is proportional to the absorbed intensity I_0A with a proportionality factor C^{-1} depending on the specific heat of the sample material, the sample thickness and the heat conductivity. The cooling conditions are described by an annihilation term with a thermal relaxation time τ which depends on the geometrical conditions, and is in our case of the order of some hundred μs up to a few ms [30,31,33,35,25].

Assuming steady-state conditions in (5) and rearranging we get:

$$T(T_L) = 1 - (\Delta T_L)C/(\tau I_0). \tag{6}$$

This is as a function of T_L a straight line with a slope proportional to I_0^{-1}. The solutions of (4) and (5) are the intersections of both curves (Fig. 2b). If we start at low I_0, i.e. a steep slope, we get only one solution. With increasing I_0 a second solution appears for $I_0 = I_\uparrow$ which splits by further increasing I_0 into two. The system remains, however, in the highly transmitting state until this solution disappeares for $I_0 = I_\downarrow$. Then the system jumps to the solution on the lower branch, where it remains for all $I_0 > I_\uparrow$.

A linear stability analysis shows that in the case of three solutions the middle solution is unstable and can be observed experimentally only under special conditions as will be demonstrated later. It is obvious from the construction of Fig. 2b that a certain steepness of $A(T)$ is necessary to obtain three solutions and thus optical bistability. If the intrinsic feedback by increasing absorption is not steep enough, e.g. for $\vec{E} \perp \vec{c}$ or for too high starting temperatures T_{L0}, one obtains a single solution for all incident intensities, and thus an $I_t(I_0)$ characteristics as is shown in Fig. 3b.

Room-temperature photothermal induced absorptive optical bistability has been observed for the first time in CdS [26] and has been investigated later on by various research groups [25;30-33].

2.2. Measurement of the Unstable Branch

A challenge in OB is to measure explicitly the unstable branch, which is the separatrix between the basins of attraction for the upper and lower branch, respectively. The first successful attempt for induced absorptive OB has been reported in [30], where the bistable element has been placed in a hybrid ring resonator as the one described in section 3. Another possible technique is the following. The system is kept by a holding beam I_h on the upper branch in the bistable region of induced absorptive OB (see Fig. 4a). Then an additional pulse of a duration $\tau_P \ll \tau$ is applied to heat the sample further (see the peak in Fig. 4b). This pulse can be e.g. a UV-N_2-laser pulse [34] or a pulse produced by a modulator in the Ar^+ laser beam [35]. This pulse heats the sample instantaneously in the illuminated area, consequently the transmission decreases. After some thermal relaxation time, a temperature profile and a transmission builds up, which are situated above or below the unstable branch.

If the energy of the additional pulse is properly chosen, the system can be prepared on the unstable branch. Due to finite fluctuations of the laser and

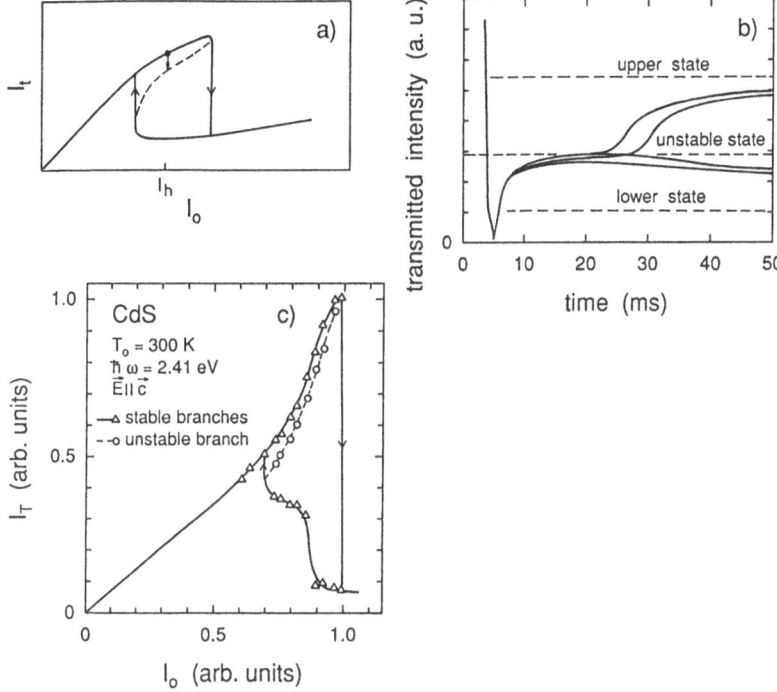

Fig. 4: Principle of how to measure the unstable branch in absorptive OB (a), temporal evolution of the transmission after application of a short additional pulse to prepare the system on the unstable branch (b), measured unstable branch (c). From [35]

of the experimental exactness, the unstable branch can be reached only with a relative precission $0 <| \epsilon |\ll 1$, so the system decays after some time either to the upper or lower stable branch as can be seen in Fig. 4b. This time increases with decreasing $| \epsilon |$. The point on the unstable branch is determined by the plateau in Fig. 4b. A variation of the working point I_h allows to determine the whole unstable branch. A result of such a measurement is shown in Fig. 4c. The stable branches have been determined by just measuring the stationary hysteresis loop. The fact that the relation $I_t(I_0)$ on the upper branch is slightly superlinear can be attributed to weak influences of a Fabry-Perot mode structure due to the sample surfaces. The unstable branch, determined by the method outlined above, lies close to the upper branch. This occurs in induced absorptive OB, depending on sample thickness and steepness of the increase of absorption, and is known also for photo-electronic optical nonlinearities [36].

2.3. Switching Dynamics

Optical bistability is a first-order phase transition in a driven, dissipative system, as can be seen e.g. from the appearance of hysteresis [36]. In contrast

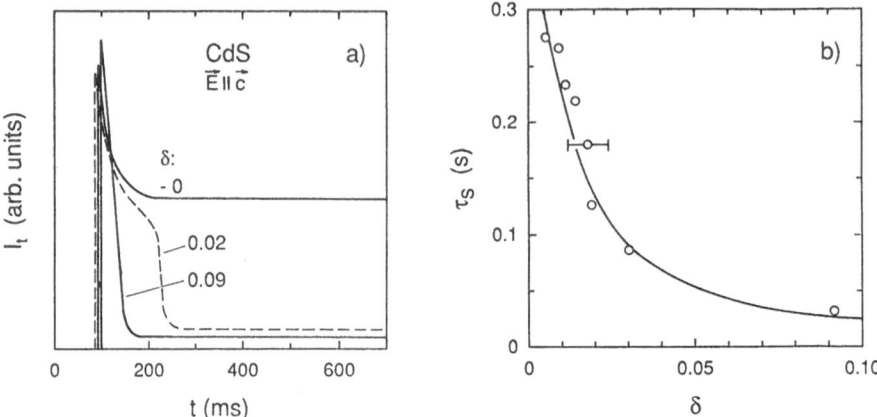

Fig. 5: The transmitted intensity as a function of time when the incident intensity I_0 is switched on in a steplike fashion from 0 to $I_\downarrow(1+\delta)$ (a), the delay between the step in I_0 and the switching process as a function of δ (b). From [30]

to e.g. the gas-liquid transition of a real or van der Waals gas, the transition from one phase to the other does not follow a Maxwell-construction through the coexistence region. The transitions start from the points which separate the unstable from the metastable regime. Consequently, critical slowing down has to be expected in the transition from one branch to the other [37,38]. This phenomenon has been investigated for the first time for induced absorptive OB independently by [30,33] and subsequently by other authors [25,35]. The method is the following: $I_0(t)$ is switched on at $t = 0$ in a step-like fashion from a value below $I_\downarrow$ to a value slightly above by a relative amount $\delta = (I_0 - I_\downarrow)/I_\downarrow$. Fig. 5a shows an example for $I_0(t) = 0$ for $t < 0$. After a short transient feature, the system remains for a certain time τ_d on the upper branch until the switching sets in. This delay time τ_d diverges for $\delta \to 0$.

Fig. 5b shows τ_d as a function of δ and a calculated curve exhibiting the logarithmic singularity expected for critical slowing down. The agreement between experiment and theory is good, but the value of the thermal relaxation time τ has to be fitted to the experimental data, giving $\tau \approx 20$ ms. On the other hand, $\tau \approx 300$ μs fits the data from the stationary measurements. Two attempts have been made until now to explain this discrepancy: In one case the heat-flow in the sample during the switching process [39], in the other case the dependence of the absorption $A(T_L, t)$ [25] have been taken explicitly into account.

Recently the critical slowing down has been investigated for the switching-back process as well as for crystals attached to different substrates [35]. Here it turned out, that the delay times are changed by an order of magnitude depending on the heat conductivity of the substrate. In contrast, the delay time depends only weakly on the incident intensity for $t < 0$, and has particularly no diverging behaviour as a function of this quantity.

2.4. Some Further Aspects of Photo-Thermal Optical Bistability

Here we briefly to mention some experiments that have been made using the mechanisms of OB described above. The interested reader is referred to the literature cited.

Samples with dielectric mirrors evaporated onto the surfaces have an increased finesse of the resonator, so that the system shows dispersive OB or various mixed forms between dispersive and induced absorptive OB. Examples for this behaviour can be found in [16,30].

If noise is applied to the input signal $I_0(t)$, the hysteresis loops get narrower if the frequency spectrum of the noise has sufficiently strong components below the inverse thermal relaxation time [35].

By evaporating metal contacts onto the crystal one can apply an electric field in the area of the light spot. This leads to a current and thus to an additional heating of the crystal. So an opto-electric or electro-optic connection is possible with this so-called photo-thermal SEED-element (Self-Electrooptic-Effect-Device) [40,41].

By changing the environmental temperature T_{L0} in (5) one can change various quantities of the hysteresis loop, like the bistable width $I_\downarrow - I_\uparrow$ or the contrast between the two stable states of transmission. This gives the possiblity to build an all-optical temperature sensor using photo-thermally induced OB [42].

Finally it should be mentioned that induced absorptive OB may occur due to its built-in feedback mechanism also repeatedly because of partial switching into the depth of the sample, if the (heat-)diffusion is sufficiently small like e.g. in glasses [36,43,27]. The appearance of lateral structures in the switching process has been described in [44,45].

3. Optical Resonators

In this section we investigate the scenarios which can be observed and calculated, if we bring an induced absorptive bistable or monostable device into a (hybrid) ring resonator. These investigations complement those for the Ikeda resonator, which is basically a dispersive nonlinear element in an optical ring resonator [46]. The main difference is that the dispersive nonlinear element yields OB only in connection with a Fabry-Perot or ring resonator. In contrast, the induced absorptive element used here can be optically bistable already without external feedback.

3.1. The Hybrid Ring Resonator with a Bistable Element

First we assume that we use the polarization $\vec{E} \parallel \vec{c}$ for the ring resonator, i.e. photo-thermal induced absorptive OB with a hysteris-loop of the type shown in Fig. 3a.

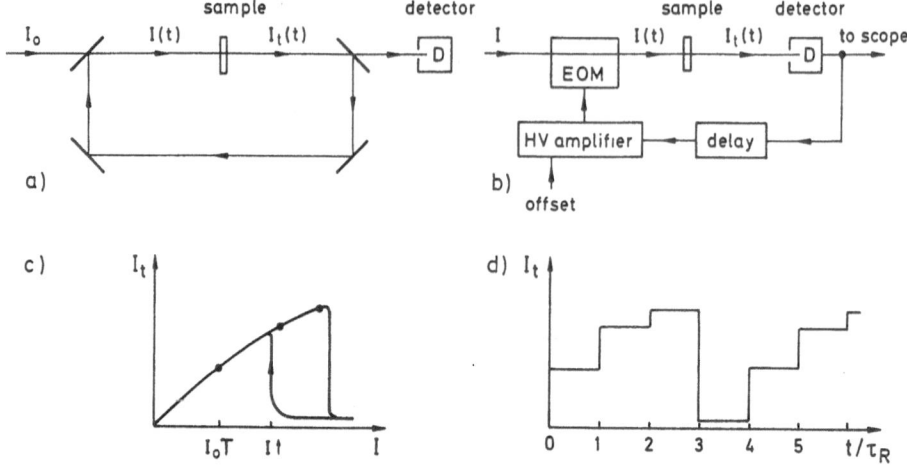

Fig. 6: Basic scheme for an optical ring resonator containing an optically non-linear sample (a), realization by a hybrid opto-electronic ring resonator (b), the bistable hysteresis loop of the sample (c), the temporal evolution of the self-oscillations (d). From [30]

In Fig. 6 we explain schematically what we expect by placing such a device in a ring resonator, assuming that the round trip time τ_R is long compared to the characteristic switching time τ which is some ms. Fig. 6a shows the ring resonator. To realize $\tau_R \gg \tau$ experimentally we use a hybrid resonator as shown in Fig. 6b, following an idea by [47] and calculations by [48,49]. Here, an electronic delay feeding a high-voltage amplifier (HVA) and a linear electro-optic modulator (EOM) allows us to simulate some 10^4 km of optical delay-length. The nonlinearity is really the optical nonlinearity of the CdS sample in the Ar^+ laser beam. The choice of the amplification of the electric signal in the delay-line and/or in the HVA allows to simulate various product reflectivities of the set of four mirrors and beamsplitters in Fig. 6a. So the hybrid ring resonator is a rather good simulation of the purely optical one, with two minor exceptions. We loose the phase information of the light in the electronic branch of the hybrid resonator. On the other hand, a purely optical resonator of some 10^4 km length would also significantly surpass the coherence length of most laser systems, so that adding intensities is actually in agreement with the conditions in the optical resonator. The other approximation is that we neglect the formation of lateral structures which may appear due to partial switching of the device [49].

The incident intensity I_0 is chosen so that the part transmitted through the first beam-splitter TI_0 is below the switch-back intensity $I_\uparrow$ of the bistable loop (Fig. 6c), and the sample is assumed to be in the transparent state at the beginning. This means that the intensity falling on the sample I is almost completely transmitted (for reasons of simplicity we neglect the sample reflectivity for the moment). A small fraction of the transmitted intensity is coupled

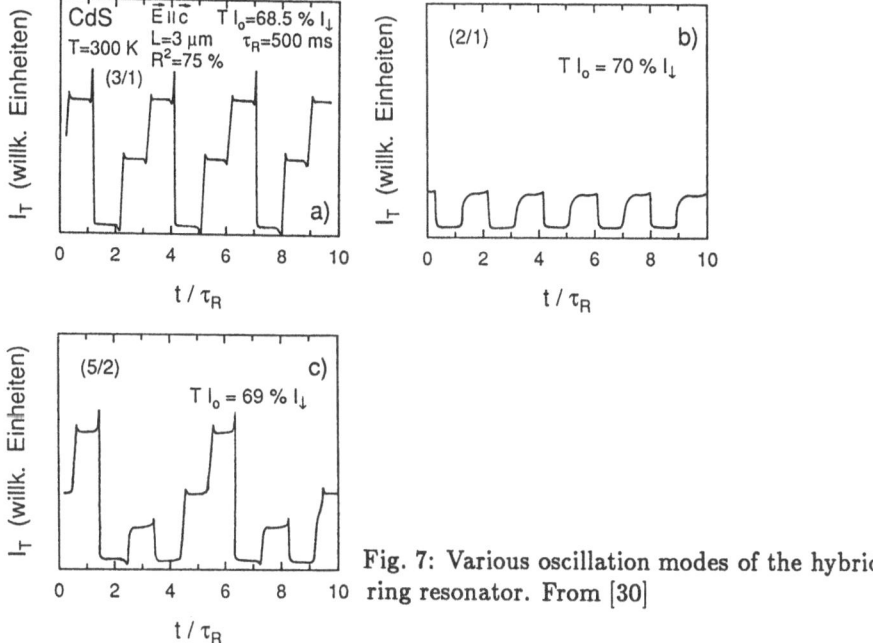

Fig. 7: Various oscillation modes of the hybrid ring resonator. From [30]

through the next mirror and detected by the detector D. After one round-trip time, the incident intensity and the transmitted one fall on the sample. Here we assume that we have to simply add the intensities. This procedure repeats itself several times (Fig. 6d) with steplike increases of the intensity impinging on the sample and being transmitted. After some round trip times I eventually exceeds $I_\downarrow$, and the crystal switches into the absorbing state. This means that almost no light is transmitted any more. Consequently, after another round trip time τ_R, only $T I_0$ falls on the sample, the sample switches back into the transparent state and the system starts again, giving rise to self-oscillations with a period which is an integer multiple of τ_R.

Our system can be described by (5) in connection with the delay equation (7), i.e. the expression for $I(t)$

$$I(T_L, t) = I_0 T + R I_t(t - \tau_R) = I_0 T + R I(t - \tau_R) \, exp[-\alpha(T_L(t - \tau_R))L] \quad (7)$$

Eq. (7) is an iteration rule that maps the value of $I(t)$ to the value of $I(t + \tau_R)$. In contrast to well-known equations like the logistic map or the circle map, (7) is not a function but shows hysteretic behaviour as can be seen in Fig. 3a.

Now we want to present some experimental results observed with the hybrid ring resonator described above.

In Fig. 7 we show various self-oscillation modes which occur when we change the control parameter input intensity I_0. We use the following nomenclature to describe the various modes: an (n/m) mode has a total period of $n\tau_R$ and m maxima. Correspondingly we see in Fig. 7 a $(3/1)$, a $(2/1)$ and a $(5/2)$ mode.

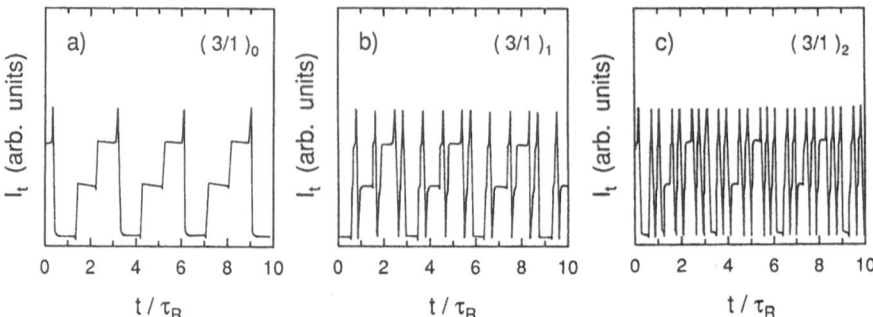

Fig. 8: Various coexisting oscillation modes of the hybrid ring resonator. From [30]

The small spikes at the end and/or the beginning of the steps are transient effects and reflect the reaction-time of the crystal. The (3/1) and (2/1) modes are fundamental modes, the (5/2) one belongs to the next generation. According to theoretical predictions and experimental investigations [30,31,48], various generations of modes can be constructed according to a Farey-tree structure, i.e. by adding numerators and denominators independently. We expect thus between an $(n1/m1)$ and an $(n2/m2)$ mode an $(n1+n2/m1+m2)$ mode with a smaller stability range against variations of the control parameter.

When we reduce the round trip-time, the oscillations become smeared out, the strict locking of the oscillation periods into multiples of τ_R is relaxed, and finally the oscillations disappear and the system remains for constant I_0 in a temporarily constant state, i.e. we have a transition from a limit cycle to a fixed point. Under these conditions the unstable branch can be measured [30].

When we briefly interrupt the incident beam, we can produce various oscillation modes under the same system conditions. An example is given in Fig. 8. A small additional spike is introduced in the system, which has basically the same shape as the temporally long structures before the perturbation. This coexistence of various modes at the same excitation conditions defines a new type of bi- or multistability. An (n/m) mode with i additional spikes is called an $(n/m)_i$ mode. So Fig. 8 shows a $(3/1)_0$, a $(3/1)_1$ and a $(3/1)_2$ mode. The basic reason for the occurence of these higher modes is the rather high dimensionality of the phase space of our system. For constant I_0 we can choose independently the intensity falling on the sample and its temperature (or absorption coefficient). These are two independent variables, which can be chosen during one round trip τ_R/τ times giving a dimensionality $d \approx 20$.

If we interrupt I_0 for a time τ_S with $\tau < \tau_S < \tau_R$ during an $(n/m)_0$ oscillation, the crystal switches to the upper branch but no light is transmitted. After one round-trip, the incident intensity during τ_S is only TI_0 and so forth, producing the self-similarity of the additional spikes stated above.

Until now I_0 was kept constant. An external harmonic modulation of I_0 with a frequency f leads to new phenomena [31]. The most interesting regime is one where f is comparable with τ_R^{-1}. We find under all conditions periodic oscillations. The oscillation period locks either into integer multiples of τ_R for $A < 0.4$ or in multiples of f^{-1} for $A > 0.4$. No non-periodic oscillations have been found so far, even when τ_R and f^{-1} form a rather irrational relation close to the golden mean.

The periodic oscillations show again a Farey-tree structure if we consider $\bar{T}f$, where $\bar{T}$ is the total oscillation period divided by the number m of maxima. The stability range of the various modes, if investigated either as a function of f for constant A or as a function of A for constant f, yields again a devil's staircase [30].

To conclude this subsection, we want to demonstrate two cases where temporally irregular oscillations appear.

3.2. The Hybrid Ring Resonator with a Monostable Element

Here we use the same hybrid ring resonator as above, but with a monostable optical device as explained in section 2.1. (Fig. 3b). Doing so we again observe oscillations for constant I_0, which are for a certain range of I_0 periodic with a period corresponding to integer multiples of τ_R. For another range of I_0 values, we get irregular oscillations. Actually we see in Fig. 9a-c a sequence of period doublings leading to deterministic chaotic behaviour. For a review of this topic see e.g. [50].

With the procedure of Grassberger and Procaccia [51] applied to the signal of Fig. 9c we found for the Haussdorff dimension D and the Kolmogorov entropy K, and for the correlation dimension D_2 and correlation entropy K_2

$$D \approx D_2 = 2.6 \pm 0.3 \quad \text{and} \quad K > K_2 = (100 \pm 25)\tau^{-1} \tag{8}$$

respectively, for embedding dimensions d above 6, giving thus evidence for the deterministic chaotic behaviour of the system. The reason for the appearance of chaos is the parabolic maximum in the iteration represented by Fig. 3b.

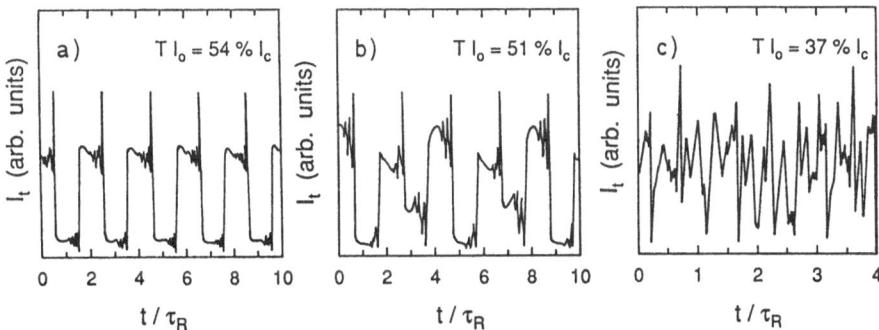

Fig. 9: Oscillation modes of the hybrid ring resonator with a monostable CdS crystal for various control parameters I_0. From [30]

101

3.3. The All-Optical Ring Resonator

The last experiment which we want to present in this section are self-oscillations in a purely optical resonator [35]. The concept is shown in Fig. 10a. We use three photo-thermally induced absorptive optically bistable devices. Each one has a holding beam, which is situated slightly below $I_\uparrow$, and the transmission of one element is the switching beam of the next. One bistable element acts as an inverter, i.e. it has high transmission T for low input intensity $I_0 < I_\uparrow$ and low transmission for high input intensity $I_0 > I_\downarrow$. Two of them in series restore the logic level, however, delayed by two switching times. Consequently, two of the elements produce an optical delay while the third one produces the oscillation. In Fig. 10b we show the temporal output of two of the three elements. The third one looks similar, with a corresponding shift of the logic levels. It is obvious that the oscillation is not strictly periodic in time. In contrast, computer simulations yield periodic oscillations. The origin of this discrepancy is not yet clear. It might be due to small temporal fluctuations of the laser beam, or the irregular behaviour might be an intrinsic effect of the system. As shown in connection with Fig. 5, the switching time is not a constant, but is determined by the critical slowing down. Small differences in the characteristics of the three devices may add up to produce after several periods an irregularity in the oscillation. In this case, we would have another type of deterministic chaotic behaviour, namely intermittency.

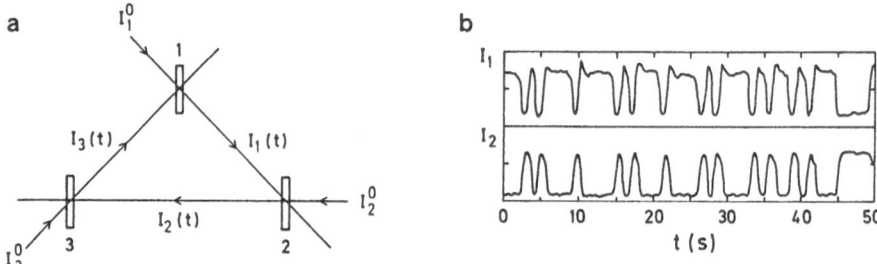

Fig. 10: Setup of an all-optical ring resonator (a), temporal oscillation of two of the transmitted beams (b). From [35]

4. Summary and Outlook

We have shown that photo-thermal optical nonlinearities caused by exciton-phonon interaction and nonradiative recombination give rise to a rich variety of optical and electrooptical effects, and to devices such as mono- and bistable switches, memories, oscillators, etc.

   This group of aspects is strongly connected with nonlinear dynamics and synergetics [4,9,50,52]. We observe the formation of temporal structures and of collective behaviour appearing in the systems. Optics is here especially valuable

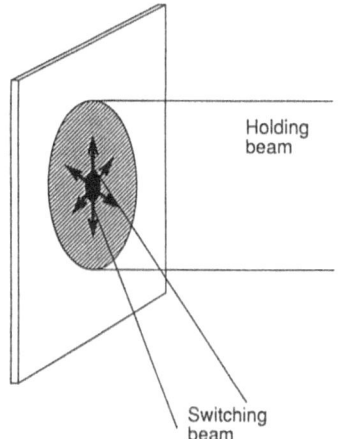

Fig. 11: Expansion of an area in the spot of a bistable element switched by a   spatially and temporally short pulse.

Holding beam

Switching beam

because it is often possible to study and to influence the nonlinearity and the feedback mechanism separately.

Future investigations may be concerned with the formation of spatial structures similar to the ones observed in the time domain. These effects should be especially distinct in the case of large spot sizes, giving rise to spatio-temporal pattern formation if connected with a suitable feedback. An example would be the situation indicated in Fig. 11. A sample showing thermally induced OB is illuminated by a constant holding beam between $I_\uparrow$ and $I_\downarrow$. When the center of the spot of the holding beam is "vaccinated" with a spatially and temporally short laser pulse, the sample will switch into the highly absorbing state in the center area. Due to the intrinsic feedback of absorptive OB and to the heat diffusion in the sample, the area of high absorption will spread all over the holding beam spot. Here it may be possible to observe the evolution of regular or irregular spatial structures, and the speed at which the switched area expands depends eventually critically on the holding beam intensity with respect to $I_\uparrow$ or $I_\downarrow$.

Another group of experiments dealing with such spatial phenomena are measurements using laser-induced gratings. Here the spatial inhomogenity is induced by the interference pattern of two laser beams that create areas of different indexes of absorption and refraction, and thus lead to (self)-diffraction of the incident light.

Further investigations concerning the hybrid ring resonator as described in section 3.1. and 3.2. will be the transition from the mode-locking behaviour to the period-doubling route into chaos that is caused by the transition from the bistable to the monostable input-output characteristics of the nonlinear crystal. This bistable→monostable transition can be induced by changing several control parameters e.g. by an increase of the environmental temperature T_{L0} or by varying the polarization of the incident light, as shown in section 2.1.

In conclusion, we hope that the reader of this contribution has gained an insight into the interesting field of optical nonlinearities in solids, and especially into its possibilities to study universal physical properties of nonlinear systems.

Acknowledgements: Most of the results presented here have been found in the research group of one of the authors (CK). We therefore thank all coworkers for their excellent work in this field done e.g. by Dr. K. Bohnert, C. Dörnfeld, M. Graf Lambsdorff, A. Witt and A. Schmidt. Thanks are also due to many colleagues all over the world for stimulating and fruitful discussions.

Financial support is acknowledged from various institutions, especially from the Deutsche Forschungsgemeinschaft through the Sonderforschungsbereich 185 Nichtlineare Dynamik, from the Materialforschungsschwerpunkt of the Land Rheinland-Pfalz, the European Community through the Stimulation Action and the Stiftung Volkswagenwerk.

The high-quality CdS samples used in most of the experiments have been grown by Dr. G. Müller-Vogt at the Kristall- and Materiallabor of the University Karlsruhe.

Bibliography

[1] Optical Bistability, Dynamical Nonlinearity and Photonic Logic, B.S. Wherrett and S.D. Smith eds., Phil. Trans. R. Soc. London A 313, 191-451 (1984)

[2] Optical Bistability: Controlling Light with Light, H.M. Gibbs, Acad. Press (1985)

[3] Excitonic Optical Nonlinearities, D.S. Chemla ed. special issue, JOSA B 2, 1135-1243 (1985)

[4] Optical Chaos, J. Chrostowski and N.B. Abraham eds., SPIE Proc. 667 (1986)

[5] Optical Bistability III, H.M. Gibbs, P. Mandel, N. Peyghambarian and S.D. Smith eds., Springer Proceedings in Physics 8 (1986)

[6] Laser-Induced Gratings, H.J. Eichler, D. Pohl and P. Günter, Springer series in Opt. Sci. 50 (1986)

[7] From Optical Bistability Towards Optical Computing, P. Mandel, S.D. Smith, B. Wherrett eds., North Holland (1987)

[8] Optical Bistability IV, W. Firth, N. Peyghambarian and A. Tallet eds., J. Physique, 49 C2 (1988)

[9] Optical Nonlinearities and Instabilities in Semiconductors, H. Haug ed, Acad. Press (1988)

[10] Laser Spectroscopy of Solids II, W.M. Yen ed., Topics in Applied Physics, 65, Springer (1989)

[11] Optical Nonlinearity and Bistability of Semiconductors, F. Henneberger ed. phys. stat. sol. b 150, 347-919 (1988)

[12] C. Klingshirn in "Advances in Nonradiative Processes in Solids", NATO ASI Series 249, 529 (1991), B. DiBartolo ed.

[13] H. Haug and C. Klingshirn, Physics Rep. 70, 315 (1981)

[14] B. Hönerlage, R. Levy, J.B. Grun, C. Klingshirn and K. Bohnert, Physics Rep. 124, 161 (1985)

[15] S. Schmitt-Rink, D.S. Chemla and D.A.B. Miller, Advances in Physics 38, 89 (1989)

[16] M. Wegener, C. Klingshirn, S.W. Koch and L. Banyai, Semiconductor Science and Technology 1, 366 (1986)

[17] Excitons, E.I. Rashba and M.D. Sturge eds., Modern Problems in Condensed Matter Sciences 2, North Holland (1982)

[18] C. Klingshirn in Energy Transfer Processes in Cond. Matter, B. DiBartolo ed., NATO ASI series B 144, 285 (1984) Plenum

[19] F. Urbach, Phys. Rev. 92, 1324 (1953)

[20] W. Martienssen, J. Phys. Chem. Sol. 2, 257 (59) and ibid. 8, 294 (59)

[21] D. Dutton, Phys. Rev. 112, 785 (1958); M. Kurik, phys. stat. sol. a 8, 9 (1971); S. Spiegelberg, E. Gutsche and J. Voigt, phys. stat. sol. b 77, 233 (1976)

[22] H. Sumi and Y. Toyozawa, J. Phys. Soc. Japan 31, 342 (1971)

[23] J. D. Dow and D. Redfield, Phys. Rev. 85, 94 (1972)

[24] J.G. Liebler, S. Schmitt-Rink and H. Haug, J. Luminesc. 34, 1 (1985)

[25] J. Gutowski, J. Hollandt and I. Broser, Z. Physik B, 76, 547 (1989)

[26] C. Klingshirn, M. Wegener, C. Dörnfeld, M. Lambsdorff, J.Y. Bigot and F. Fidorra in Ref. 5, 129; M. Lambsdorff, C. Dörnfeld and C. Klingshirn, Z. Physik B 64, 409 (1986)

[27] C. Klingshirn, M. Wegener, C. Dörnfeld and M. Lambsdorff in Ref. 7, 252

[28] D.A.B. Miller, JOSA B 1, 857 (1984)

[29] D. Jäger, F. Forsmann and B. Wedding, IEEE JQE, 1453 (1985)

[30] M. Wegener and C. Klingshirn, Phys. Rev. A 35, 1740 and 4247 (1987); M. Wegener, Ph.D. thesis, Frankfurt (1987)

[31] M. Wegener, C. Klingshirn and G. Müller-Vogt, Z. Physik B 68, 519 (1987)

[32] F. Henneberger, phys. stat. sol. b 137, 371 (1986)

[33] I. Haddad, M. Kretzschmar, H. Rossmann and F. Henneberger, phys. stat. sol. b 138, 235 (1986)

[34] M. Wegener, C. Klingshirn, A. Daunois, J.-Y. Bigot, N. Cherkaoui Eddeqaqi and J.B. Grun, Appl. Phys. Lett. 52, 685 (1988)

[35] J. Grohs, A. Schmidt, M. Kunz, C. Weber, A. Daunois, A. Rupp, W. Dötter, F. Werner and C. Klingshirn, Proc. SPIE 1127, 39 (1989)

[36] Dynamics of First-Order Phase Transitions in Equilibrium and Nonequilibrium Systems, S.W. Koch, Lecture Notes in Physics 207, Springer (1984)

[37] Nonequilibrium Phase Transitions, E. Schöll, Springer (1987)

[38] J.A. Goldstone, E.M. Garmire, IEEE JQE 17, 366 (1981)

[39] M. Kretzschmar, F. Henneberger, H. Rossmann and I. Haddad, phys. stat. sol. b 143, K71 (1987)

[40] A. Witt, M. Wegener, V.G. Lyssenko, C. Klingshirn, G. Wingen, Y. Iyechika, D. Jäger, G. M ller-Vogt, H. Sitter, H. Heinrich and H.A. Mackenzie, IEEE JQE 24, 2500 (1988)

[41] V. Kazukauskas, J. Grohs, C. Klingshirn, G. Wingen and D. Jäger, Z. Physik B 79, 149 (1990)

[42] J. Grohs, M. Müller, A. Schmidt, A. Uhrig, C. Klingshirn, H. Bartelt, Optics Commun. 78, 77 (1990)

[43] G.R. Olbright, H.M. Gibbs, N. Peyghambarian, H.E. Schmidt, S.W. Koch and H. Haug in Ref. 5, 186

[44] S.W. Koch and E.M. Wright, Phys. Rev. A 35, 2542 (1987)

[45] H.X. Nguyen, V.D. Egorov, A. Harendt and E.V. Nazanova, phys. stat. sol. b 148, 407 (1988)

[46] K. Ikeda, H. Daido and O. Akimoto, Phys. Rev. Lett. 45, 709 (1980) and 48, 617 (1982); J.V. Moloney, Phys. Rev. A 33, 4061 (1986)

[47] Klaus Bohnert, private communications

[48] M. Lindberg, S.W. Koch and H. Haug, JOSA B 3, 751 (1986)

[49] I. Galbraith and H. Haug, JOSA B 4, 1116 (1987)

[50] Deterministic Chaos, H.G. Schuster, VCH (1988)

[51] P. Grassberger and I. Procaccia, Phys. Rev. Lett. 50, 346 (1983), Phys. Rev. A 28, 2591 (1983) and Physica D 9, 189 (1983)

[52] Synergetics, an Introduction, H. Haken Springer Series in Synergetics 1

Chaos and Nonlinear Effects in Josephson Junctions and Devices

U. Krüger[1], *J. Kurkijärvi*[2], *M. Bauer*[1], *and W. Martienssen*[1]

[1]Physikalisches Institut der Universität Frankfurt am Main
und Sonderforschungsbereich Nichtlineare Dynamik,
W-6000 Frankfurt am Main 1, Fed. Rep. of Germany
[2]Åbo Akademi, Institute of Physics,
Porthansgatan 3, SF-20500 Åbo/Finnland

1. Introduction

The planar Josephson tunnel junction driven by both ac- and dc-current sources is a relatively simple physical system which displays a wide range of nontrivial nonlinear phenomena. Chaotic behaviour in this system was first revealed in simulations by *Huberman* et al. [1] and by *Kautz* et al. [2]. Subsequently the existence of chaos in Josephson junctions has been verified by theory [3 − 10] and simulations [11 − 20].

Most effort has been concentrated on the so-called "Resistively shunted junction" (RSJ)-model [21]. This model is known to provide a qualitatively correct description of the behaviour of the real junction in many relevant cases. It assumes that the underlying differential equation governing the dynamics of the Josephson junction is isomorphic to that of a driven pendulum which is one of the most simple equations of motion exhibiting chaos and describing an interesting physical system. The same model seems relevant also in other contexts such as charge density waves and various electronic circuits. Electronic circuits indeed are often used in analog simulations of the properties of the Josephson junction. This allows chaos to be studied through a full interplay between theory, simulation and experiment.

A great deal of insight into phenomena of chaotic behaviour and the various routes to chaos has been achieved through the use of simple models or model maps. The hope is that very simple models may share universal properties in common with "real" chaotic systems with a large number of degrees of freedom. Nevertheless, the link between simple dynamical systems and real physical systems remains to be found. No practicable recipe is known, say, for a systematic reduction of the latter to the former.

A first step in this direction could be a study of the rf-driven Josephson junction as a system for which one can hope to compare the experimental behaviour to numerical and analog simulations. The reason is that it contains the lowest number of parameters among the experimental systems that have been found to exhibit chaotic behaviour. Such a comparison should not only provide information about the transition to chaos and the chaotic state in Josephson junctions, but might also yield information about the adequacy of models of Josephson junctions.

Apart from the inherent interest in chaotic behavior, the rf-biased junction is also of practical importance. It defines, for example, the standard of voltage, an application for which chaotic behavior must be avoided. What is more, many other technical devices have been developed which rely on the extreme nonlinearity of Josephson junctions, including magnetometers based on Superconducting Quantum Interference Devices (SQUID's) [22], digital circuits [23] and parametric amplifiers [24]. In spite of extensive analysis, it is only recently that the possibility of chaotic behaviour in Josephson devices designed for technical applications has been considered. These studies have been practically motivated as chaos has been suggested as the source of the large levels of excess noise observed in Josephson parametric amplifiers.

In the present paper we will try to give an overview of the different theoretical and experimental studies of chaos and nonlinear effects in Josephson junctions and devices. Since the amount of theoretical and especially numerical work on Josephson systems has been enormous in recent years, a fair account of all contributions within a limited amount of space is impossible. We therefore place main emphasis on experiments and theoretical investigations which directly rely on experimental studies. Despite the significant consequences with regard to technical applications, experimental evidence for chaos in Josephson junctions has not been extensive, and progress is not yet as spectacular as in the numerical work. This is merely due to the fact that the relev ant dynamical behaviour of Josephson junctions emerges at very high frequencies (GHz-region) where experimental data analysis is not easily accessible. In addition, period doublings do not lead to large changes in the voltage across a junction, and noise can have other reasons than chaos. A better understanding, on the other hand, is extremely important for identifying the role of deterministic noise in real physical systems where also thermal noise is present.

The paper is organized in the following way: Section 2 reviews the RSJ-model and defines the experimental circumstances for its validity. Section 3 sketches briefly how specified Josephson junctions are manufactured. Section 4 characterizes parameter-regions of the model which are experimentaly accessible in order to investigate nonlinear and chaotic behaviour. Section 5 describes experimental evidence for chaos in Josephson junctions revealed by inspecting the junction's I/V-characteristic, frequency spectrum and noise temperature. Also a number of experimental results on modifications of the RSJ-junctions are explored, and finally a short outlook is given in section 6.

2. The RSJ-Model

Solid-state theory cannot give an exact and universal relation between the current and voltage of an arbitrary designed and manufactured junction. The study of dynamical properties of various junctions requires models. Josephson tunnel junctions of the SIS-type are the best studied and most important for the investigations reported in this article. They consist of two superconductors (S) separated by a thin insulating (I) layer. In most cases, thin vacuum-deposited superconducting films serve as electrodes. The oxide of the lower (base) electrode plays the role of the insulator. The oxide layer thickness d_I is of the order of ten to thirty atomic sizes $(10 - 30\text{Å})$ so that pairs of electrons (Cooper pairs) have a small but nonvanishing probability $(p \sim 10^{-5} - 10^{-3})$ of penetrating from one electrode into the other via quantum tunneling through the energy barrier created by the insulator [25].

Such penetrations result in a nonvanishing normal conductance σ_N in the normal state $(T > T_c)$, and in the Josephson effect in the superconducting state $(T < T_c)$. The supercurrent I_s through the junction is given by the relation:

$$I_s = I_0 \sin \phi \tag{1}$$

where I_0, the critical current, is the maximum supercurrent that the junction can sustain without developing a voltage, and ϕ is the phase difference between the complex order parameters in the two superconductors. For applied currents larger than I_0, part of the current must flow through a resistive element (intrinsic to the junction or added externally) so that a voltage V develops across the junction. In this state, equation (1) is still valid, but $\phi(t)$ evolves with time according to the relation

$$V = (\phi_0/2\pi)\dot{\phi} \tag{2}$$

where $\phi_0 \equiv h/2e$ is the flux quantum and the dot denotes differentiation with respect to time. The two other intrinsic parameters of the junction are its self-capacitance C, brought about by the overlap of the two superconductors, and its nonlinear quasiparticle conductance

$$\sigma_{qp}(V) = I_{qp}/V \tag{3}$$

where I_{qp} is the quasiparticle current.

The Resistively shunted junction (RSJ) model, shown in Fig.2, introduced by *McCumber* and *Stewart* [21], accepts the simplest expressions for the current components: it describes an ideal Josephson junction (Eq.1) shunted by a resistance R (Eq. 3) and its self-capacitance C and driven by a current source which includes a dc component of amplitude I_0 and an rf component of amplitude I_1 and frequency ω_1. The equations relating these idealized parameters are written as

$$\frac{dV}{dt} + R^{-1}V + I_c \sin \phi = I_{dc} + I_{rf} \sin(\omega t) \tag{4a}$$

$$2eV = \hbar\frac{d\phi}{dt} \tag{4b}$$

where V is the voltage across the junction, e is the electron charge and $\hbar$ is Planck's constant divided by 2π.

According to these basic equations, if R is large, the I/V curve consists of three separate branches as depicted in Fig.1. For currents below the critical value, not only the superconducting state but also the state with finite resistance is possible. The junction, revealing the same potential as a periodically driven damped pendulum, behaves like a particle with a large effective mass, so that with appropriate initial conditions it can move along the energy profile even at $|I| < I_c$, leaping over the

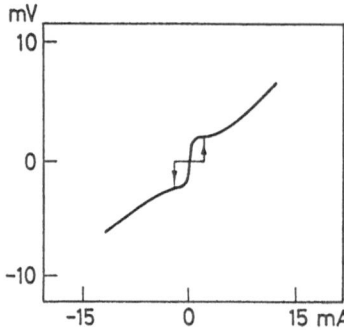

Fig. 1. The dc I/V curve of the Josephson junction with low damping.

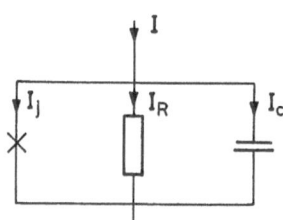

Fig. 2. Circuit diagram of the rf-biased Josephson junction

energy barrier by inertia. As a result of this multiple valued V(I) dependence, the I/V curve has an hysteretic shape.

If we measure all currents in units of I_0 and time in units of ω_p^{-1}, where $\omega_p = (2eI_c/\hbar C)^{1/2}$ is the plasma frequency, Eqs. (4a) and (4b) may be rewritten as

$$\frac{d^2\phi}{d\tau^2} + \beta_c^{-1/2}\frac{d\phi}{d\tau} + \sin\phi = i_{dc} + i_{rf}\sin(\omega\tau) \tag{5}$$

Here, $\beta_c = 2eI_0R^2C/\hbar$ is the McCumber parameter. It should be noted that some authors [3, 4, 7] use a normalization where time is measured in units of ω_c^{-1}, and $\omega_c = 2eI_cR/\hbar$. In comparing literature citations it is useful to notice that $\omega_c/\omega_p = \beta_c^{-1/2}$. With this normalization to ω_c, Eqs. (4) have the form

$$\beta_c\frac{d^2\phi}{d\tau^2} + \frac{d\phi}{d\tau} + \sin\phi = i_{dc} + i_{rf}\sin(\omega\tau) \tag{6}$$

Indeed, since there are three important frequencies in the case of the RSJ-model, it is also possible to normalize the time unit to ω_a^{-1}, where $\omega_a = (RC)^{-1}$. These three frequencies arise naturally in connection with the $\omega - \beta_c$ diagram which will be discussed later. We note that $\omega_a\omega_c = \omega_c^2$. It is also useful to note that $\beta_c = Q^2$, where Q is the quality factor of the oscillator. In most cases we shall make use of the plasma frequency normalization in this work. Equation (6) has the minimum number of three degrees of freedom $d\phi/d\tau$, ϕ, $\omega\tau$ required for chaotic solutions.

In spite of the simplicity of the RSJ-model one should keep in mind that it is only valid in a narrow temperature region just below the critical temperature T_c of the superconducters involved [26, 27]. At the usual operation temperature, $T \leq T_c/2$ real junctions show several important deviations from the RSJ-model; the most substantial ones are the voltage (frequency) dependance of the quasiparticle conductance $\sigma_{qp}(V)$ (Eq.3), the inhomogeneous current distributions in the tunnel barrier due to the finite London-penetration depth λ_J [28] and geometrical self-resonance effects (Fiske-Resonances [29]) which may interfere with the tunnel current I_s.

3. Fabrication technology

Several excellent review articles are available on the fabrication of Josephson junctions and devices [30]. Only a brief description will be presented here.

Figure 3.a shows a sketch of a typical tunnel junction structure. To fabricate the junction, a thin film of a superconducting material with a thickness of a few hundred nanometers is deposited over a clean surface of an insulating barrier (1). After the film is patterned using photolithography to form the "base electrode" (2) of the desired shape, it is covered with a relatively thick insulating layer (3). Then photolithography is used again to form a "window" in the insulator with the area A defining the area of the junction (from $\sim 10^{-8}$ to $\sim 10^{-2} cm^2$). After cleaning the surface of the base electrode inside the window, it is oxidized to form a very thin $(d_I \sim 2 - 5nm)$ insulating barrier (4) of the tunnel junction. The fabrication is then completed by deposition of the upper superconducting layer, and its patterning to form the "counter-electrode" (5).

In the beginning of the Josephson effect research in the early 1960s, the tunnel junctions were fabricated using the so called "soft" superconductors (Sn,In,Pb) and thermal oxidation to form the barrier. With such relatively simple technology, a reproducible critical current density j_c did not exceed $\sim 10^2 A/cm^2$, and the dimensionless capacitance parameter β, which in device variables is defined by

$$\beta = \frac{2e}{\hbar} V_c^2 \left(\frac{\varepsilon_r \varepsilon_0}{d_I} \right) j_c^{-1} \tag{7}$$

was therefore large: $\beta > 10^3$.

As we will see later, β should be on the order of $\beta \sim 1$ in nearly all practical devices based on the Josephson-effect. In addition, reproducible measurements and reliable device applications strongly depend on the thermal recycling from room temperature ($\sim 300K$) to the temperature of liquid helium ($\sim 4.2K$). There are different ways to meet these requirements by fabrication technology. One possibilty is due to the development of the "lead-alloy technology" by *Matisoo et al.* [31] in the 1970s. In this technology, the base electrodes are fabricated of the alloy of Pb with Au and In, and oxidized in the plasma of an rf glow discharge, while alloys of Pb with Au or Bi are used for the counter electrodes. This technology provides reproducible junctions together with a great variability of barrier thickness and hence of j_c. The parameter β can thus be reduced to the desired values in the range of unity.

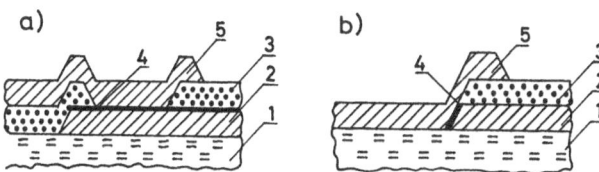

Fig. 3. Schematic side view on typical tunnel junction structures. (a) planar type and (b) edge type. Notation: 1, dielectric substrate; 2, superconducting base electrode; 3, thick insulating layer; 4, thin insulating layer forming the tunnel barrier; 5, superconducting counter-electrode.

Another fabrication technology makes use of "rigid" superconductors like Nb at least for the base electrode. Evaporation or sputter processes with Niobium, however, require good vacua. In addition, oxide barriers on niobium have very complex substructures and relatively high specific capacitance, so that considerable advantage is achieved by replacing the natural oxide by artificial barriers. Such barriers can be made either by the deposition of an insulator, or a semiconductor, or some other metal, e.g. Al, with its consequent oxidation. In addition to niobium, several other rigid superconducting compounds are under study to serve as electrode materials, including NbN, Nb_3Sn, V_3Si and Nb_3Ge. With the exception of NbN for those materials the use of artificial barriers is almost inevitable.

We should mention three other types of junction configurations. The edge configuration (Fig. 3b) is appropriate for the design of very small area junctions without making use of submicron techniques like electron-beam-lithography or X-ray-lithography. In the case of edge junctions ion milling is used to form a step edge on the base electrode (2) and the thick insulator (3). After oxidation of the edge area (4), the counter electrode (5) is deposited, so that the vertical junction size is defined by the base electrode thickness and can be as small as $\sim 0.1\,\mu m$. The junction area can be reduced from ~ 0.1 to $0.01\,\mu m^2$. For junctions like these the values of β can be less than unity.

Noticeable attention in the context of chaos in Josephson junctions has been payed to "weak-links" (Fig.4). In these so-called SNS-structures the weak electrical contact between the electrodes is provided by some link made of a normal metal (N) or superconductor (S). The fabrication technology for these junctions can be very similar to that of planar SIS-type junctions but their electrical properties, e.g. the I/V-curve, are rather different. The Josephson current in the SNS weak links is due to the proximity-effect: the ability of a superconducting condensate of Cooper-pairs to conserve the amplitude and phase of its order-parameter $\phi(r,t)$ even in a normal metal at distances of the order of the coherence length. For typical superconductors and normal metals, the coherence length is tens and even hundreds nanometers. The thickness d_N of a normal layer can be much greater than that of an oxide layer.

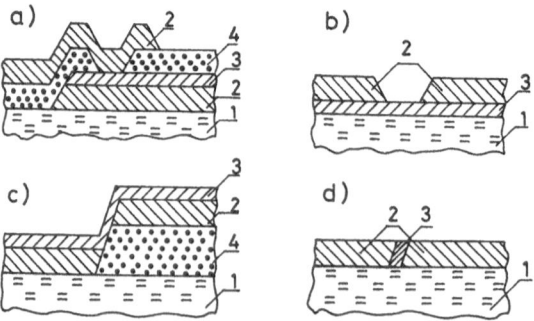

Fig. 4. Schematic side view on typical weak link structures. (a) SNS sandwich, (b) planar type variable-thickness bridge, (c) edge type variable thickness-bridge, (d) combined sandwich-bridge structure. Notation: 1, dielectric substrate; 2, superconducting electrodes; 3, normal metal layer; 4, thick insulating layers.

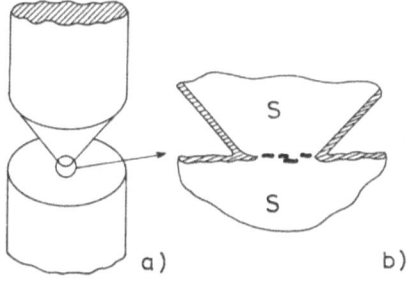

Fig. 5. (a) Schematic general view and (b) microstructure of a point contact junction.

Finally a very common configuration of Josephon junctions is the point contact (Fig.5). The most attractive feature of "point-contacts", which played an important role in the history of the field and which has brought them back in the context of high-T_c-research, is the extreme simplicity of their fabrication: it is sufficient to press slightly a sharply pointed superconducting needle to a plane surface of another superconducting electrode with the help of some adjusting material. While the pressure is weak, the electrodes are separated by the thin oxide layer which inevitably covers their surface. Of course, the characteristics of these contacts can differ by a large extent from SIS and SNS junctions due to microshorts and granular capacity.

4. Choice of Parameters

The first step towards investigations of chaos and nonlinear effects in Josephson junctions is choosing suitable junction parameters. A great difficulty in the description of the nonlinear solutions of the RSJ equations arises from the multitude of a four-dimensional parameter space $(i_{dc}, i_{rf}, \beta_c, \omega)$. Varying only a single parameter may lead to an infinite number of different periodic and quasiperiodic plus chaotic solutions [32] . In addition, several solutions exist with interwoven, fractal basins of attraction meaning that initial conditions within different basins entail different asymptotic solutions.

Altogether the behaviour of Eq. (6) has been treated in an enormous amount of numerical and analog simulations. It seems to be a completely hopeless task to draw a comprehensive guide to the resulting types of solutions the harmonically driven RSJ equation can assume in all corners of the parameter space. Therefore, typically the system has been investigated in the i_{rf}, ω plane for a fixed damping parameter β_c. A particularly thorough investigation was carried out by Kautz and Monaco [33]. Related model systems with a $\cos\phi$-term [34] and a nonlinear quasiparticle I/V-curve [35] have also been considered. In order to introduce some structure into the problem we subdivide the discussion into two sections according to whether the junction is driven by an ac current only or a combined ac and dc current [20]. For both cases we sketch a number of phase diagrams for different parameter values. In order to emphasize systematic tendencies, a great amount of detail has been omitted.

4.1. Simulations of ac driven Josephson junctions without a dc bias

In the literature the case of ac driven Josephson junctions without a dc bias appears to be variously shaped. Nevertheless a contour of a structured picture has arisen out of the large amount of work invested in the problem. *MacDonald et al.* [35] performed numerical simulation for $\beta_c = 4$. Fig.6 sketches areas in the i_{rf}, ω plane where more complicated than just periodic or steady solutions have been found. The RC cut-off frequency ω_{RC} is indicated in order to clarify the role of the characteristic frequencies.

Fig. 7 sketches the same phase portrait for the value $\beta_c = 16$ according to analog simulation of Eq. 6 by *d'Humières et al.* [3]. In order to give an impression of the whole phase space being displayed, the range of frequencies on the horizontal axis is the same in the two figures.

In a digital simulation *Pedersen et al.* [4] studied an equivalent phase portrait for $\beta = 25$ which resulted in the picture sketched in Fig. 8. In this case, the RC cut-off frequency lies well below the lowest frequency studied. Apart from differences in detail this picture again corresponds to an analog simulation with the same β_c performed by *Huberman et al.* [1]. With regard to experiments on real Josephson-junctions the question arises, however, whether there may be complex behaviour also below ω_{RC} which would reduce the bandwith required for a direct experimental observation of chaos. More information on this point has been gained by *Kautz et al.* [3] who concluded, from numerical investigations, that chaotic behaviour mainly

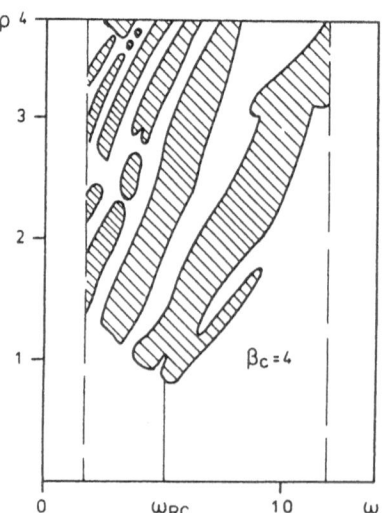

Fig. 6. The phase portrait of the RSJ equation in the ω, i_{ac} plane for $\beta_c = 4, i_{dc} = 0, \rho = i_{ac}$. The hatching indicates areas with multiply periodic or chaotic solutions (simplified from MacDonald et al. [35].

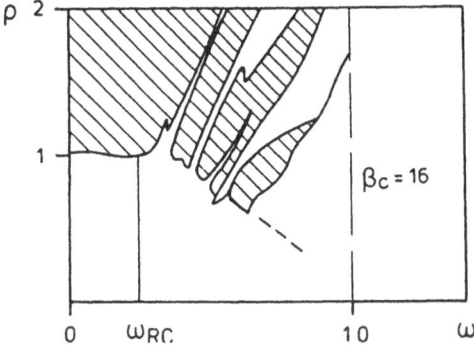

Fig. 7. Same phase portrait as above but for $\beta_c = 16, i_{dc} = 0$ and $\rho = i_{ac}$. (simplified from d'Humières et al. [3].

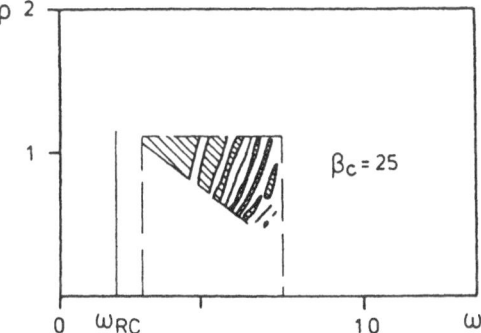

Fig. 8. Same phase portrait as above but for $\beta_c = 25, i_{dc} = 0$ (simplified from Pedersen et al. [4]).

occurs between the plasma frequency ω_p and the RC cut-off frequency. This result has been confirmed by studies performed by *Cirillo et al.* [36] who mapped the lower border of the occurence of complex behaviour in the ω, i_{ac} plane for a number of different β_c values.

Another question of principal importance is what kind of behaviour one first encounters when the driving amplitude is increased for fixed ω, β_c. In general all authors agree that period doubling is first to appear at frequencies on the high side of the minimum of the ω, i_{ac} boundary of the hatched area of complicated solutions. At higher driving amplitudes successively chaos and odd periodicities, as well as periodic regions can be resolved. On the lower side, the intermittency route to chaos has been found. On the basis of a very careful analog simulation of the complex phase portrait at the value $\beta_c = 4$, *Yeh et al.* [37] suggest the following mechanism for the onset of intermittency on the low frequency side: when the driving amplitude is just large enough to get the phase over the first sinusoidal potential hump, a dephasing mess occurs and the amplitude is suddenly reduced. The amplitude then begins to grow slowly and finally goes overboard again after a long tranquil period. This picture is consistent with the reconstructed return map and measurements of the scaling behaviour of the laminar length, indicating that the observed intermittency is of type I. Above the minimum of the phase boundary, *Yeh et al.* [37] observe period doubling again and establish agreement with the universal features of the period doubling transition such as Feigenbaum's constants δ and α.

4.2. Simulations of ac driven Josephson junctions with a dc bias

In contrast to the former case the nonlinear behaviour of a junction with a finite dc drive is theoretically much better understood at least to the extent that the average voltage across the junction is different from zero. This is mainly due to the pioneering work by *Bak et al.* [38] and *Jensen et al.* [39] on the universal properties of circle maps. Using stroboscopic Poincaré-map techniques, *Bak et al.* pointed out that the appropriate return map of the RSJ-system (Eq. 6) with $i_{dc} \neq 0$ is the one-dimensional circle map. In addition they predicted global universal properties of the mode-locking phenomena [39]. Formally *Bak et al.* state in two steps that first the two dimensional Poincaré section turns into a dissipative two-dimensional return map

$$\phi_{n+1} = f(\phi_n, \dot{\phi}_n) \tag{8a}$$

$$\dot{\phi}_{n+1} = g(\phi_n, \dot{\phi}_n) \tag{8b}$$

which in turn reduces into a one-dimensional circle map:

$$\phi_{n+1} = [\phi_n + 2\pi\Omega + g(\phi_n)](mod2\pi) \tag{9}$$

where the function $g(\phi_n)$ is a periodic function of the phase ϕ_n . The detailed shape of the function $g(\phi_n)$ is an indication for the type of the nonlinearity in the system. Once the return map has been identified as the skeleton of the different kinds of nonlinear behaviour which might occur in the RSJ-system with a given set of parameters, it is no longer necessary to integrate the differential equation. One can use the one-dimensional return map instead which is a great deal faster to iterate than to solve the original Eq. 6.

Circle maps are the simplest representation of a dynamical system showing the interplay between periodic, quasiperiodic and chaotic behaviour. The fundamental quantity describing the dynamics of such a map is the winding number W, the average rotation per iterate. The iterate of a given phase ϕ_n in this sense is the sequence $\phi_0, \phi_1 \ldots$ where $\phi_1 = f(\phi_0)$, $\phi_2 = f(f(\phi_0))$ etc. The winding number W is defined as the limit:

$$W = \lim_{n \to \infty} \frac{f^n(\phi) - \phi}{n} \tag{10}$$

As long as f is a *diffeomorphism*, strong theorems hold: The winding number is then well defined and neither depends on the choice of an initial point nor on the detailed form of $g(\phi)$. The winding number can take either a rational value $W = P/Q$, $P, Q \in \mathbb{Z}$ or an irrational value. For the two-parameter family of maps, e.g. the "sine-circle-map" sketched in Fig.9

116

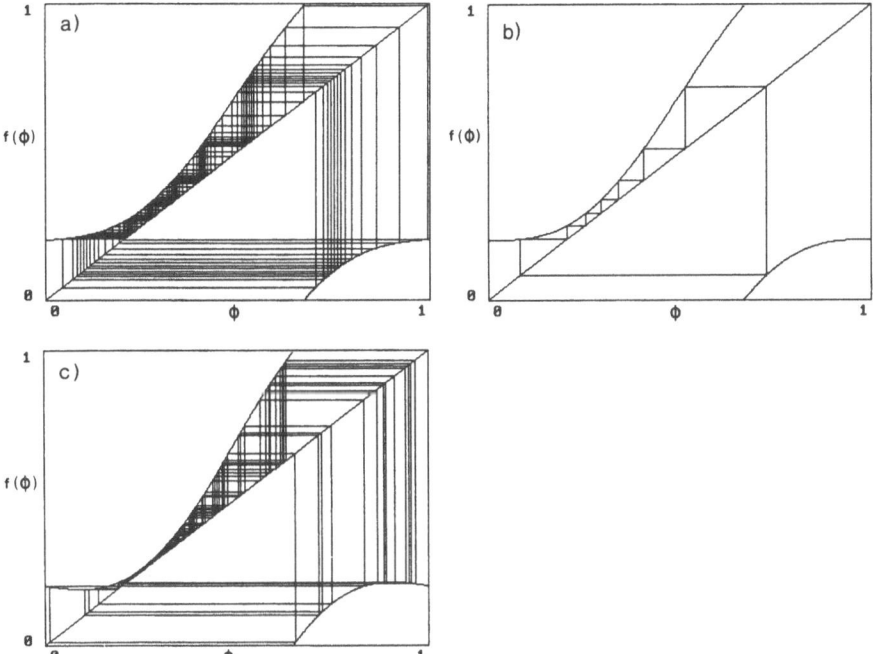

Fig. 9. Iterations for the circle map for $\Omega = 0.2$ and (a) $K = 0.9$, (b) $K = 1$, and (c) $K > 1$ the map develops local extrema and chaotic behaviour may occur.

$$f_{\Omega,K}(\phi) = \phi + \Omega + \frac{K}{2\pi} \sin 2\pi \phi \qquad (11)$$

we can distinguish the following cases.

For $0 < K \leq 1$, where the map is invertible, the iterations locks on a finite interval of Ω to every rational number P/Q. The motion is therefore called "mode-locked". For each rational W there exists a whole interval, which we shall denote $\Delta\Omega(W)$, where that winding number persists. Nevertheless, these intervals summed over all the rational numbers do not fill the Ω axis, and there is room for irrational winding numbers. This means that the relation $W = W(\Omega)$ is highly non-trivial. It is continuous and non-decreasing and has a constant plateau for each rational W. Such an object is called a "Devil's staircase" [32]. In contrast to periodic motion, quasiperiodic motion only occurs for specific $\Omega(W)$. If the given irrational W is approximated to higher and higher order by rationals, the corresponding mode-locked intervals $\Delta\Omega$ will shrink to a point. The probability of quasiperiodic intervals therefore depends strongly on the function g.

Even more complicated behaviour can be encountered if we allow the map to be noninvertible by varying the parameter $K \geq 1$. Now, in the special case of $K = 1$ the locking intervals do fill the whole axis, i.e. the Devil's staircase of steps becomes complete. It can be shown numerically that the fractal dimension of the remaining set of unlocked intervals with irrational winding numbers is $d_H = 0.87$, i.e. smaller than one. In other words: the set defined by removing the mode-locked

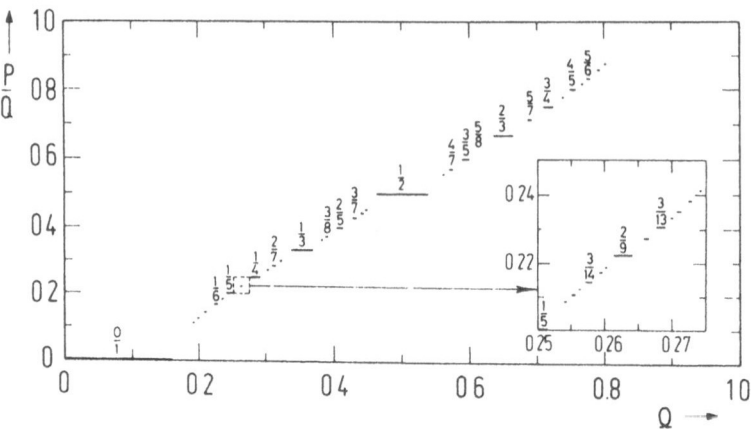

Fig. 10. The Devil's staircase formed by mode locked intervals calculated at $K = 1$. The self-similar structure is emphasized by magnification.

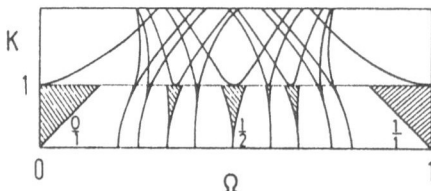

Fig. 11. Phase diagram of the circle map. As K increases, the tongues become wider, and eventually they overlap for $K > 1$. (from Jensen et al. [39]).

intervals is a Cantor set with the fractal dimension d_H and the self-similar structure sketched in Fig.10 found to be universal, i.e. independent of the particular choice of the function $g(\phi_n)$. Thus at $K = 1$ the mode-locked intervals have measure 1 and the quasiperiodic motion zero measure, in contrast to the situation at $K = 0$ where the quasiperiodic motion has measure 1 and the periodic motion measure 0. When $K > 1$ the sine-circle-map becomes noninvertible, and one finds chaos generated by period-doublings or intermittency or bistable responses where the winding number depends on the initial condition.

The phase diagram for the sine-circle map is shown in Fig.11. Each of the "tongues" coming ot of the Ω axis corresponds to a mode locked region with some rational winding number. They start out with zero width. As K increases they get wider. Above $K = 1$ they start overlapping. In this region f has two extrema, and period doublings take place inside the tongues.

In principle these results carry over to Josephson junctions. In practice there is the difficulty that there is no simple relationship between K and Ω and the junction parameter. Indeed, both K and Ω are functions of all four parameters of the Josephson junction. It is therefore very difficult to determine the critical value $K = 1$ in terms of real junction parameters in order to check the fractal dimension of the set of unlocked intervals. As a consequence, confirmed results on the problem have been rather scarce. *Yeh et al.* [40] performed analog and numerical simulations by just varying the current in a small interval. They obtained the result $d_H = 0.91 \pm 0.4$ which is slightly higher than the above quoted prediction by *Jensen et al.* for the sine-circle map. *Alstroem et al.* [41] made a similar determination. They first searched one complete set of critical parameters $(\beta_c, i_{ac}, i_{dc}, \omega)$ and then followed the critical line by changing the current and β_c simultaneously. Their result agrees exactly with the prediction $d_H = 0.87$. It should be noted, however, that this technique can not be used in a real experiment on Josephson junction as the parameter β_c is fixed by the fabrication process. Other investigations of the same problem, involving studies of prominent types of nonlinear behaviour of the system such as instabilities between competing locked steps and period doubling sequences as well as chaos on steps and simulations of the resulting voltage behaviour, have been successively performed by different groups. A more complete review of these results can be found in *Kurkijärvi* [20], *Christiansen et al.* [42] and *He et al.* [43].

Finally, in the context of Josephson junctions, noise and chaotic behaviour is often described in terms of a noise temperature, i.e., the temperature in the dissipative element necessary to generate the observed noise thermally at the given frequency range. Noise temperatures are usually measured at low frequencies, < 100 kHz. All the above mentioned authors find that the effective noise temperature resulting from deterministic chaos reaches its highest value in the hesitation between locked steps. Such chaos is intermittent as identified with the aid of a Poincaré map by *Seifert et al.* [44] and by *Kautz et al.* [45]. *Goldhirsch et al.* [46] tried to simulate the corresponding power spectrum with models of hopping between locked states. These studies have been extended in recent times by different extensions and modifications of the RSJ model, which have turned out to be more easily accessible in experiment. This will be discussed in the second part of chapter 5.

5. Experiments on real Josephson junctions

Here we try to describe what experiments have been done so far on real Josephson junctions. As already mentioned, it has proven rather difficult to perform controlled experiments which can be modelled by Eq.6, using conventionally fabricated Josephson junctions, while still operating at frequencies at which the lumped circuit model is appropriate. In principle the experimental difficulties arise from the fact that the characteristic voltage oscillations have their fundamental frequency, or lower order subharmonics (with subharmonics number between, say, 1 and 5), at very high frequencies in the range of 5 to 200 GHz. Since the investigation of distinct routes to chaos, or of scaling behaviour at the critical line etc., requires digitized data, thorough experimental analysis of nonlinear effects in Josephson junctions is mainly restricted to qualitative methods and statements. In general it is necessary to use elaborate microwave technology at least to detect frequency spectra and low noise spectra of the junction dynamics. Parasitic effects usually increase with frequency. They are of even greater concern in this case as one must attempt to match the low-impedance ($\leq 1\Omega$) of the tunnel junction.

Another main disadvantage is that thermal noise, most often not taken into account in simulations, has a major effect on the outcome of the experiments. This is because the energy of the thermal oscillations may very well be of the same order of magnitude as the intrinsic energy levels in the Josephson junction, and complicated nonlinear interactions are to be expected. On the other hand, any Josephson circuit possessing enough reactive elements to make the governing equation third-order should exhibit chaos and thus provide a vehicle for the study of its appearance in a real physical system. Therefore also a variety of Josephson circuits and devices have been investigated besides the "pure" junction, as we shall see below.

5.1. Experiments on Josephson parametric amplifiers

In the end of the seventies a large amount of experimental work was done on the externally pumped Josephson parametric amplifier [47] . In some cases low noise temperatures were found. Most often, however, experiments showed considerable excess noise [48][49]. Two types of amplifiers have been investigated, the four photon amplifier, which is pumped at approximately the plasma frequency, and the three photon amplifier for which the pump frequency is twice the plasma frequency. In the four photon amplifier the dc bias current is typically zero. In the language of dynamical systems the gain in the three photon, dc current biased mode is due to the presence of a nearby period doubling instability [50]

For experimentalists the observed noise rise has been a major puzzle, and a large number of theoretical and numerical papers [51] have dealt with the problem. The conclusion is that the very large noise temperatures often observed cannot reasonably be explained by traditional noise sources such as Johnson noise, shot noise, or quantum noise. Due to the pioneering work by *Huberman et al.* [1] and by *Pedersen et al.* [4] the appearance of chaos was taken into consideration. It was suggested that chaos was the origin of the excess noise in the four photon amplifier.

More recently the performance of the three-photon amplifier has been reexamined in the light of the phenomena of noise-induced transitions near bifurcations [52]. The authors state that chaos in this case is most probable not the reason for the excess noise. The reason is that in order to achieve gain, the amplifier must be operated at a stable bias point below the threshold for a period-doubling bifurcation of the pump signal and thus below the threshold of chaos. Their other conclusion - based on analog measurements - was that the observed noise rise required the presence of thermal noise. They suggested that the noise rise was due to thermally induced hopping between a biased point in the high gain region (that would be stable in the absence of thermal noise) and an unstable point in the bifurcated region.

Since the amplification process in the four photon Josephson parametric amplifier is quite different from that of the three photon device, a firm conlusion on the excess noise in the latter has yet to be made.

5.2. Measurements of I/V-characteristics

Before the term chaos was connected to Josephson junctions, researchers sometimes encountered irregular shaped I/V-characteristics in samples subjected to strong applied rf signals. In many cases such junctions were discarded because of assumed defects during fabrication. It is now known that such irregular I/V curves may be a signature of chaos. Therefore the most simple way to measure chaos is to compare experimentally obtained I/V curves with numerically calculated ones. An example of such an experiment is sketched in Fig.12 which shows an experimentally obtained I/V curve with loss of phaselock on the rf induced step.

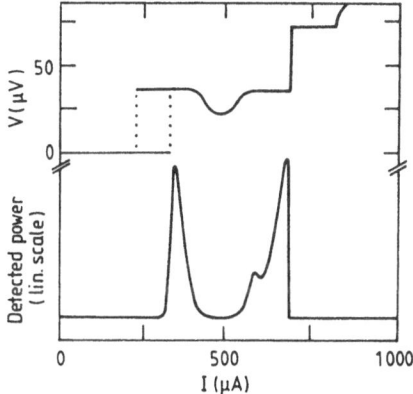

Fig. 12. Upper curve: rf induced step with loss of phase lock. Applied frequency 17.6 GHz. Lower curve: Detected half harmonic power at 8.8 GHz. (from Pedersen et al. [53]).

Also shown in Fig.12 is the spectrum of half harmonic generation as measured with a sensitive microwave spectrometer. These experimental curves which contain two period doubling bifurcations and a chaotic region on an rf induced step can be claimed as among the first pieces of experimental evidence for chaos in Josephson junctions. Other indications of chaos at the I/V curve are negativ differential resistance [55][56] and hopping between metastable states [57]. Especially for the hopping phenomena a general correlation between substructure in the I/V characteristic and enhanced noise has been reported [58]. However, in experiments most of the complicated structure is typically washed out because of thermal noise. By comparing experiments with calculations that includes thermal noise, the existence of chaos has sometimes been demonstrated indirectly [55].

5.3. Low frequency measurements of chaos in Josephson junctions

Gubankov et al. [59] were among the first who have studied chaos experimentally in the ac driven Josephson tunnel junction. Their junctions had an area on the order of $20\mu m^2$ and a critical current of typically $0.2mA$. They monitored output signals

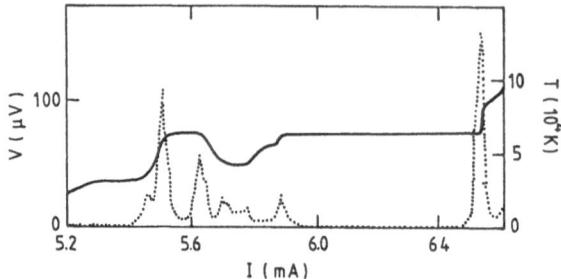

Fig. 13. Experimental I/V curve showing loss of phase lock on an rf induced step. Lower curve is the corresponding low frequency noise temperature (from Octavio et al. [60]).

from the junction at 1 kHz with a 100 Hz band, thus at frequencies much less than either the applied frequency or the plasma frequency. In order to compare the experimental results with theoretical predictions, they inspect the ω, i_{ac} plane of the devices looking for those regions where large values of low frequency noise could be measured. Altogether the results agree quite nicely with the numerical investigations cited above.

Their technique was later improved by *Octavio et al.* [60] who reported noise temperatures of the order 10^6 K or more. High noise levels are accompanied by period doubling effects on voltage steps. *Octavio et al.* worked with Pb-Te-Pb junctions with a 40 and 75 nm tellurium barrier thickness, junction areas of the order $10^4 \mu m^2$ and critical current densities between 10 and $400 A/cm^2$. The junctions were exposed to 35 GHz radiation. Noise was recorded at frequencies of $50 - 100$ kHz. Fig.13 shows an experimental curve from Ref.[60]. However, such high noise temperatures are not only due to chaos since it can be shown from numerical simulations of Eq.6 that only noise levels of order 10^3 can be expected from complex dynamical behaviour in this type of junc tion. The most probable reason for very high noise temperatures is the fractal basin boundary of different rf induced steps which serve as an attractor of the junction's dynamics. Thus several solutions exist for a given bias condition, and thermal noise causes stochastic switching between steps [61] . This behaviour is not deterministic chaotic since no universal scaling laws for the switching phenomena are valid.

On the other hand, *Octavio et al.* also reported that a very small change of parameters may change the shape of the observed I/V curve significantly from a nice set of discrete steps to an almost smooth curve. When similar junction parameters were simulated with an analog computer, they saw a large number of small steps resembling the Devil's staircase, indicating that for this junction the experimentally used parameter set ($\omega = 0.16, \beta_c = 1.1$) was quite close to the critical line of the corresponding circle map.

The switching effect between different rf induced steps has also been investigated by an experiment on Nb-aSi-Nb junctions by *Hu et al.* [62]. They exposed the junction to far infrared laser radiation at 419 GHz, which corresponds roughly to the plasma frequency of the junction. The junction had an area of about $5 \mu m^2$ with the parameters: $I_{ac} = 490 \mu A$ at 1.4 K, $\beta_c \sim 4$. The authors measured enormous noise temperatures up to 10^{11} K which were recorded at about 100 Hz. The observed I/V curves could be reproduced by numerical simulations yielding chaotic solutions for different values of the driving amplitude.

Davidson et al. [63] also investigated the low frequency noise from a junction irradiated by a microwave source. In particular they reported experiments on the nature of the excess noise that occurs between rf induced steps where two steps have an overlapping current range, i.e. coexist. They observed a significant thermal influence, which - depending on bias current - appears to be either thermally induced hopping or noise induced intermittency. Besides these investigations on "pure" junctions, also microbridges [66], two-dimensional junction arrays, rf-SQUIDS [67] and dc-SQUIDS [68] have been investigated in order to find chaos and nonlinear effects.

Although the reported experiments have revealed great insight in the nonlinear behaviour of Josephson junctions, it is obvious with regard to the numerical and analog simulations that the conclusive chaos experiment on Josephson junctions still has to be performed. It would consist of applying an ac drive with a frequency comparable to the plasma frequency of the junction and in measuring directly the voltage response. This experiment would require specially designed Josephson junctions with plasma frequencies well below 10 GHz together with an elaborated filter and impedance matching and very fast digitizing techniques. In addition, the whole bandwidth from zero to roughly the plasma frequency should be recorded in order to identify period doubling sequences and other routes into chaos by analysing the frequency spectra. Such an experiment has been tried by several laboratories, but no reproducible results have been reported so far.

5.4. Experiments in other Josephson junction systems

Josephson junctions need not to be ac driven in order to exhibit nonlinear or chaotic behaviour. As demonstrated by *Marcus et al.* [64] and by *Miracky et al.* [57][58] a voltage biased junction with a series inductance reveals a third order system that may exhibit chaos. This device generates the following third-order differential equation:

$$\beta_L \beta_c \frac{d^3\phi}{d\tau^3} + \beta_c \frac{d^2\phi}{d\tau^2} + \frac{d\phi}{d\tau}(1 + \beta_L \cos \phi) + \sin \phi = i \qquad (12)$$

where $\beta_L = 2\pi L I_c 2e/h$ is the inductance parameter. The LC-resonance frequency $\Omega_{LC} = 1/\sqrt{\beta_L \beta_c}$. The system is seen to depend on three dimensionless parameters: β_L, β_c and i. The first two are fixed with respect to a given junction, but the continuously variable third one serves as the control parameter. *Miracky et al.* realized this model by a junction configuration of an area $350 \times 400 \mu m^2$ with typically $I_c = 1mA, C = 6nF, R = 2m\Omega$, and $L = 3pH$. The I/V curves of these devices exhibit stable regions of negative differential resistance as sketched in Fig.14.

Very large increases in the low-frequency voltage noise with equivalent noise temperatures of 10^6 K or more, observed in the vicinity of these regions, arise from switching, or hopping, between subharmonic modes. Moderate increases in the noise, with temperature of about 10^3 K, arise from chaotic behaviour. Both of these conclusions are substantiated by analog simulations. Measurements of the low frequency

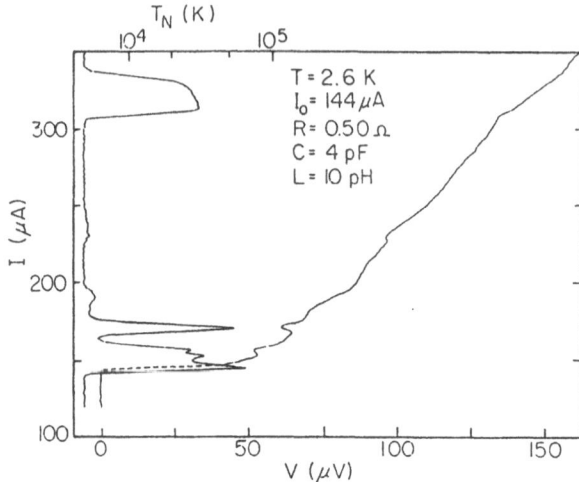

Fig. 14. I/V curve and noise-temperature T_N vs. current I at 97 kHz for a small inductively shunted junction: $\beta_c = 0.4, \beta_L = 4$. (from Miracky [57]).

spectrum of the hopping noise in one type of junction show a $1/f^2$-dependence, independent of both bias current and temperature. A simple amplifier utilizing the negative differential resistance is found to exhibit a "noise rise". Analog and digital simulations on this system [57] indicate that a purely deterministic hopping between two unstable modes, accompanied by excess low-frequency noise, is possible. The differential equation (Eq.12) describing the junction system can be reduced to a one-dimensional map in the vicinity of one of the unstable modes.

A similar system with a large inductance ($L \sim 40nH$) has been studied by *Bauer et al.* [65]. This configuration was especially designed to lower the intrinsic system frequencies in order to obtain phase portraits, return maps and further system characteristics directly derived from the recorded time signals. The system exhibits low frequency (kHz-region) relaxation-oscillations of the junction voltage when it is biased with constant current. When the bias is modulated by an ac-current the authors find mode locking behaviour as well as quasiperiodicity at distinct parameter values. Analyzing the data, they tried to model the dynamical behaviour of the system by a 1D circle map as sketched in Fig.17. In that context also questions concerning the global properties of the mode-locking behaviour of relaxation oscillators appear [73]. Josephson-junctions with critical currents in the order of $I_{max} = 100mA$ are used. The junctions consist of PbInAu base electrodes and PbAu top electrodes separated by a thin oxide layer. These junctions are shunted by an inductivity of $40nH$ with a resistance of approx. $1m\Omega$ .When a constant current I is supplied to this system, the measured junction voltage shows sharp periodic pulses. The pulses are interpreted as relaxation oscillations due to hysteretic switching between two different states of the nonlinear I/V characteristic displayed by the system (Fig.15).

A superimposed ac-current leads to a more complicated behaviour as sketched in Fig.16.

124

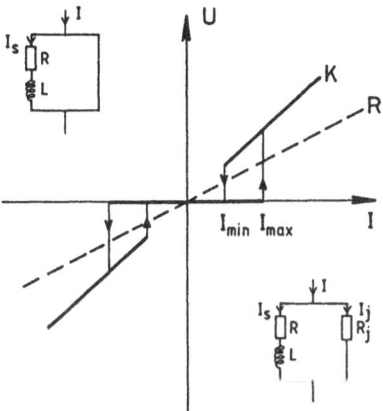

Fig. 15. Schematic representation of the I/V curve and the equivalent circuit diagram for the different states of the junction system.

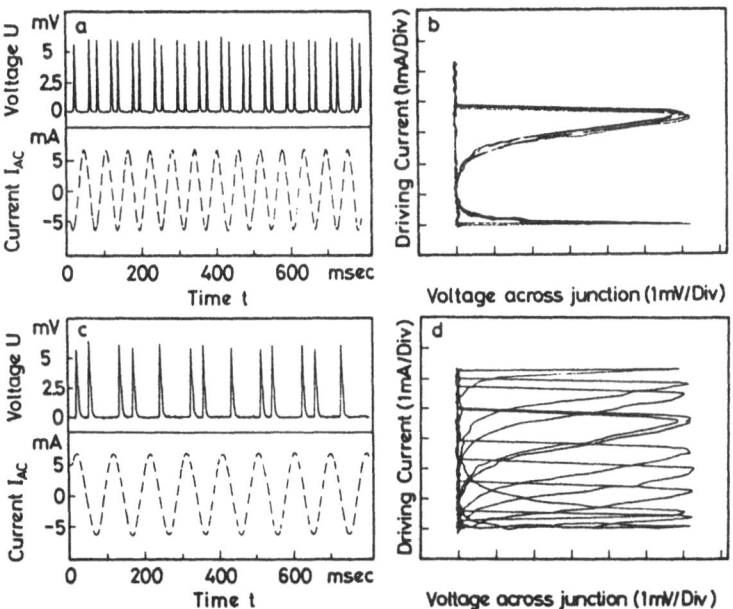

Fig. 16. Depending on the parameter values the system reveals periodic (mode-locked) (a,b) and quasiperiodic behaviour (c,d). Phase portraits are obtained by plotting the junction voltage versus the driving current. In the mode-locked state the corresponding phase-portrait exhibits a closed track indicating the periodicity of the signal (b). In the quasiperiodic state, on the other hand the track fills up the whole region (c).

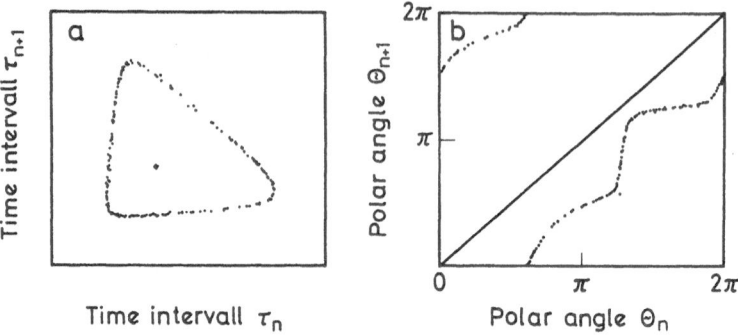

Fig. 17. Reconstruction of the circle map from a measured time signal. (a) the time interval between subsequent pulses is plotted as τ_{n+1} vs. τn. (b) The corresponding return map is obtained by a transformation to polar coordinates revea ling a $1D$ circle map.

In order to investigate the mode-locking behaviour in more detail, *Bauer et al.* measured the pulse frequency in dependence of the driving frequency which reveals a Devil's staircase structure. By numerical investigations on the basis of the above mentioned circuit diagram, the fractal dimension is determined to be $d_H = 0.75 \pm 0.03$ which is slightly lower than the value found for the circle map. The deviation in the dimension of the set of unlocked intervals from the theoretical value found for the circle map may result from the fact, that the steps are studied along a line of constant value for the amplitude I_{ac}, which is only an approximation of the more complicated line of complete mode locking. Another explanation could be the deviation from universality that has been reported in recent papers [71][72] for driven relaxation oscillators.

6. Outlook

Among the systems we did not discuss in detail here, especially the long rf-driven Josephson-junctions has become a very active field of research in recent years. In the absence of an rf-signal, the long Josephson junction may support solitons, which is a very ordered spatial mode originating in the nonlinear nature of the Josephson junctions. The applied rf-signal, on the other hand, may introduce chaos into the problem, as we have seen above. Thus, in such a system we may see a competition between space-time coherence and chaos, and immensely complicated space-time structures will arise. In this case we must talk about both, chaos in time and chaos in space and also about the interactions between the two. Such investigations typically require very big computers and are somewhat at the forefront of chaos research. It should be noted that we have been discussing the interplay between temporal chaos and thermal noise, and temporal chaos and spatial chaos. So far only a few reports have been published on the interaction of all three ingredients [69]; this in spite of the fact that they are obviously present in many real Josephson junctions. The topic of chaos and spatial coherence in long Josephson junctions is very fascinating also because of the very fundamental questions it raises concerning the organisation scheme of high-dimensional space-time order. Very recent papers by *P. Bak et al.* [70]on self organized critical phenomena create the impression that a great deal of effort in the field of nonlinear dynamics has to be expected from the investigation of new spatiotemporal scaling laws in dissipative nonequilibrium systems.

References

1. B.A. Huberman, J.P. Crutchfield, N.H. Packard: Appl.Phys.Lett. **37** 730 (1980)
2. R.L. Kautz: J.Appl.Phys **52** 3528 (1981)
   R.L. Kautz: IEEE Trans.Magn. **25** 1399 (1989)
3. D. D'Humières, M.R. Beasley, B.A. Huberman, A. Libchaber: Phys.Rev. A26 3483 (1982)
4. N.F. Pedersen, A. Davidson: Appl.Phys.Lett. **39** 830 (1981)
5. A.H. MacDonald, M. Plischke: Phys.Rev.B **27** 201 (1983)
6. M. Octavio: Phys.Rev.B **29** 1231 (1984)
7. M. Cirillo: J.Appl.Phys. **60** 338 (1986)
8. R.L. Kautz: J.Appl.Phys. **62** 198 (1987)
9. A. Davidson, N.F. Pedersen: Phys.Rev.A **36** 2455 (1987)
10. Y.J. Yeh, Y.H. Kao: Appl.Phys.Lett. **42** 299 (1983)
11. G. Cicogna: Phys.Rev. A **42** 1901 (1990)
12. X. Yao, J.Z. Wu, C.S. Ting: Phys.Rev. B **42** 244 (1990)
13. J. Kutnik, M. Odehnal: J.Low.Temp.Phys. **65** 353 (1986)
14. J.A. Ketoja, J. Kurkijärvi: Phys.Lett. A **105** 425 (1984)
15. J. Blasing, T. Holz, J. Schreiber: Physica Status Solidi B **157** 109 (1990)
16. P. Alstroem, M.T. Levinsen: Phys.Rev. B **31** 2753 (1985)
17. V.K. Kornev, K.Yu. Platov, K.K. Likharev: IEEE Trans.Magn. **MAG-21** 586 (1985)
18. M. Octavio: Physica A **163** 248 (1990)
19. H.D. Jensen, A. Larsen, J. Mygind: IEEE Trans.Magn. **25** 1412 (1989)
20. for a review see: J. Kurkijärvi: SQUID 85, (Eds. H.D. Hahlbohm, H. Lubbig), de Gruyter, Berlin (1985).
21. W.C. Steward: Appl.Phys.Lett. **12** 277 (1968)
    D.E. McCumber: J.Appl.Phys. **39** 3113 (1968)
22. for a review see: SQUID 85, (Eds. H.D. Hahlbohm, H. Lubbig), de Gruyter, Berlin (1985).
23. see e.g.: A. Barone, G. Paterno: "Physics and Applications of the Josephson effect", Wiley, New York (1982)
24. see e.g.: K.K. Likharev: "Dynamics of Josephson Junctions and Circuits", Gordon and Breach, New York (1986)
25. B.D. Josephson: Phys.Lett. **1** 251 (1962)
    P.W. Anderson, J.M. Rowell: Phys.Rev.Lett. **10** 230 (1963)
26. L.G. Aslamasov, A.I. Larkin: JETP Lett. **9** 150 (1969)
27. L.G. Aslamasov, A.I. Larkin, Yu.N. Ovchinnikov: Sov.Phys.-JETP **55** 323 (1968)
28. S.N. Artemenko, A.F. Volkov, A.V. Zaitsev: Sov.Phys.-JETP **76** 1816 (1979)
29. M.D. Fiske: Rev.Mod.Phys. **36** 221 (1964)
30. L.D. Jackel: SQUID 80, (Eds. H.D. Hahlbohm, H. Lubbig), de Gruyter, Berlin (1980) and references therein.
31. J. Matisoo: Phys.Lett. A **29** 473 (1969)
32. for a review see: H.G. Schuster: "Deterministic Chaos", VCH-publishers, Weinheim, 2.Aufl. (1988)
33. R.L. Kautz, R. Monaco: J.Appl.Phys. **57** 875 (1985)
34. K. Sakai, Y. Yamaguchi: Phys.Rev. B **30** 1219 (1984)
35. A. Aiello, A. Barone, G.A. Ovsyannikov: Phys.Rev. B **30** 456 (1984)
36. M. Cirillo, N.F. Pedersen: Phys.Lett. A **90** 150 (1982)
37. W.J. Yeh, Y.H. Kao: Phys.Rev.Lett. **49** 1888 (1982)
38. P. Bak, M.H. Jensen, P.V. Christiansen: Solid State Commun. **51** 231 (1984)
39. M.H. Jensen, P. Bak, T. Bohr: Phys.Rev. A **30** 1960, 1970 (1984)
40. J.W. Yeh, D.R. He, Y.H. Kao: Phys.Rev Lett. **52** 480 (1984)
41. P. Alstroem, M.H. Jensen, M.T. Levinsen: Phys.Lett. A **103** 171 (1984)

42. P.L. Christiansen, N.F. Pedersen: International Conference on Singular Behaviour and Nonlinear Dynamics, (Eds. S. Pnevmatikos, T. Bountis), World Scientific, Singapore (1989)

43. D.R. He, J.W. Yeh, Y.H. Kao: Phys.Rev. B **31** 1359 (1985)

44. H. Seifert: Phys.Lett. **42** 213 (1983)

45. R.L. Kautz: J.Appl.Phys. **52** 6241 (1981)

46. I. Goldhirsch, Y. Imry, G. Wasserman, E. Ben-Jacob: Phys.Rev. **29** 1218 (1984)

47. for a review: N.F. Pedersen: Physica Scripta. Vol.**T13** 129 (1986)

48. R.Y. Chiao, M.J. Feldman, D.W. Pederson, B.A. Tucker, M.T. Levinsen: Future Trends in Superconductive Electronics, AIP Conference Procedings **44** 259 (1978)

49. M.J. Feldman, M.T. Levinsen: Appl.Phys.Lett. **36** 854 (1980)

50. K. Wiesenfeld, B. McNamara: Phys.Rev.A **33** 629 (1986)

51. see e.g.: N.F. Pedersen: SQUID 80 (Eds. H.D. Hahlbohm, H. Lubbig), de Gruyter, Berlin (1980).

52. R.F. Miracky, J. Clarke: Appl.Phys.Lett. **43** 508 (1983)

53. N.F. Pedersen, O.H. Soerensen, B. Dueholm, J. Mygind: J.Low Temp.Phys. **38** 1 (1980)

54. K. Okuyama, H.J. Hartfuss, K.H. Gundlach: J.Low Temp.Phys. **44** 283 (1981)

55. D.C. Cronemeyer, C.C. Chi, A. Davidson, N.F. Pedersen: Phys.Rev. **B32** 2667 (1985)

56. U. Krüger: Diplomarbeit, Univ.Frankfurt (1988)
    U. Krüger, W. Martienssen: Workshop, Kryoelektronische Bauelemente, Braunschweig (1987)

57. R.F. Miracky: Ph.D. Thesis, Lawrence Berkeley Laboratory, Univ. of California, Berkeley (1984)

58. R.F. Miracky, J. Clarke, R.H. Koch: Phys.Rev.Lett. **50** 856 (1983)

59. V.N Gubankov, K.I. Konstantinyan, V.P. Koshlets, G.A. Ovsyannikov:
    IEEE Trans.Magn.**MAG-19** 637 (1983)

60. M. Octavio, C. Readi Nasser: Phys.Rev. **B30** 1586 (1984)

61. E.G. Gwinn, R.M. Westervelt: Phys.Rev.Lett. **54** 1613 (1985)
    R.L. Kautz: J.Appl.Phys. **58** 424 (1985)

62. Qing Hu, J.U. Free, M. Iansiti, O. Liengme, M. Tinkham: IEEE Trans.Magn. **MAG-21** 590 (1985)

63. A. Davidson, B. Dueholm, M.R. Baesley: Phys.Rev.B **32** 154 (1986)

64. P.M. Marcus, Y. Imry, E. Ben-Jacob: Solid State Commun. **41** 161 (1982)

65. M. Bauer, U. Krüger, W. Martienssen: Europhys. Lett. **9** 191 (1989)

66. C. Noeldecke, R. Gross, M. Bauer, G. Reiner, H. Seifert: J.Low.Temp.Phys. **64** 235 (1986)

67. I.M. Dimitrenko, D.A. Konotop, G.M. Tsoi, V.I. Shnyrkov: J.Low.Temp.Phys. **9** 340 (1983)

68. E. Ben-Jacob, Y. Imry: SQUID 85, (Eds. H.D. Hahlbohm, H. Lubbig), de Gruyter, Berlin (1985) and references therein.

69. see e.g.:S. Pnevmatikos, T. Bountis (Eds.):International Conference on Singular behavior and Nonlinear Dynamics, World Scientific, Singapore (1989)

70. P. Bak, C. Tang, K. Wiesenfeld: Phys.Rev. A **38** 364 (1988)

71. A. Cumming, P.A. Linsay: Phys.Rev.Lett. **59** 1633 (1987)

72. P. Alstroem, B. Christiansen, M.T. Levinsen: Phys.Rev.Lett. **61** 1679 (1988)

73. S. Habip, M. Bauer, D.R. He, W. Martienssen, Preprint

74. B.B. Mandelbrot: "The fractal geometry of nature", Freeman, San Francisco (1982)
    P. Cvitanovic: Phys.Scr. **9** 202 (1985)

Chaotic Dynamics in Spin-Wave Instabilities

H. Benner, F. Rödelsperger, and G. Wiese

Institut für Festkörperphysik, Technische Hochschule Darmstadt,
W-6100 Darmstadt, Fed. Rep. of Germany

Ferromagnetic samples excited by strong microwave fields show a variety of nonlinear phenomena. We report on magnetic resonance experiments in yttrium iron garnet (YIG) probing spin-wave instabilities above the first-order Suhl threshold. A very complex multistable behaviour and various types of auto-oscillations and sequences of bifurcations are observed, ending up in chaos. The experimental behaviour can be interpreted in terms of a multimode-model taking into account both spin waves and magnetostatic modes.

1. Introduction

Magnetic insulators are very interesting objects for studying nonlinear spin dynamics, since even their simplest equation of motion

$$\dot{\vec{M}}(\vec{r},t) \; = \; - \, \gamma \, \vec{M}(\vec{r},t) \times \vec{H}_{\text{eff}}(\vec{r},t) + dissipation \tag{1a}$$

contains intrinsically nonlinear terms. Considering non-fluctuating terms only and neglecting dissipation for a moment, the time evolution of magnetization is described by the torque of an effective field

$$\vec{H}_{\text{eff}}(\vec{r},t) \; = \; \vec{H} + \vec{h}\cos\omega t + \overleftrightarrow{A}\cdot\vec{M}(\vec{r},t) + D\nabla^2\vec{M}(\vec{r},t)$$

$$+ \vec{\nabla}\{\vec{\nabla}\cdot\int d^3r'\,\frac{\vec{M}(\vec{r}\,',t)}{|\vec{r}-\vec{r}\,'|}\} \quad . \tag{1b}$$

For the systems of interest $\vec{H}_{\text{eff}}$ is composed of the external dc and ac magnetic fields, of demagnetizing and single-site anisotropy fields, of an exchange field, and of a dipolar field, depending themselves on the magnetization and giving rise to nonlinearities. Usually, this equation is discussed only

129

for weak excitation ($h \ll H$, AM, etc.) by linearizing its r.h. side with respect to deviations of $\vec{M}(\vec{r}, t)$ from thermal equilibrium $\vec{M}_0$, and this simplified form is familiar from most of the present textbooks. However, even if confining ourselves to uniform magnetization, the exact solution of (1) shows a bistability at moderate excitation amplitudes (the so-called "foldover effect" [1]). Moreover, the additional effect of the non-local exchange and dipolar fields may result in more complicated threshold phenomena indicating self-induced formation of spatial and temporal structures.

Part of these phenomena have already been observed more than thirty years ago in high-power ferromagetic resonance (FMR) experiments [2-5], and have theoretically been explained by Suhl [6] to result from the parametric excitation of spin waves through transverse pumping of the uniform mode. Suhl considered two cases, where the excitation of spin waves should be most efficient: The *first-order Suhl instability* results from the excitation of spin waves at half the pumping frequency, $\omega_k = \omega/2$, the *second-order instability* from spin waves at $\omega_k = \omega$. There exists a third type, the *parallel-pumping instability*, where standing spin waves at $\omega_k = \omega/2$ are directly excited by a uniform microwave field polarized parallel to $\vec{H}$ [4,5]. The specific properties of these instabilities have extensively been studied since a long time and have been reviewed in different articles [1,7,8].

The recent progress of nonlinear dynamics has stimulated the re-examination of high-power FMR experiments under the aspect of "routes to chaos". Recent experiments, mostly performed on yttrium iron garnet (YIG), have exhibited a variety of nonlinear phenomena, e.g. low-frequency auto-oscillations, period doublings, quasiperiodic behaviour, mode-locking, irregular periods, and chaos [9-18]. The experiments were mainly discussed in terms of models considering only a minimum number (2 or 3) of coupled spin-wave modes [17-21]. More realistic descriptions have been obtained by taking into account the entire degenerate spin-wave manifold [22,23] or by direct integration of the equation of motion for $\vec{M}(\vec{r}, t)$ [24].

Here, in an exemplary way, we focus on recent experimental results obtained on spheres of yttrium iron garnet (YIG) within the *coincidence regime* of the first-order Suhl instability (Sect. 3). Several new and interesting phenomena have been observed: the occurrence of auto-oscillations starting directly at the Suhl threshold, sudden jumps of the magnetization amplitude, and the occurrence of multistability accompanied by a very complex sequence of bifurcations ending up in chaos [25]. We report on techniques how to characterize chaotic systems, and present analyses of our data. Finally, we discuss our experimental results in terms of a multi-mode model, based on Suhl's theory, but taking into account the specific properties of long-wavelenght modes (Sect. 4). A new nonlinear coupling mechanism based on indirect excitation of the magnetostatic (4,3,0) mode is derived. Numerical simulations show that both the multistable behaviour and the complex time dependences can be understood to arise from such indirect couplings.

2. Spin-Wave Instabilities

2.1 Suhl Instabilities

Suhl's theory [6] was based on the idea of weakly coupled eigenmodes with nonlinear couplings becoming efficient only at high amplitudes. Expanding the magnetization $\vec{M}(\vec{r}, t)$ into Fourier components, i.e. into a uniform mode $\vec{m}_0(t) \sim \exp(i\omega_0 t)$ and spin waves $\vec{m}_k(\vec{r}, t) \sim \exp(i\vec{k}\cdot\vec{r} - i\omega_k t)$, the derivation proceeds in two steps: (i) to find the eigenmodes of the linearized form of (1), and (ii) to determine their nonlinear couplings from a higher-order expansion.

Considering a spin system with isotropic ferromagnetic exchange and dipolar interactions exposed to a static magnetic field $\vec{H} = \hat{z} H$, the dispersion relation for spin waves reads [26]

$$\omega_k = \gamma \sqrt{(H - 4\pi M^z N^z + Dk^2)(H - 4\pi M^z N^z + Dk^2 + 4\pi M^z \sin^2\theta_k)} \quad . \quad (2)$$

Here, $\gamma = g\mu_B/\hbar$ denotes the gyromagnetic ratio, N^x, N^y and N^z the demagnetization coefficients, D the stiffness constant which is proportional to the exchange integral, and θ_k is the angle between the wave vector $\vec{k}$ and $\vec{H}$. The dipolar anisotropy gives rise to a band of excitation energies depending on θ_k. The frequency ω_k is lowest for $\vec{k}$ parallel to $\vec{H}$ and highest for perpendicular orientation, as shown in Fig. 1. The frequency of the uniform mode is located within this band or slightly above. Its exact position depends sensitively on sample shape and is given by [26]

$$\omega_0 = \gamma \sqrt{[H + 4\pi(N^x - N^z)M][H + 4\pi(N^y - N^z)M]} \quad . \quad (3)$$

For vanishing single-site anisotropy and spherical samples ($N^x = N^y = N^z = \frac{1}{3}$),

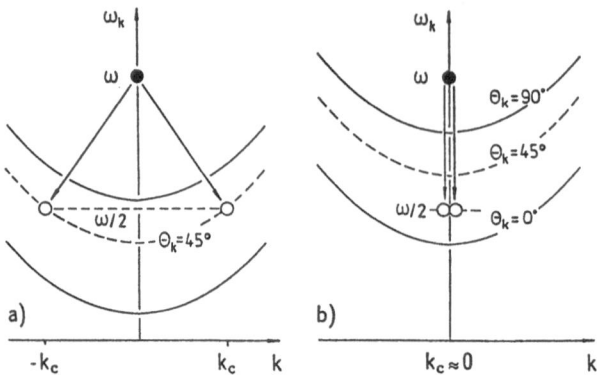

Fig. 1. Parametric excitation of non-uniform modes in case of the first-order Suhl instability ($\omega_k = \omega/2$). a) Excitation of spin waves ($k_c \simeq 10^4 \text{-} 10^5 \text{cm}^{-1}$ and $\theta_k \approx 45°$) in low magnetic field. b) Excitation of long-wavelength modes ($k_c \to 0$ and $\theta_k < 45°$) in high magnetic field.

ω_0 simplifies to $\omega_0 = \gamma H$ and, therefore, remains unaffected by demagnetizing fields and by *foldover* mechanisms [1,27] resulting from the temperature or power dependence of magnetization.

Since the spin-wave approximation remains no longer valid in the limit $k \to 0$, we should mention a different type of non-uniform eigenmodes, the *magnetostatic* or *Walker modes* [30], which have not been considered in the original theory of Suhl. For wavelengths of the order of sample dimension, the dipolar field becomes sensitive to the sample shape, but the exchange field may be totally neglected. Therefore, the spatial distribution of magnetization may be calculated from the magnetostatic Maxwell equations [26]. The corresponding set of eigenmodes is characterized by the indices (n,m,r) and shows a discrete spectrum located inside the spin-wave band. To give an example for their rather complicated structure, the $(4,3,0)$ mode is presented in Fig. 2.

The second step of the derivation is significantly simplified by taking into account that in FMR only the uniform mode is directly excited by the transverse microwave field $\vec{h}(t) = \hat{x}\, h \cos \omega t$, and that only resonant terms can efficiently couple to it. Hence, parametric excitation of spin waves occurs only under the following conditions: (i) In the first-order instability one photon of frequency ω excites a spin-wave pair with opposite wavevectors $(\vec{k}, -\vec{k})$ – i.e. a *standing* spin wave – of frequency $\omega_k = \omega/2$ (Fig. 1). Experimentally, this instability shows up as a *subsidiary absorption* generally occurring in lower

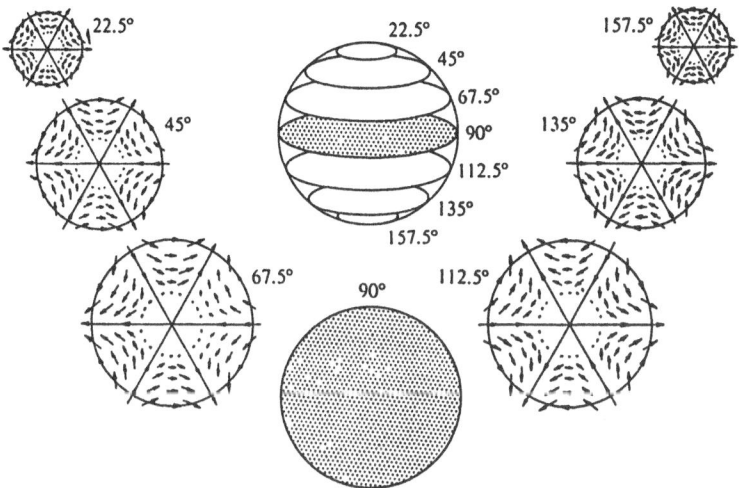

Fig. 2. Magnetostatic $(4, 3, 0)$ mode in a sphere showing the instantaneous positions of the rotating magnetization vectors in planes transverse to the dc magnetic field (from [26]).

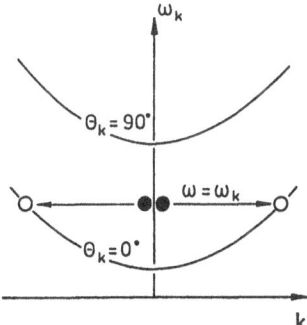

Fig. 3. Parametric excitation of spin waves in case of the second-order Suhl instability ($\omega_k = \omega$).

field than the FMR signal of the uniform mode [2]. (ii) In the second-order instability two photons of frequency ω are necessary to excite a standing spin wave of frequency $\omega_k = \omega$ (Fig. 3). In experiment *premature saturation* of the common FMR signal is observed [3]. Either instability only occurs above a critical value of the amplitude b_0 of the uniform mode, which means, above a certain threshold of the irradiated microwave power. For the first-order instability this threshold reads

$$h_{c1} = \frac{2\omega_k \sqrt{\eta_k^2 + (\omega/2 - \omega_k)^2} \sqrt{\eta_0^2 + (\omega - \omega_0)^2}}{\gamma^2 \, 4\pi M \sin\theta_k \cos\theta_k \, (\omega_k + \gamma H - \gamma 4\pi N^z M + \gamma D k^2)} \quad , \tag{4}$$

where η_k and η_0 denote the damping of the spin waves and of the uniform mode.

For a limited range of magnetic fields (and pumping frequencies) it is possible to satisfy the first-order instability condition $(\omega_k = \omega/2)$ simultaneously with the resonant excitation of the uniform mode $(\omega_0 = \omega)$. That means, subsidiary absorption can directly occur at the ferromagnetic resonance field, giving rise to an extremely small pumping threshold. For spherical samples this *coincidence regime* is limited by the conditions $4\pi M/3 \le H \le 8\pi M/3$; the lower limit results from the formation of domains, above the upper limit $\omega_0/2$ remains below the spin-wave band. As long as $H \le 0.41 \cdot 4\pi M$, the threshold $h_{c1} \approx \eta_k \eta_0 (\gamma^2 4\pi M \sin\theta_k \cos\theta_k)^{-1}$ shows a minimum at $k_c \gg \pi/d$ and $\theta_k \simeq 45°$ (Fig. 1 a). For larger fields $\omega_0/2$ no longer meets the 45°-branch, and the minimum threshold is expected to occur for $\theta_k < 45°$ and $k \to 0$ (Fig. 1 b). However, one has to keep in mind that the spin-wave approximation then becomes dubious.

2.2 Parallel-Pumping Instability

The elliptic Larmor precession of the spins in presence of dipolar or single-site anisotropy gives rise to a periodic variation of the longitudinal magnetization. Thus, not only for the uniform mode, but also for standing spin waves — in contrast to travelling waves — there occurs an oscillating *uniform* contribution to M^z, which can *directly* be excited by a longitudinal pumping field $\vec{h}(t) = \hat{z}\, h\cos\omega t$. Since M^z oscillates with twice the precession frequency, resonance is obtained for $\omega = 2\omega_k$.

The direct excitation of standing spin waves by parallel pumping represents an inherently nonlinear mechanism, which becomes efficient only above a certain threshold of h. Confining ourselves to resonance, this threshold is given by [4,5]

$$h_{c3} = \frac{\omega\,\eta_k}{\gamma^2\,2\pi M \sin^2\theta_k} \quad . \tag{5}$$

The $\sin^2\theta_k$ dependence corresponds to the fact that due to the dipolar anisotropy the ellipticity is strongest and the threshold lowest for perpendicular orientation of the wavevectors $\pm\vec{k}$ with respect to $\vec{H}$. Apart from the different pumping mechanisms, the scheme of parametric excitation resembles to that of the first-order Suhl instability (Fig. 1). There, however, the lowest threshold occurred for $\theta_k \simeq 45°$. Moreover, parallel pumping experiments need a much stronger excitation than transverse pumping; within the coincidence regime the threshold h_{c3} exceeds the threshold h_{c1} by a factor ω/η_0, which is typically in the order of 1000.

Parallel pumping has been established as a standard technique for the investigation of nonlinear spin dynamics. A review of recent results is given in [8,18].

3. Experiments in YIG Spheres

3.1 Experimental Set-up

A prototype ferromagnet used most often for the investigation of nonlinear spin dynamics is yttrium iron garnet (YIG) $Y_3Fe_5O_{12}$ [28]. To be exact, YIG represents a ferrimagnet of cubic symmetry. Its magnetic Fe^{3+} ions ($^6s_{5/2}$) are located on tetrahedral sites (3 per unit cell) and octahedral sites (2 per unit cell) (Fig. 4). Magnetic order occurs below 550 K. The magnetic moments on either site form ferromagnetic sublattices of antiparallel orientation with weak preference of the <111> direction. The excitation of "antiferromagnetic" modes occurs at much higher frequencies than applied in the present FMR investigations, so one merely probes "ferromagnetic" properties. Since the net magnetization is rather strong at room temperature, there is no need for low-temperature experiments. Moreover, high-quality single crystals with a very precise shape are commercially available. The most important advantage of YIG, however, is its extremely narrow resonance line with a width of less than 1 Oe ($\Delta H_0 \equiv 2\eta_0/\gamma$), resulting in a very small threshold for spin-wave

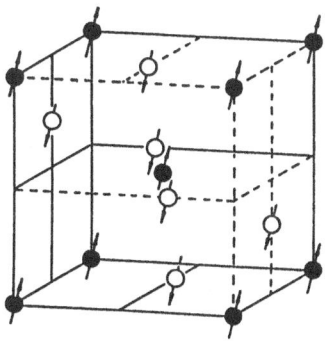

Fig. 4. Magnetic structure of yttrium iron garnet $Y_3Fe_5O_{12}$; only tetrahedral ($\circ$) and octahedral ($\bullet$) sites of Fe^{3+} are shown.

135

Table 1. Properties of yttrium iron garnet

magnetization at 300 K	$4\pi M_s$	1760 G
gyromagnetic ratio	γ	$2\pi \cdot 2.798$ MHz/Oe
stiffness constant	D	$4.48 \cdot 10^{-9}$ Oe cm^2
coincidence regime (for spheres)	$H_l - H_u$	$590 - 1170$ Oe [*)]
	$\omega_l - \omega_u$	$1.65 - 3.27 \cdot 2\pi$ GHz [*)]
FMR linewidth at 300 K	ΔH_0	$\simeq 1$ Oe
linewidth of spin waves	$\Delta H_{k \to 0}$	0.3 Oe

[*)] for "30° direction"; these limits vary with orientation by up to 10 % owing to crystal-field anisotropy.

---

instabilities. Note that the threshold amplitude in the coincidence regime only amounts to $h_{c1} \approx \Delta H_0 \Delta H_k / 8\pi M < 1$ mOe, corresponding to an irradiated power of some $10\,\mu$W.

In view of such small thresholds, "high power" FMR experiments could in principle be performed on every conventional ESR spectrometer. For various reasons, however, we preferred a broad-band (1 - 4 GHz) transmission-type set-up (Fig. 5). For transverse excitation and detection of the uniform mode, instead of a microwave cavity, we used two micro-coils with perpendicular orientation in order to prevent disturbations by the radiation field. The squared amplitude of the driving field h at sample position is proportional to the input power P_{in} supplied by the microwave source. The signal transmitted to the pick-up coil was amplified and detected by a diode. Within the quadratic regime of the diode, the rectified signal is proportional to the squared amplitude $|h_0|^2$ of the uniform mode, i.e. to the transmitted power P_{tr}. By means of a digital oscilloscope and an integrating voltmeter we recorded both the time dependence of $P_{tr}(t)$ and its average value $\bar{P}_{tr}$ with respect to the input power P_{in}.

The presented data were obtained at room temperature on a highly polished sphere of pure YIG with 0.71 mm diameter. We generally used the "30° sample orientation", where the crystal is rotated by 31.7° from the <100> to the <111> direction. For this special orientation the shift of ω_0 by a weak cubic crystal field becomes exactly zero [29].

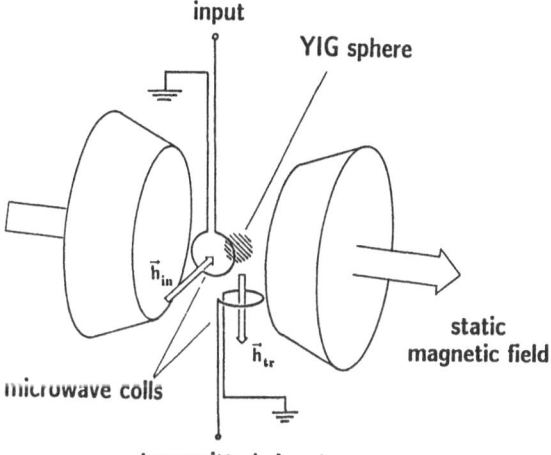

input

YIG sphere

$\vec{h}_{in}$

static
magnetic field

$\vec{h}_{tr}$

microwave coils

transmitted signal

Fig. 5. Experimental set-up. The driving coil (1 - 4 GHz) is directly fed by a very stable microwave oscillator. The signal transmitted to the pick-up coil is amplified, rectified, and recorded by a digital oscilloscope.

3.2. Observed Phenomena

In conventional FMR experiments the Suhl thresholds were reported to show up by saturation of the resonance line [3] or by additional absorption in lower field [2]. In our experiments, instead, we observed a very sharp and asymmetric break at the top of the line, becoming broader with increasing input power and showing "noisy" oscillations (Fig. 6). The coincidence condition means that ω and H are locked to resonance and can only be varied simultaneously.

The coincidence regime of the investigated YIG sphere was ranging from 640 to 1210 Oe (1.8-3.4 GHz). Below 640 Oe the FMR signal vanishes due to the occurrence of magnetic domains. Above 1210 Oe a drastic increase of the threshold shows up, indicating the changeover from first- to second-order Suhl instability. The intermediate field range may be divided into three regimes differing by their characteristic behaviours:

(i) *Regime A* (640-680 Oe): Up to the Suhl threshold $\bar{P}_{tr}$ increases linearly with P_{in}, indicating that only the uniform mode is excited. Above the threshold $\bar{P}_{tr}$ remains constant for a range of nearly 10 dB (Fig. 7a). No oscillations are observed, until finally a sudden decrease of $\bar{P}_{tr}$ occurs,

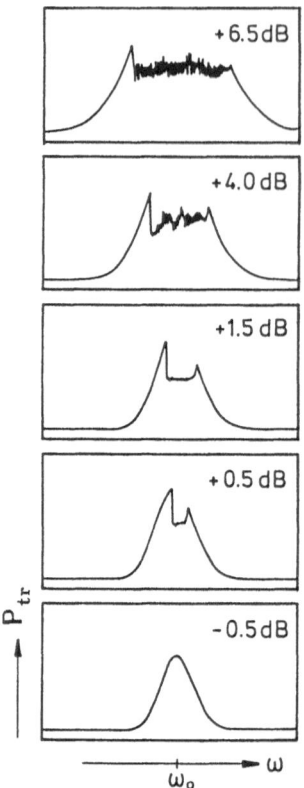

Fig. 6. Shape of the FMR line below and above the Suhl threshold (denoted by "0 dB"). Frequency scale: $\pm 10\,\text{MHz} \cdot 2\pi$. The asymmetric breaks show hysteresis effects and time-dependent oscillations.

accompanied by a transition from constant to chaotic time dependence of $P_{tr}(t)$. On variation of P_{in} the system shows a hysteresis indicating a bistability: The upper branch corresponds to constant time dependence and the lower branch to chaotic behaviour.

(ii) *Regime B* (680-950 Oe) is characterized by a variety of sudden jumps of $\bar{P}_{tr}$ starting directly above threshold, accompanied by the occurrence of a complicated multistability (Fig. 7b). If P_{in} is again decreased after a jump, the system remains on the new level. Variation of P_{in} yields a number of new levels, i.e. for the same input power there exist several stable states, sometimes up to ten. Simultaneously, the system shows very complex types of time dependence. We observed constant behaviour, periodic and quasiperiodic oscillations, intermittency and chaos. Generally, the oscillation amplitudes are

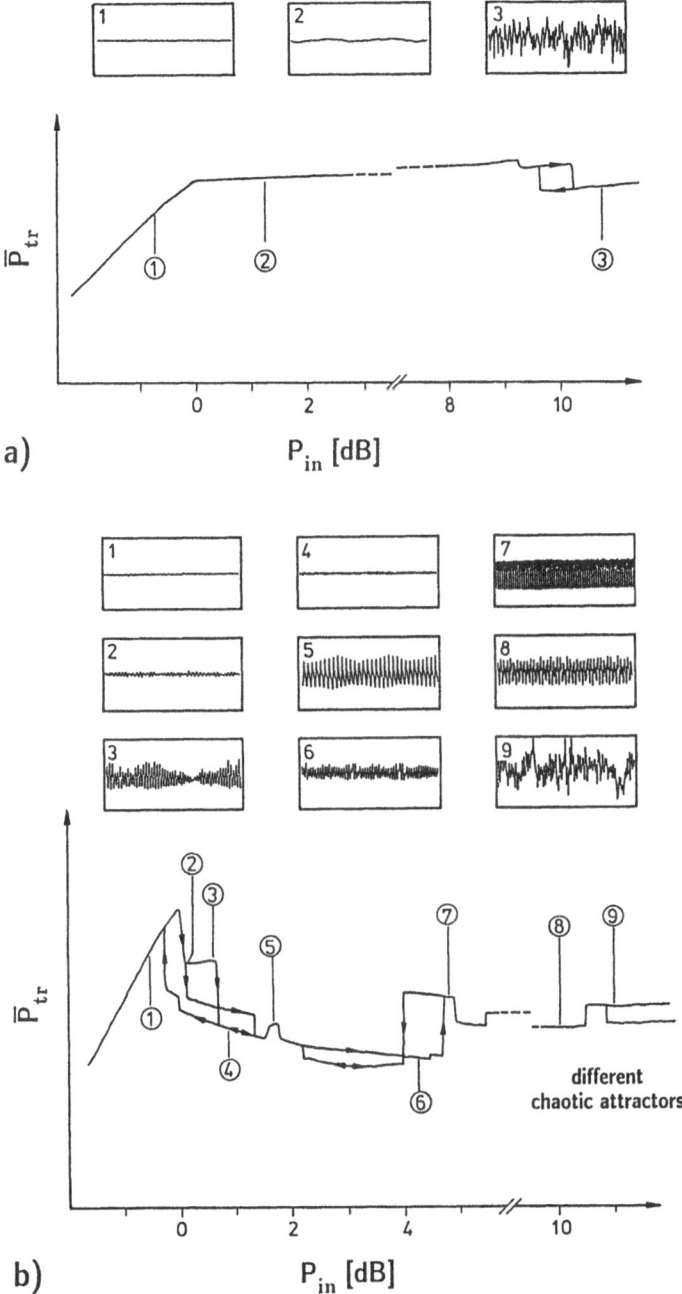

Fig. 7. Average transmitted FMR signal $\bar{P}_{tr}$ vs. input power P_{in}. a) Regime A ($H = 668$ Oe; $\omega_p/2\pi = 1.90$ GHz; 30°-orientation). b) Regime B ($H = 838$ Oe; $\omega_p/2\pi = 2.33$ GHz; 30°-orientation). Insets show time dependences of P_{tr} at certain points.

less than 10% of $\bar{P}_{tr}$, and frequencies occur between 10 and 600 kHz, most often between 30 and 200 kHz, apart from higher harmonics. The sudden jumps of $\bar{P}_{tr}$ are always accompanied by an abrupt change of the observed time dependence. As long as $\bar{P}_{tr}$ remains on the same level, $P_{tr}(t)$ shows only continuous variations of the oscillations or bifurcations without any hysteresis (e.g. transitions from constant to periodic, from periodic to quasiperiodic, locking phenomena, period doublings). About 5 - 10 dB above threshold the system becomes chaotic, showing various kinds of irregularity at higher input power. Moreover, the corresponding chaotic levels sometimes appear already at lower power within the regime of regular behaviour. The experiment turns out to be very sensitive to the control parameters ω and H; a detuning of H by only 0.2 Oe was already sufficient to change the level pattern. Nevertheless, the levels could be reproduced even after several weeks.

(iii) *Regime* C (950-1210 Oe): Qualitatively, the behaviour resembles very much that of regime A. $\bar{P}_{tr}$ remains constant above the threshold, and oscillations occur only at higher excitation. The transition to chaos also occurs at higher input power (20-30 dB above the Suhl threshold).

For other sample orientations ($\vec{H}$ parallel to the <100>, <110> and <111> directions) we found very similar behaviour, but the three regimes were slightly shifted in field (by +90 Oe, −15 Oe and −50 Oe, respectively).

We interpret these different regimes to be related to the different types of parametrically excited modes. It is obvious to identify A with the regime of spin waves with $k \simeq 10^4 - 10^5 \mathrm{cm}^{-1}$ and $\theta_k \approx 45°$, whereas in B and C we only expect long-wavelenght modes ($k \to 0$) to be excited. Apparently, the observed discontinuities are related to the discreteness of these modes. The different behaviour of B and C, however, cannot be understood in terms of the conventional Suhl theory, but requires a more detailed discussion given in Section 4.

3.3 Scenarios

From the aspect of nonlinear dynamics the most interesting phenomena occur in regime B. To get a more systematic impression of the observed oscillations, a large number of time series of $P_{tr}(t)$ — up to 16 K data points each — was recorded on variation of P_{in} or some other control parameter. The corresponding power spectra were obtained by Fourier transformation, and their strongest spectral components were plotted versus the parameter under variation. This way, one obtains a number of "maps" visualizing the dependence of the oscillation frequencies on various control parameters. Such maps are useful for classifying the observed routes to chaos.

As a general result, we found that there is no global correspondence to one of the well-known scenarios of *Feigenbaum, Ruelle-Takens-Newhouse* or *Pomeau-Vidal* [31], but a variety of parts from all of them. This might correspond to the fact that the nonlinearities of our real system are more complicated and are based on a larger number of internal degrees of freedom (which are probably related to certain modes) than in other systems showing the standard routes above. However, restricting the control parameters to certain regimes where $\overline{P}_{tr}$ remains on the same level, such correspondence does exist.

Very often, quasiperiodicity was observed with up to three fundamental frequencies. A typical example, suggesting an interpretation in terms of the Ruelle-Takens scenario, is shown in Fig. 8. Very close above threshold (denoted by "0 dB") the system starts oscillating at about 130 kHz, that means, a first Hopf bifurcation changes the *fixed point* into a *limit cycle*. At 2.5 dB a second fundamental frequency of 40 kHz occurs – corresponding to a second Hopf bifurcation – together with several sum and difference frequencies of harmonics. Note that no jump is observed in $\overline{P}_{tr}$. With increasing microwave power both oscillation frequencies seem to vary independently [1], indicating that the attractor is a *2-torus*. This quasiperiodic oscillation remains stable for about 1 dB. Then, according to Ruelle and Takens, one would expect a third Hopf bifurcation and the immediate break of the arising unstable *3-torus* to chaos. The different behaviour observed in our experiment can be interpreted by the spin system switching over to a coexisting stable attractor. Nevertheless, we also found experimental examples where a third fundamental frequency occurred for an extended parameter range. There are other levels where the different frequencies tend to lock. At 5 dB, for instance, instead of quasiperiodicity a period-5 (P 5) oscillation takes place. Low-period oscillations, such as P 2, P 3, P 4 or P 6 were observed rather often, but sometimes also higher periods of 11 or even 25. The changeover to chaos is generally accompanied by a jump of $\overline{P}_{tr}$. Since in most cases this changeover does not start from a 2-torus, it cannot be related to the third bifurcation of a Ruelle-Takens

---

[1] Experimentally, one cannot exclude a very fine-graduated locking to rational frequency ratios similar to a *devil's staircase*, but this would be rather unlikely.

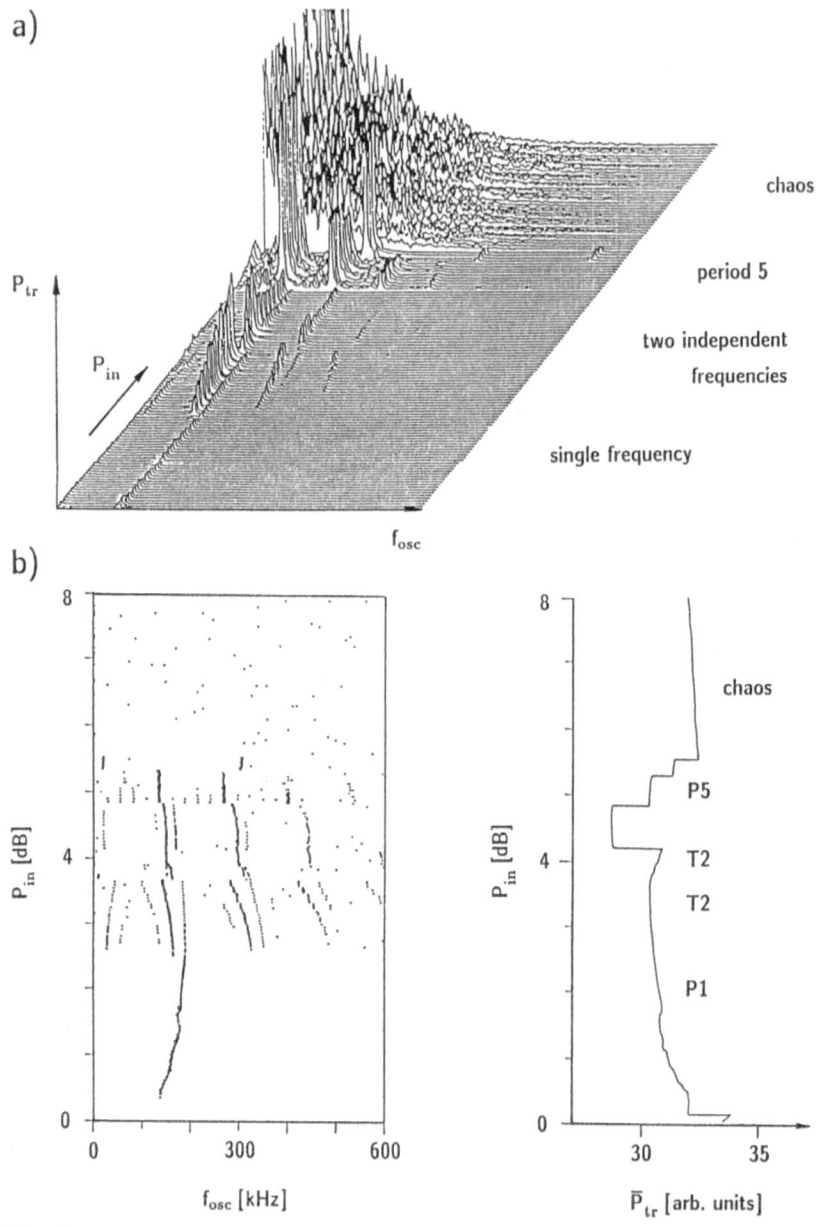

Fig. 8. Power spectra of auto-oscillations in Regime B with respect to P_{in} (input power is scaled to the Suhl threshold). a) Three-dimensional "landscape" of spectral components. b) "Map" of oscillation frequencies; the corresponding level of $\bar{P}_{tr}$ is presented in the r.h.s. diagram. The system shows quasiperiodic behaviour and mode-locking.

scenario. We rather suppose that the chaotic behaviour results from a sudden increase of the number of coupled modes and does not follow one of the standard routes.

A period-doubling route as reported by other authors from both transverse and parallel pumping experiments [9,12] was observed up to period 8 but occurred rather seldom. More often, only a single period doubling occurred, remaining stable for an extended range of P_{in} and then changing directly over to chaos. Though the Feigenbaum route is known to be very sensitive to noise which might suppress the subsequent period doublings, we rather interpret the observed behaviour to present an independent route.

We also observed intermittency starting either from a fixed point, from a limit cycle or from a 2-torus. Figure 9 shows the breaking of a period-2 limit cycle. By analyzing the distribution of the lengths of the laminar phases [32], the observed signals could clearly be attributed to each of the universal types I, II or III or to crises [31,33]. Surprisingly, we found that the signal became more irregular and the duration of the laminar phases became shorter for decreasing excitation P_{in}.

3.4 Fractal dimensions

The chaotic behaviour of a dissipative system can be attributed to the existence of a strange attractor. Specific techniques have been developed to determine its characteristic properties such as dimensions, entropies or Lyapunov exponents. For details see e.g. the second chapter of this book.

It has been shown [34,35] that certain properties of an attractor $\vec{U}(t)$ can be reconstructed from a time series of experimental data representing a single component $U_1(t)$. In a first step, one has to construct d-dimensional vectors $\vec{x}$ of a new phase space e.g. by means of the *time delay method*:

$$\vec{x}(t_0) = \{ U_1(t_0), U_1(t_0+\tau), \ldots, U_1(t_0+(d-1)\cdot\tau) \} \quad . \tag{6}$$

In a second step, from N of such vectors $\vec{x}_i \equiv \vec{x}(t_i)$ the correlation integral [2]

$$C_d^2(\ell) = \lim_{N\to\infty} \frac{1}{M} \sum_{i=1}^{M} \frac{1}{N} \sum_{j=1}^{N}{}' \Theta\left[\ell - \| \vec{x}_i - \vec{x}_j \| \right] \quad , \tag{7}$$

---

[2] Following the work of Grassberger and Procaccia [36] we confine ourselves to estimates of D_2 and K_2.

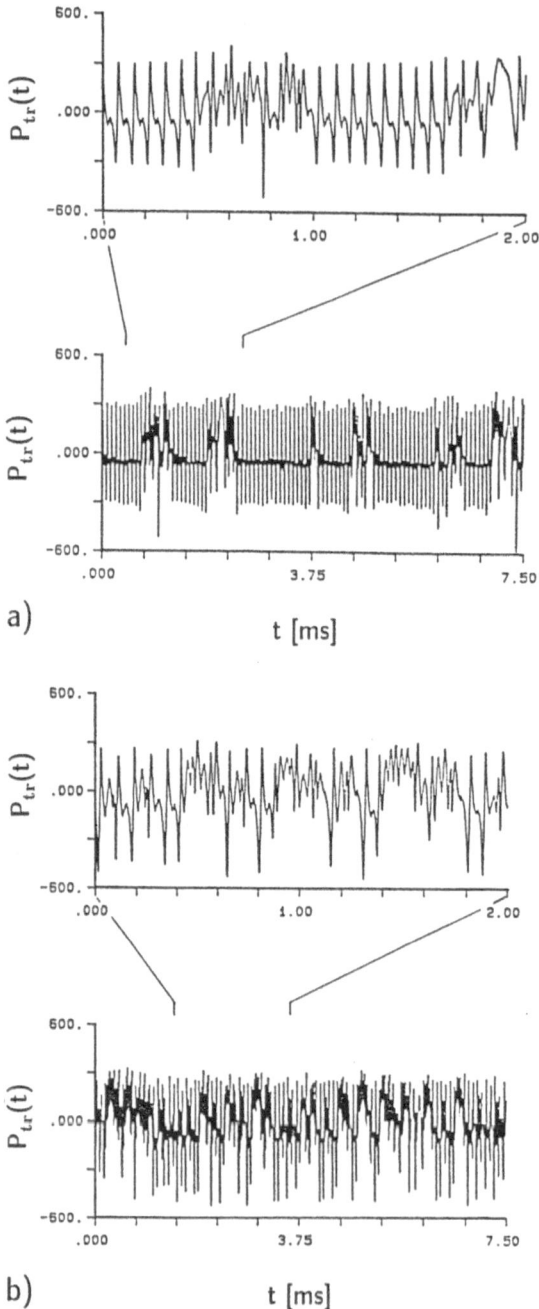

Fig. 9. Breaking of a period-2 limit cycle by intermittency ($\omega/2\pi = 2.51\,\mathrm{GHz}$, $H = 980\,\mathrm{Oe}$). Input power is a) 4.8 dB, b) 4.6 dB (!) above the Suhl threshold. Upper pictures show parts of both signals with higher time resolution.

144

is calculated, where N denotes the total number of data points, M the number of reference points, $\Theta(x)$ the Heaviside function, ℓ a distance parameter and $\|\vec{x}\|$ some arbitrary norm, e.g. Euclidian norm. The correlation integral $C_d^2(\ell)$ counts the number of pairs of d-dimensional vectors with a distance $|\vec{x}_i - \vec{x}_j|$ smaller than a given value ℓ. [3] Grassberger and Procaccia [36] have shown that $C_d^2(\ell)$ can be used to estimate the correlation dimension:

$$D_2 \simeq \lim_{\ell \to 0} \lim_{d \to \infty} \frac{\log C_d^2(\ell)}{\log \ell} \quad . \tag{8}$$

To this end, $\log C_d^2(\ell)$ is plotted vs. $\log \ell$. In practice, since the limit $\ell \to 0$ cannot be reached, the slope of the curve is taken instead. The limit $d \to \infty$ is not essential and is taken only to guarantee a proper embedding. In addition, the correlation entropy K_2 can be obtained from the vertical distance of two neighbouring curves $\log C_d^2(\ell)$ and $\log C_{d+1}^2(\ell)$ at fixed ℓ [36]:

$$K_2 \simeq -\lim_{\tau \to 0} \lim_{\ell \to 0} \lim_{d \to \infty} \tfrac{1}{\tau} \cdot \left(\log C_{d+1}^2(\ell) - \log C_d^2(\ell) \right) \quad . \tag{9}$$

Here, the limit $d \to \infty$ results from the original definition of K_2 [31] and is necessary to obtain the correct asymptotic behaviour.

In order to get some idea of the minimum number of relevant degrees of freedom involved in the time evolution, we have systematically analysed the correlation dimension of our data. To this end we have evaluated a large number of time series of $P_{tr}(t)$, up to 16 000 points each, taken from the whole investigated parameter range. For regular signals we found values of D_2 close to 1 or 2, in some cases even close to 3, as expected from the power spectra. Chaotic signals, in general, showed higher fractal dimensions ranging from 3 to larger than 10. It is interesting to note, that multistability also occurs in the chaotic regime, indicating the coexistence of several strange attractors (Fig. 10). At 10 dB there occur e.g. two separate levels of chaotic oscillations differing markedly in amplitude, but also in fractal dimension. We even observed a different tendency of variation with respect to the control parameter; while on the lower level D_2 was found to increase with P_{in}, it was decreasing on the upper one.

In total, this analysis supports our earlier impression that jumps of $\overline{P}_{tr}$ are related to discontinuous changes of relevant degrees of freedom. As long as

---

[3] By the restriction $|i-j| > 5 \ldots 20$ we exclude the comparison of vectors that follow immediately after each other, which might result in artifacts.

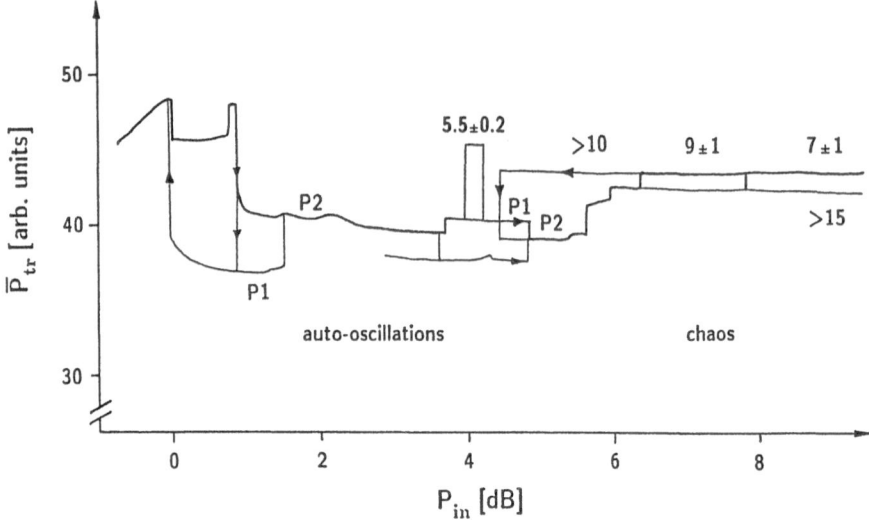

Fig. 10. Correlation dimensions estimated by the method of Grassberger and Procaccia (regime B, $H = 839\,\text{Oe}$). Two chaotic levels occur above $7\,\text{dB}$, differing in amplitude and dimension.

the system remains on the same level, only slight and continuous changes of D_2 occur on variation of P_{in}, whereas D_2 changes drastically at every jump. It is tempting to ascribe such behaviour to some nonlinear mechanism "switching certain modes on or off" with respect to the control parameters and to the previous state of the system.

4. Theoretical Model

Since Suhl's theory only considers the excitation of short-wavelength standing spin waves, this theory seems inadequate to explain the more interesting effects of regimes B and C, which we attribute to the excitation of modes with wavelengths in the order of sample dimension. Therefore, we have extended Suhl's treatment of the first-order instability to account as well for the specific properties of magnetostatic modes and long-wavelength modes from the crossover regime to spin waves.

4.1 Spin waves and Magnetostatic Modes

Following previous works we expand the magnetization into eigenmodes of the linearized equations of motion. The transverse components are given by

$$\vec{m}_\perp(\vec{r}, t) = a_0(t)\, \vec{m}_0(\vec{r})\, e^{-i\omega_0 t} + \sum_\kappa a_\kappa(t)\, \vec{m}_\kappa(\vec{r})\, e^{-i\omega_\kappa t} \quad . \tag{10}$$

$\vec{m}_0(\vec{r})$ and $\vec{m}_\kappa(\vec{r})$ describe the magnetization fields of the uniform mode and of the relevant non-uniform eigenmodes. κ is a generalized mode index denoting the wave-vector $\vec{k}$ in case of spin waves or the indices (n, m, r) in case of magnetostatic modes. The ω_κ are the respective eigenfrequencies. Due to the presence of weak nonlinear couplings between these modes, the complex amplitudes $a_0(t)$ and $a_\kappa(t)$ are slowly varying functions of time. We prefer a complex notation, keeping in mind that only the real part has physical meaning. For simplicity, the magnetization is normalized to M_s. Then, $\vec{m}_0(\vec{r})$ can be written as $\vec{m}_0(\vec{r}) = \hat{x} + i\, \hat{y}$, where $\hat{x}$ and $\hat{y}$ denote unit vectors in the respective directions. The z-component of the normalized magnetization is given in lowest order by

$$m_z(\vec{r}, t) = 1 - \tfrac{1}{2}\left[\mathrm{Re}\{\vec{m}_\perp(\vec{r}, t)\}\right]^2 \quad . \tag{11}$$

We consider an effective field $\vec{H}_{\mathrm{eff}}(\vec{r}, t)$ composed of contributions from the external field $H \cdot \hat{z}$, from the demagnetizing field $-\tfrac{4}{3}\pi M_s \cdot \hat{z}$ of the sphere, from the driving microwave field $h\cos\omega t \cdot \hat{x}$ and from dipolar and exchange fields arising from all non-uniform modes. All these terms have to be inserted in (1). In order to derive equations of motion for the amplitudes $a_0(t)$ and $a_\kappa(t)$ we make use of the following relations of orthogonality

$$\int_{\mathrm{sample}} d^3r\, (\hat{z} \times \vec{m}_\kappa^*(\vec{r})) \cdot \vec{m}_{\kappa'}(\vec{r}) \equiv (\vec{m}_\kappa(\vec{r}) | \vec{m}_{\kappa'}(\vec{r})) \sim \delta_{\kappa, \kappa'} \quad , \tag{12}$$

which first have been derived for magnetostatic modes [37], but also hold for spin waves or modes from the crossover regime [38].

In our further derivation we confine ourselves to *three-magnon processes*, that means we only consider nonlinear couplings of lowest order in a_0 and a_κ (bilinear terms in the equations of motion). This restriction allows us to describe the couplings by exact analytic expressions and can be justified as follows: Terms of higher order are only relevant at higher input power far above threshold. In our experiment, however, the most interesting effects (regime B) occur directly above threshold where the three-magnon approximation should be valid. Next, we neglect all rapidly oscillating terms, since only the coupling between resonant terms will be effective. Introducing the phenomenological dampings $-\eta_0\,a_0$ and $-\eta_\kappa\,a_\kappa$, we finally arrive at

$$\dot{a}_0 = -\eta_0\,a_0 \; - \sum_{\kappa\kappa'} \hat{\rho}^*_{\kappa\kappa'}\,a_\kappa\,a_{\kappa'}\,e^{-i\,(\omega_\kappa+\omega_{\kappa'}-\omega_0)\,t} + \gamma\,\frac{h}{2}\,e^{-i\,(\omega-\omega_0)\,t}$$

$$\dot{a}_\kappa = -\eta_\kappa\,a_\kappa \; + \sum_{\kappa'} \rho_{\kappa\kappa'}\,a_0\,a^*_{\kappa'}\,e^{-i\,(\omega_0-\omega_{\kappa'}-\omega_\kappa)\,t} \;. \qquad (13)$$

The dipolar coupling coefficients $\rho_{\kappa\kappa'}$ and $\hat{\rho}_{\kappa\kappa'}$ are defined by

$$\rho_{\kappa\kappa'} = \hat{\rho}_{\kappa\kappa'} + \hat{\rho}_{\kappa'\kappa} \; ; \qquad \hat{\rho}_{\kappa\kappa'} = \frac{\gamma}{4iV} \int_{\text{sample}} d^3r \; \vec{m}^*_\kappa \cdot \vec{m}_0 \; H^z_{\text{dip}}[\vec{m}^*_{\kappa'}] \; , \qquad (14)$$

and $\vec{H}_{\text{dip}}[\vec{m}_\kappa]$ denotes the integral operator generating the dipolar field of the mode in brackets. These equations of motion differ from Suhl's model only by the more general types of coupling. Confining ourselves to standing spin waves with wavevectors $\pm\vec{k}$ and $\pm\vec{k}\,'$, then only the diagonal elements of $\rho_{kk'}$ are non-zero, owing to the orthogonality of Fourier components with different wavevectors [4]. We obtain

$$\rho_{kk'} = \frac{\gamma\,4\pi M\,\sin\theta_k\,\cos\theta_k\,(\omega_k + \gamma H - \gamma 4\pi M/3 + \gamma D k^2)}{2\,\omega_k}\,\delta_{kk'} \quad , \qquad (15)$$

which is consistent with [6] and yields the threshold amplitude h_{c1} of (4) describing the excitation of a standing spin wave. For magnetostatic modes, however, the dipolar field of $\vec{m}_\kappa$ remains no longer orthogonal to $\vec{m}_{\kappa'}$, and the excited modes are mutually coupled by nondiagonal coefficients $\rho_{\kappa\kappa'}$.

4.2 Excitation of the (4,3,0) Mode

We now want to proceed beyond the Suhl mechanism and show that, apart from the non-diagonal coefficients $\rho_{\kappa\kappa'}$, there exist even more important mutual couplings between the spin waves, too. So far, we have admitted several modes at $\omega/2$, but only the uniform mode at ω. Within the same formalism as presented above one can easily show that there also exist

---

[4] Note that the dipolar field of a certain mode must have the same spatial periodicity as the mode itself.

couplings to other non-uniform modes, provided that their frequencies are close to ω (because of the resonance approximation). These modes cannot be directly excited by the uniform pumping field but only indirectly through the modes at $\omega/2$. Their couplings are described by coefficients similar to (14) replacing $\vec{m}_0$ by the mode in question. The detailed discussion [25] shows that only the magnetostatic $(4,3,0)$ mode (Fig. 2) which is degenerate with the uniform mode for *arbitrary fields* can efficiently be excited by such an indirect mechanism and has to be treated as an additional degree of freedom. Including the $(4,3,0)$ mode in (13) and eliminating the explicit time dependences by means of the transformations

$$b_0 \quad = \quad a_0 \, e^{-i(\omega_0 - \omega)t}$$

$$b_{430} \quad = \quad a_{430} \, e^{-i(\omega_{430} - \omega)t}$$

$$b_\kappa \quad = \quad a_\kappa \, e^{-i(\omega_\kappa - \omega/2)t} \tag{16}$$

we finally arrive at the following set of autonomous equations of motion:

$$\dot{b}_0 \quad = \quad - \, (\eta_0 + i\Delta\omega_0) \, b_0 \; - \tfrac{1}{2} \sum_{\kappa\kappa'} \rho^*_{\kappa\kappa'} \, b_\kappa b_{\kappa'} + \gamma \, \tfrac{h}{2}$$

$$\dot{b}_{430} \quad = \quad - \, (\eta_{430} + i\Delta\omega_{430}) \, b_{430} \; - \tfrac{1}{2} \sum_{\kappa\kappa'} \sigma^*_{\kappa\kappa'} \, b_\kappa b_{\kappa'}$$

$$\dot{b}_\kappa \quad = \quad - \, (\eta_\kappa + i\Delta\omega_\kappa) \, b_\kappa \; + \sum_{\kappa'} (\, \rho_{\kappa\kappa'} \, b_0 b^*_{\kappa'} + \sigma_{\kappa\kappa'} \, b_{430} b^*_{\kappa'} \,) \quad . \tag{17}$$

Here, $\Delta\omega_0 = \omega_0 - \omega$, $\Delta\omega_{430} = \omega_{430} - \omega$, $\Delta\omega_\kappa = \omega_\kappa - \omega/2$, and the coupling coefficients are defined by (14) and by

$$\sigma_{\kappa\kappa'} \quad = \quad \hat{\sigma}_{\kappa\kappa'} + \hat{\sigma}_{\kappa'\kappa} \quad ;$$

$$\hat{\sigma}_{\kappa\kappa'} \quad = \quad \frac{\gamma}{4iV} \int_{\text{sample}} d^3r \left\{ \, \vec{m}^*_\kappa \cdot \vec{m}_{430} \, H^z_{\text{dip}}[\vec{m}^*_{\kappa'}] + \tfrac{1}{2} \, \vec{m}^*_\kappa \cdot \vec{m}^*_{\kappa'} \, H^z_{\text{dip}}[\vec{m}_{430}] \, \right\} \quad . \tag{18}$$

According to (17) the indirect excitation of the $(4,3,0)$ mode acts as an additional mutual coupling between the excited modes. The following discussion will show that most of the observed phenomena can be explained by this more complex type of equations of motion.

4.3 Discussion

In our present understanding the observed multistability and complicated forms of time dependence are closely related to the types of excited modes.

In regime A the excited spin-wave modes are dense in ω. Therefore, the modes with the lowest threshold satisfy exactly $\omega_\kappa = \omega_{\kappa'} = \omega/2$. Since they are degenerate, the couplings through the $(4,3,0)$ mode tend to synchronize them. We have simulated this situation by considering several spin waves of equal frequency and threshold, mutually coupled by the $(4,3,0)$ mode. Using realistic parameters for YIG, the numerical solutions of (17) were found to be in agreement with the experimental behaviour: Above threshold, the squared amplitude of the uniform mode $|b_0|^2$ (which is proportional to the transmitted power P_{tr}) remains constant and does not show any oscillations within a wide range of P_{in} (Fig. 12 a).

In regimes B and C the modes with the lowest thresholds are long-wavelength modes, formally described by $\vec{k} \to 0$. Since the threshold increases when changing over from spin waves to magnetostatic modes [39], the pairs of modes with the lowest thresholds can arise only from the crossover regime between them. It can be shown that these modes are no longer dense in ω and

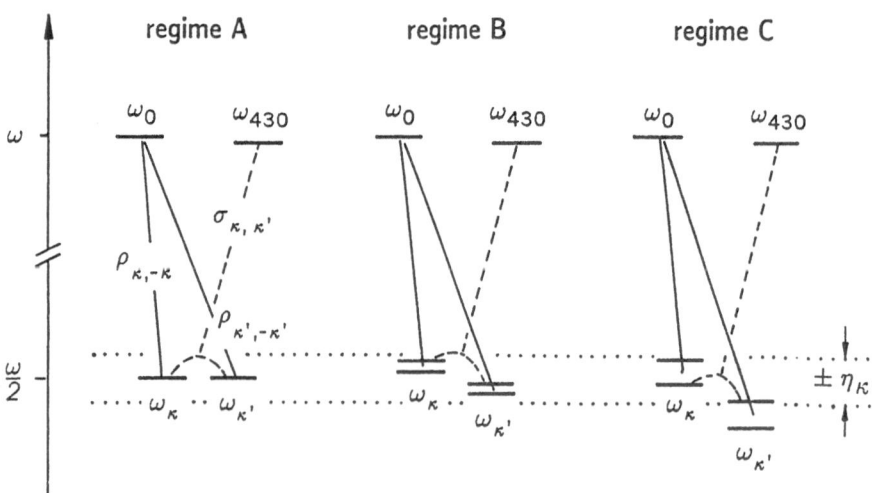

Fig. 11. Indirect couplings through the $(4,3,0)$ mode. The average detunings of the coupled modes determine the type of behaviour. Dotted: bandwidth of parametric exitation by the uniform mode. The lower pair of modes in regime C is not excited.

that their frequency splittings increase with magnetic field [25]. Then, the resonance condition for parametric excitation should rather be written as [5] $\omega_{+\kappa} + \omega_{-\kappa} = \omega$; owing to the discreteness of these modes, in general this condition is not exactly satisfied. So, in regime B we have pairs of modes with slightly different centre frequencies $\omega_\kappa = (\omega_{+\kappa} + \omega_{-\kappa})/2$ and $\omega_{\kappa'} = (\omega_{+\kappa'} + \omega_{-\kappa'})/2$ which are mutually coupled by the $(4,3,0)$ mode (Fig. 11). Hence, several pairs with centre frequencies close to $\omega/2$ are excited and take part in a collective oscillation. The sensitive dependence on control parameters, especially on the detuning beween ω and γH, can be understood to result from the discreteness of the eigenfrequencies of the relevant modes: Even small changes of H lead to the excitation of different modes.

We have simulated this situation by considering several pairs mutually coupled by the $(4,3,0)$ mode. Their frequency splittings and the detunings of their centre frequencies from $\omega/2$ were chosen to be of the order of η_κ. Indeed, we found periodic, quasiperiodic, and chaotic oscillations of $|b_0|^2$ (Fig. 12 b). As in experiment, we also observed period doublings, locking phenomena, and intermittency. The oscillations of $|b_0|^2$ vanished for very small detunings from $\omega/2$, which corresponds to the absence of oscillations in regime A. The simulations also showed sudden jumps and bistable behaviour of $|b_0|^2$, while the type of oscillation changed at every jump.

A more intuitive explanation of these jumps can be given as follows: The adiabatic elimination of the $(4,3,0)$ mode from the equations of motion ($\dot{b}_{430} = 0$) gives rise to additional terms of third order in the equations of motion, particularly to a renormalization of the eigenfrequencies:

$$\tilde{\omega}_\kappa = \omega_\kappa + \sum_{\kappa'} T_{\kappa\kappa'} |b_{\kappa'}|^2 . \tag{19}$$

Therefore, the frequencies of the modes are shifted with increasing excitation. Since only pairs within a small range of detuning from $\omega/2$ can be excited by the uniform mode, the renormalization of frequency can act as a (self-amplifying!) mechanism "switching on" certain new pairs or "switching off" already excited ones. This is indicated by a sudden jump of $|b_0|^2$ and by a change of time dependence and fractal dimension.

In regime C the frequency distances of the pairs coupled by the $(4,3,0)$ mode have become rather large. If a certain pair with a centre frequency ω_κ

---

[5] The notation is chosen in imitation of the case of standing spin waves. If κ represents the indices (n,m,r), then $-\kappa$ corresponds to $(n, 1-m, r)$ [25].

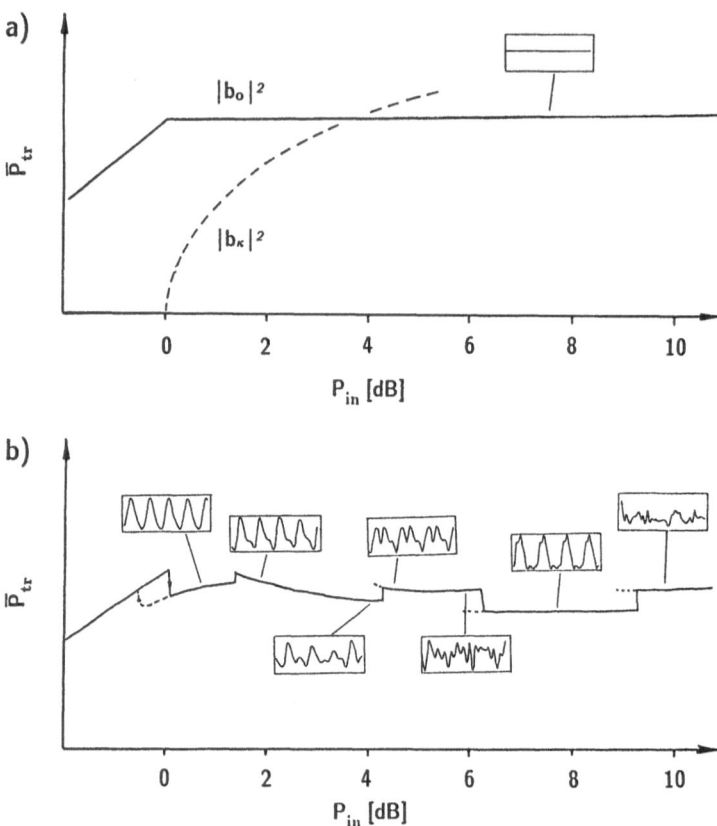

Fig. 12. Simulated FMR signal P_{tr} with respect to the input power. a) Regime A: excitation of several degenerate spinwaves ($\Delta\omega_\kappa = 0$). P_{tr} remains constant and time-independent. b) Regime B: 10 coupled modes were considered with detunings of the order of their linewidths ($\Delta\omega_\kappa \simeq \eta_\kappa$). P_{tr} shows sudden jumps in amplitude, hysteresis effects, and a complicated time dependence.

close to $\omega/2$ is excited, the average detuning of other pairs, coupled through the $(4,3,0)$ mode, is too large to be excited by the uniform mode (Fig. 11). Then, the couplings through the $(4,3,0)$ mode are no longer effective and can be neglected. Therefore, only uncoupled modes close to the lowest threshold are excited, showing a similar behaviour as in regime A.

5. Conclusions

The experimental results presented in this chapter indicate that transverse-pumped YIG spheres in the coincidence regime of the first-order Suhl instability are of particular interest for studying nonlinear phenomena by FMR. The experimental conditions differ from the more often investigated second-order Suhl instability or from parallel pumping by the variety and complexity of the observed phenomena, and also by more complicated nonlinear mechanisms. Most of the effects reported in previous investigations, such as period doublings, quasiperiodicity, or mode-locking, have been observed in this system, but also new phenomena were found: sudden jumps in the transmitted FMR signal and a well reproducible multistability. We have tried to classify the conditions of their occurrence and to understand their basic mechanisms.

Altogether, we have seen that the investigation of spin systems represents a very interesting topic both from the viewpoint of solid-state physics and of nonlinear dynamics. The reasons are manifold: (i) They represent intrinsically nonlinear systems with nonlinearities originating from well-known interactions. (ii) Their nonlinearities can partly be controlled by external fields – as in the present example. (iii) Their nonlocal couplings allow the formation of spatial structures. Nevertheless, in experiment one meets the problem that most of the interesting phenomena in spin systems occur on rather inconvenient time and length scales. While the time scale of auto-oscillations – typically a microsecond – remains accessible by modern electronics, it has been impossible so far to resolve dynamic magnetic patterns of micrometer size. Conventional magnetic resonance only probes the uniform mode, optical scattering experiments [40] suffer from too low resolution. Very recently, new technics for recording the local magnetization have been developed [41,42]. If they could be improved to probe standing spin waves, then again the coincidence regime would offer favourite conditions, because of the long-wavelength modes involved. Such investigations could be of crucial importance for confirming (or rejecting) details of the presented model.

Acknowledgements

We thank Prof. Dr. W. Tolksdorf from Philips Research Laboratory, Hamburg, for supplying us with high quality samples. This project of SFB 185 "Nichtlineare Dynamik" was partly financed by special funds of the Deutsche Forschungsgemeinschaft.

References

[1] R. W. Damon in: Magnetism, Vol. 1, ed. by G. T. Rado and H. Suhl
(Academic, New York 1963) pp. 551

[2] R. W. Damon: Rev. Mod. Phys. 25, 239 (1953)

[3] N. Bloembergen, S. Wang: Phys. Rev. 93, 72 (1954)

[4] F. R. Morgenthaler: J. Appl. Phys. 31, 95S (1960)

[5] E. S. Schlömann, J. J. Green, U. Milano: J. Appl. Phys. 31, 386 S (1960)

[6] H. Suhl: J. Phys. Chem. Solids 1, 209 (1957)

[7] V. E. Zakharov, V. S. L'vov, S. S. Starobinets: Sov. Phys.- Usp. 17, 896 (1975)

[8] V. S. L'vov, L. A. Prozorova in: Spin Waves and Magnetic Excitations, Vol. 1, ed. by
A. S. Borovik-Romanov and S. K. Sinha (Elsevier, Amsterdam 1988) pp. 233-285

[9] G. Gibson, C. Jeffries: Phys. Rev. A 29, 811 (1984)

[10] F. Waldner, D. R. Barberis, H. Yamazaki: Phys. Rev. A 31, 420 (1985)

[11] H. Yamazaki, M. Warden: J. Phys. Soc. Jpn. 55, 4477 (1986)

[12] F. M. de Aguiar, S. M. Rezende: Phys. Rev. Lett. 56, 1070 (1986) ;
F. M. de Aguiar, A. Azevedo, S. M. Rezende: Phys. Rev. B 39, 9448 (1989)

[13] T. L. Caroll, L. M. Pecora, F. J. Rachford: J. Appl. Phys. 64, 5396 (1988) ;
Phys. Rev. A 42, 377 (1989)

[14] H. Benner, F. Rödelsperger, H. Seitz, G. Wiese: J. de Physique, Colloque C8,
49, 1603 (1988)

[15] M. Warden, F. Waldner: J. Appl. Phys. 64, 5386 (1988)

[16] P. E. Wigen, H. Dötsch, M. Ye, L. Baselgia, F. Waldner:
J. Appl. Phys. 63, 4157 (1988)

[17] P. H. Bryant, C. D. Jeffries, K. Nakamura: Phys. Rev. A 38, 4223 (1988)

[18] H. Yamazaki, M. Mino: Progr. Theor. Phys. Suppl. 98, 400 (1989)

[19] K. Nakamura, S. Ohta, K. Kawasaki: J. Phys. C 15, L 143 (1982)

[20] X. Y. Zhang, H. Suhl: Phys. Rev. A 32, 2530 (1985)

[21] S. M. Rezende, O. F. de Alcantara Bonfim, F. M. de Aguiar: Phys. Rev. B 33,
5153 (1986)

[22] X. Y. Zhang, H. Suhl: Phys. Rev. B 38, 4893 (1988)

[23] S. P. Lim, D. L. Huber: Phys. Rev. B 37, 5426 (1988)

[24] F. Waldner: J. Phys. C 21, 1243 (1988)

[25] G. Wiese, H. Benner: Z. Phys. B 79, 119 (1990)

[26] see e.g. R. M. White: Quantum Theory of Magnetism (2. ed.), Springer Ser. Solid-State
Sci., Vol. 32 (Springer, Berlin, Heidelberg 1983) pp. 184- 202

[27] M. T. Weiss: J. Appl. Phys. 30, 146 S (1959)

[28] Landolt-Börnstein, New Series, Vol. III/4a (Springer, Berlin, Heidelberg 1970) pp. 315

[29] R. M. Hill, R. S. Bergman: J. Appl. Phys. 32, 227 S (1961)

[30] L. R. Walker: J. Appl. Phys. 29, 318 (1958)

[31] see e.g. H. G. Schuster: Deterministic Chaos (2. rev. ed.) (VCH, Weinheim 1988)

[32] F. Rödelsperger, T. Weyrauch, H. Benner: J. Magn. Magn. Mater. (1992)

[33] P. Bergé, Y. Pomeau, C. Vidal: Order within Chaos (Wiley, New York 1986) pp. 247-258

[34] H. Packard, J. P. Crutchfield, J. D. Farmer, R. S. Shaw: Phys. Rev. Lett. 45 (1980) 712

[35] F. Takens in: Lecture Notes in Mathematics, ed. by D. A. Rand and L. S. Young (Springer, Berlin, Heidelberg 1981) Vol. 898, p. 366

[36] P. Grassberger, I. Procaccia: Physica 9D, 189 (1983); Phys. Rev. A 28, 2591 (1983)

[37] L. R. Walker: Phys. Rev. 105, 390 (1957)

[38] E. Schlömann, R. I. Joseph: J. Appl. Phys. 32, 1006 (1961)

[39] P. H. Bryant: Phys. Rev. B 39, 4363 (1989)

[40] V. G. Zhotikov, N. M. Kreines: Sov. Phys. JETP 50, 1202 (1979)

[41] D. Rugar, H. J. Mamin, R. Erlandson, J. E. Stern, B. D. Terris: IBM Research Report RJ 6272 (1988)

[42] J. Pelzl, B. K. Bein: Phys. Bl. 46, 12 (1990)

Solitary Nonlinear Excitations in Spin Systems: Theory[1]

H.J. Mikeska

Universität Hannover, Institut für Theoretische Physik,
Appelstrasse 2, W-3000 Hannover, Fed. Rep. of Germany

We give a survey of the phenomena related to the existence and propagation of soliton–like excitations in quasi one–dimensional magnets. We introduce solitons as domain walls, mediating between equivalent ground states. Static soliton and anti-soliton solutions are given. A particular model is discussed, which is equivalent to the Sine–Gordon chain and exhibits three types of solutions: solitons/breathers/phonons. The relevance of these models for real materials is discussed with particular emphasis on the different effects of ferro-, respectively antiferromagnetic ordering. The statistical mechanics of domain walls are investigated using the equivalence to the Sine–Gordon chain. Treating the kinks as classical noninteracting particles, the soliton density at finite temperatures and their contribution to the dynamic structure factor in easy–plane ferromagnetic and antiferromagnetic spin chains are calculated. The stability of the soliton solution is analysed. The parameter dependence and influence of the out–of–plane fluctuations in the continuum and the discrete model is given. Finally, the driven Sine–Gordon chain with damping is considered and some results for the interplay between chaos and spatial ordering are described.

1. Domain Walls as Classical Excitations

1.1 The Static Soliton

Magnetic spin chains have proved to be very appropriate in describing soliton–like excitations, domain walls (DW), in solid state physics. In this contribution, we present a short survey of the basic aspects of soliton–like excitations in magnetic chain systems; for a detailed account we refer to a forthcoming review by Mikeska and Steiner [1]. The simplest 1D classical model supporting DW's is described by the following energy functional

$$E\{S_n\} = -J \sum_n S_n S_{n+1} - D \sum_n (S_n^x)^2 \quad . \tag{1.1}$$

The first term stands for the isotropic exchange energy and the second for the single–ion anisotropy. If the constants J and D are chosen as positive, then the system has the x–axis as a preferred direction in spin space. For the energy to be minimal in the

---

[1] Notes taken by: G. Ristow, N. Elstner and J. Behre

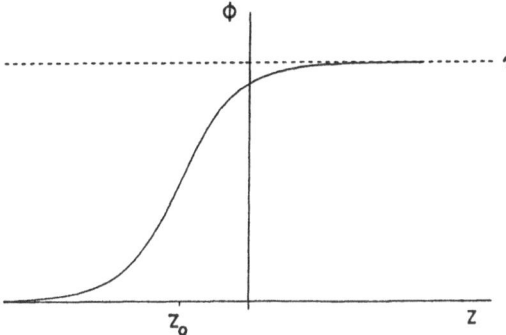

Fig. 1. soliton–like excitation

classical approach the spins should all have maximum x–components. This leads to degeneracy and therefore to two equivalent ground states:

$$|1> = |\uparrow\uparrow\uparrow\uparrow \dots \uparrow> \quad \text{and} \quad |2> = |\downarrow\downarrow\downarrow\downarrow \dots \downarrow> \ .$$

The transition between these two states is decribed by the static domain wall. Its structure is given as a compromise between the exchange interaction J ($\Rightarrow$ wide DW) and the anisotropy D ($\Rightarrow$ narrow DW). In the classical approximation the spins are treated as vectors of length S, and one works in spherical coordinates given as

$$S_n = S(\sin\theta_n \cos\Phi_n, \sin\theta_n \sin\Phi_n, \cos\theta_n) \quad . \tag{1.2}$$

The continuum approximation leads to continous variables given as $\theta_n \rightarrow \theta(z)$ and $\Phi_n \rightarrow \Phi(z)$. Thus we obtain a new energy functional where one has already subtracted the ground state energy and fixed $\theta_n \equiv \frac{\pi}{2}$

$$E\{\Phi(z)\} = JS^2 \int dz \, \{\tfrac{1}{2} \left(\frac{\partial\Phi}{\partial z}\right)^2 - \tfrac{D}{J}\cos^2\Phi\} \quad . \tag{1.3}$$

From the energy minimization ($\frac{\delta E}{\delta \Phi} = 0$) follows

$$\frac{d^2\Phi}{dz^2} = \tfrac{D}{J}\sin 2\Phi \ . \tag{1.4}$$

The soliton solution to this equation is

$$\Phi_s = 2\,\text{arctg}\ e^{\pm\sqrt{2D/J}\,(z-z_0)} \quad . \tag{1.5}$$

This solution is antisymmetric with respect to $(z_0, \frac{\pi}{2})$, and $\pm$ denotes the classical soliton and anti–soliton solutions. This is the simplest model which describes the transition between two degenerate ground states; it is sketched in Fig. 1.

1.2 The Dynamics of the Classical Sine–Gordon Chain

An expanded model which has simple domain wall dynamics [2] is given by

$$E\{S_n\} = -J \sum_n S_n S_{n+1} + D \sum_n (S_n^z)^2 - \mu B \sum_n S_n^x \quad . \tag{1.6}$$

Due to the changed sign of the second term the system shows easy–plane symmetry in the xy–plane; this symmetry is broken by the magnetic field. From the physical point of view there is only one ground state but mathematically speaking it is infinetely degenerate with respect to shifts of Φ by multiples of 2π, if the spin $S(z)$ is represented in spherical coordiantes $(\Theta(z), \Phi(z))$ as in Eq. 1.2. The energy functional can be written down in the above manner as

$$E\{\Phi(z), \theta(z)\} = JS^2 \int dz \left\{ \tfrac{1}{2} \left(\frac{\partial \Phi}{\partial z}\right)^2 + \tfrac{1}{2} \sin^2 \theta \left(\frac{d\Phi}{dz}\right)^2 + \frac{D}{J} \cos^2 \theta \right.$$
$$\left. - \frac{\mu B}{JS} \sin \theta \cos \Phi \right\} . \tag{1.7}$$

Separate energy variation in both coordinates ($\frac{\delta E}{\delta \Phi} = \frac{\delta E}{\delta \theta} = 0$) leads to the static equations of motion

$$\theta = \frac{\pi}{2} \quad \text{and} \quad \frac{d^2 \Phi}{dz^2} = \frac{\mu B}{JS} \sin \Phi \quad . \tag{1.8}$$

The first equation means that there is no deviation from the equatorial plane. If one compares the second equation with Eq. 1.4 one notices that Φ occurs instead of 2Φ. There exists a transition domain between, physical speaking, equal but mathematical different states.

In this case the approximate dynamical equations valid for $\frac{\mu B}{DS} \ll 1$ read [3]:

$$\frac{\partial^2 \Phi}{\partial z^2} - \frac{1}{c^2} \frac{\partial^2 \Phi}{\partial t^2} = m^2 \sin \Phi \tag{1.9a}$$

and

$$\theta = \frac{1}{2DS} \frac{\partial \Phi}{\partial t} \tag{1.9b}$$

where the constants are given by (the lattice constant has been set equal to unity)

$$m^2 = \frac{\mu B}{JS} \quad \text{and} \quad c^2 = 2JDS^2 \quad . \tag{1.10}$$

This is an ideal model which is also known as the *Sine–Gordon (SG) chain*. The energy scale is set by $E_0 = JS^2$.

The classical SG chain is fully integrable and has the following types of solutions:

- Solitons / Kinks

$$\Phi(z, t) = 4 \arctan \exp\left\{ \pm \frac{m}{\sqrt{1-u^2/c^2}} (z - ut - z_0) \right\}$$
$$E_{sol}(u) = \frac{8 m E_0}{\sqrt{1-u^2/c^2}} \tag{1.11}$$

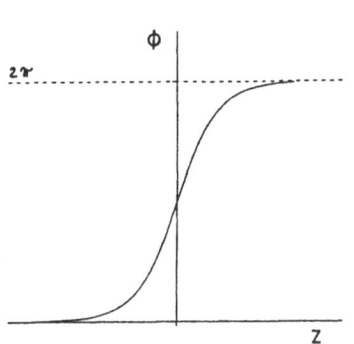

Fig. 2. Soliton in–plane angle $\Phi(z)$

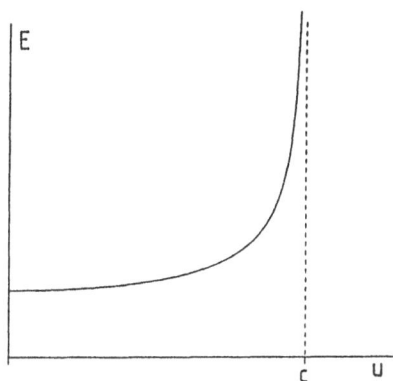

Fig. 3. Soliton energy $E(u)$

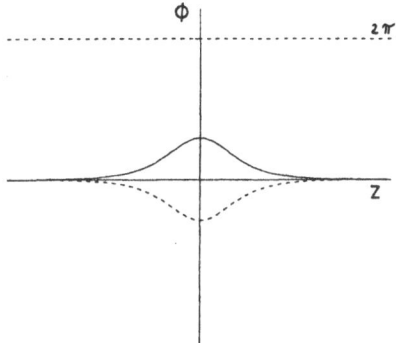

Fig. 4. Breather with low energy ($E \gtrsim 0$) and frequency $\Omega \lesssim mc$

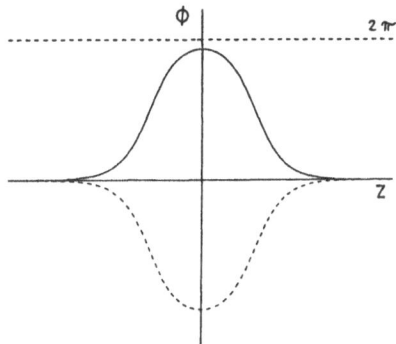

Fig. 5. Breather with high energy ($E \lesssim 2E_s$) and frequency $\Omega \ll mc$

These are moving solutions of the DW type with velocity u. One notes that the energy increases with the velocity.

- Breathers

$$\Phi_l(z,t) = 4 \arctan\left\{ \frac{1}{\sqrt{l^2 m^2 - 1}} \frac{\sin \Omega_l t}{\cosh z/l} \right\}$$

$$\Omega_l = \frac{c}{l}\sqrt{l^2 m^2 - 1}, \qquad lm = 1 \ldots \infty \tag{1.12}$$

$$E_B = \frac{2E_s}{lm}$$

The breather oscillates with a characteristic frequency Ω_l. At high energies one can interprete the breather as two solitons bound together. Here E_s denotes the static soliton energy with $E_s = 8mE_0$.

- small oscillations / phonons

$$\Phi(z,t) = A_q e^{i(qz - \omega_q t)}$$

$$\omega_q = c\sqrt{m^2 + q^2} \tag{1.13}$$

159

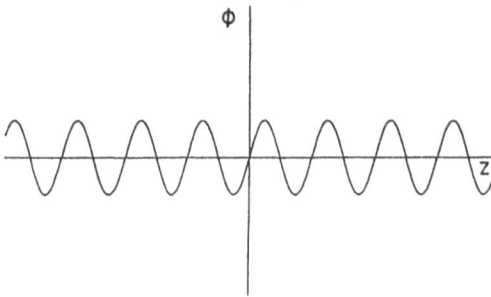

Fig. 6. Small-amplitude oscillation, i.e. $A_q \ll 1$.

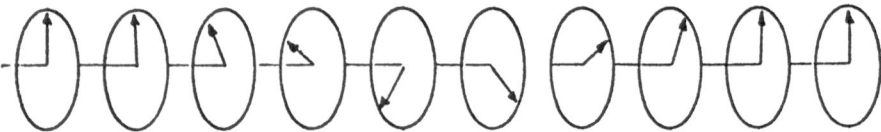

Fig. 7. Soliton configuration in easy–plane ferromagnets

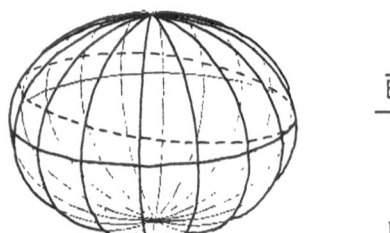

Fig. 8. Another representation of Fig. 7

In our context this "phonon" is a *Holstein–Primakoff* spin wave or magnon.

Solitons in easy-plane ferromagnets can also be viewed in a slightly different picture (see Fig. 7). Here the spins outside the localized transition region are shown to be parallel to the applied magnetic field B. An antisoliton in this picture will propagate counterclockwise. And finally in Fig. 8 the spins are shown lined up on a sphere. In this picture the small out–of–plane component of the moving soliton is visible. The directions of the spin vectors on the unit sphere are shown as one proceeds along the chain through a soliton.

1.3 Soliton-Carrying Spin Chains

1.3.1 Easy-plane Ferromagnets in an External Field

The energy functional for materials of this kind was described in section 1. The model defined by Eq. 1.7 is very valuable in describing substances like $CsNiF_3$ and

CHAB ([3,4], contribution by H. Benner to this volume). For comparison with the antiferromagnet below we note again that for the ferromagnetic chain with $J > 0$ the ground state has $\Phi = 0$, i.e. $S \parallel B$.

1.3.2 Easy–plane Antiferromagnet (AF) in an External Field

A typical example for these kind of substances is $TMMC$ (H. Benner, this volume). In this case the model of Eq. 1.7 can be used likewise, but the exchange energy J must be chosen as negative. To cope with the alternating spin pattern it is convenient to work on two sublattices. A convenient parametrization of the spin vector is [5,6]

$$\begin{pmatrix} S_{2n} \\ S_{2n+1} \end{pmatrix} = \pm S\big(\cos(\Theta \pm \theta)\cos(\Phi \pm \varphi), \dots \big) \tag{2.1}.$$

At low temperatures this leads to slow variations. The system can be treated equivalently to the SG chain, and one obtains slightly different parameters given as

$$
\begin{aligned}
E_0 &= \tfrac{1}{4}|J|S^2 \\
c &= 2|J|S \\
m &= \tfrac{\mu B}{2|J|S}
\end{aligned}
\tag{2.2}
$$

The ground state is given by $\Phi \approx \frac{\pi}{2}$, i.e. $S \perp B$ (apart from a small canting angle $\sim \frac{B}{|J|}$. For the spins it is preferable to be perpendicular to the external field as a consequence of the *spin-flop effect*. A difference important for the correlation functions to be discussed later is that now $\Phi_{SG} = 2\Phi$.

1.3.3 Ising Antiferromagnet with Weak Transversal Coupling

As an example of materials of this kind might serve $CsCoCl_3$. Due to the weak coupling these materials show nearly Ising–like behaviour. The model can only be treated with quantum mechanics, and the Hamiltonian reads

$$\mathcal{H} = 2J \sum_n S_n^z S_{n+1}^z + 2J\varepsilon \sum_n (S_n^x S_{n+1}^x + S_n^y S_{n+1}^y) \tag{2.3},$$

where $S = \frac{1}{2}$. Ising systems show well-localized quantum solitons delocalized by transverse interaction into a band of excitations [7].

2. Low-Temperature Statistical Mechanics of Classical Sine-Gordon Chains

2.1 Soliton Density at Low Temperatures

At sufficiently low temperatures only a small number of solitons exists in the Sine-Gordon chain, due to their finite excitation energy. Therefore it may be assumed that there is no overlap between kinks, e.g. they are treated as classical non-interacting excitations. Within the validity of this approximation it is possible to calculate the soliton contribution to thermodynamic quantities. So the soliton density is given by [3,8,9] :

$$n_s = n_{\bar{s}} = 4m\sqrt{\frac{mE_0}{\pi k_B T}} \exp\left(-\frac{E_s}{k_B T}\right), \qquad E_s = 8mE_0 \tag{2.1}$$

These approximations are only valid as long as the fraction $\frac{4n}{m}$ of lattice sites occupied by solitons is small. In this case eq. (2.1), which was found by phenomenological considerations, agrees with the result of transfer-matrix calculations using the full hamiltonian [9]. The following table shows how far these approximations are realized in the model systems for typical values of the parameters:

xy ferromagnet $CsNiF_3$

$$JS^2 = 23.6K \qquad T = 6K \qquad 4n/m \approx 0.024$$
$$DS^2 = 5K \qquad B = 5kG \qquad \Longrightarrow \qquad 2/m \approx 11$$

xy antiferromagnet $TMMC$

$$|J|S^2 = 128K \qquad T = 2.5K \qquad 4n/m \approx 0.043$$
$$DS^2 = 1K \qquad B = 37.5kG \qquad \Longrightarrow \qquad 2/m \approx 41$$

Allowing for finite z-components of the spins, the expressions for the energy and the density are modified in the following way:

$$E_s \rightarrow E_s(1 - \alpha T), \qquad n \rightarrow n \exp\left(3m\sqrt{\frac{J}{2D}}\right) \approx 2.5n \tag{2.2}$$

These corrections are important to achieve agreement with experimental findings, but they do not contain any qualitatively new aspects. Numerical work has been done for more quantitative results. In addition to solitons it appears that breathers should also be considered. However, the statistical mechanics including breathers is rather subtle and has been understood after much effort through the work of Johnson, Chen and Fowler [10]. The result is that it is sufficient to include in addition to solitons magnons with anharmonic terms, since breathers do not have an energy gap and their contributions can be expanded in powers of temperature.

2.2 Soliton Contribution to the Dynamic Structure Factor in Easy-plane Ferromagnets

Within the approximation of noninteracting kinks, it is possible to calculate e.g. the soliton contribution to the specific heat. Even more interesting are dynamic structure factors, since scattering experiments are sensitive to the spatial structure of excitations. In this section it will be shown how to calculate the longitudinal structure factor. Starting from the SG approximation one finds

$$S^z(x,t) = S\cos\phi(x,t) \quad . \tag{2.3}$$

The longitudinal structure factor is defined by

$$S_{\parallel}(q,\omega) = \frac{1}{(2\pi)^2} \int dt\,dx\; e^{i(qx-\omega t)} <\cos\phi(x,t)\cos\phi(0,0)> \quad . \tag{2.4}$$

Inserting the one-soliton solution (1.11) into eq.(2.3) gives:

$$\cos\phi(x,t) = 1 - 2\mathrm{sech}^2 m\gamma(x-ut) \tag{2.5}$$

$$\gamma = \frac{1}{\sqrt{1-u^2/c^2}}$$

Having many noninteracting kinks ($\frac{4n}{m} \ll 1$), this expression can be generalized to

$$\cos\phi(x,t) = 1 - 2\sum_i \mathrm{sech}^2 m\gamma_i(x-x_{0i}-u_it) + O((\frac{n}{m})^2) \quad . \tag{2.6}$$

Terms due to from the overlap of two different kinks are omitted, because they are of $O((\frac{n}{m})^2)$. Using eq. (2.6) one finds for the correlation function

$$<\cos\phi(x,t)\cos\phi(0,0)> = 1 - 2 < \sum \mathrm{sech}^2 m\gamma_i x >$$
$$+ 4 < \sum \mathrm{sech}^2 m\gamma_i(x-x_{0i}-u_it)\mathrm{sech}^2 m\gamma_i x_{0i} > \tag{2.7}$$
$$+ O((\frac{n}{m})^2) \quad .$$

The brackets stand for the thermal average over the starting positions x_{0i} and the velocities u_i of solitons and antisolitons:

$$< \sum .... > = 2 \int dx_{0i} \int du\, n(u).... \tag{2.8}$$

with

$$n(u) = n_s \frac{\exp\left(-\beta E_s\left(\frac{1}{\sqrt{1-u^2/c^2}}-1\right)\right)}{\int du \exp\left(-\beta E_s\left(\frac{1}{\sqrt{1-u^2/c^2}}-1\right)\right)}$$
$$\approx n_s\sqrt{\frac{\beta E_s}{2\pi c^2}}\exp\left(-\frac{1}{2}\beta E_s\frac{u^2}{c^2}\right) \quad . \tag{2.9}$$

All the integrals can be calculated analytically, giving:

2. term: $\quad -4\cdot 2n\int dx_0\mathrm{sech}^2 mx_0 = -\frac{16n}{m}$

$$\tag{2.10}$$

3. term: $\quad 8\int du\, n(u)\int dx_0\mathrm{sech}^2 m\gamma(x-x_0-ut)\mathrm{sech}^2 m\gamma x_0$

Using the substitution $\xi = m(x-x_0-ut)$ the Fourier transforms can be carried out very easily, yielding:

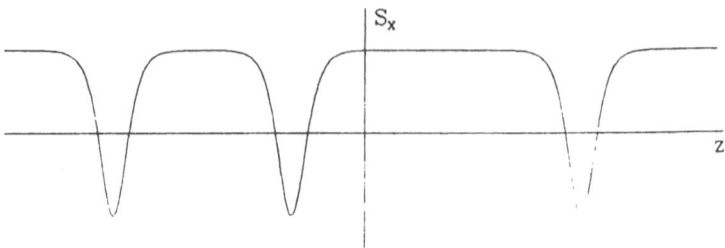

Fig. 9. Sequence of kinks in a ferromagnetic chain

$$S_\parallel(q,\omega) = (1 - 16\frac{n}{m})\,\delta(q)\,\delta(\omega)$$

$$+ \frac{8}{2\pi}\int du\, n(u)\delta(\omega - uq)\int \frac{du}{m}\frac{e^{i\frac{q\xi}{m}}}{\cosh^2\gamma\xi}\int dx_0\frac{e^{iqx_0}}{\cosh^2 m\gamma x_0}$$

This yields the final result:

$$S_\parallel(q,\omega) = (1 - 16\frac{n}{m})\,\delta(q)\,\delta(\omega)$$

$$+ 16\frac{\beta E_s}{mcq\pi^2}\,e^{-\beta E_s}\,e^{-\frac{1}{2}\beta E_s(\frac{\omega}{cq})^2}\left(\frac{\frac{\pi q}{2m\gamma}}{\gamma\sinh\frac{\pi q}{2m\gamma}}\right)^2 \tag{2.11}$$

The calculations of the transverse structure factors are now straightforward. One finds that all components of the structure factor have the general form:

$$S^{\alpha\alpha}(q,\omega) = \frac{16}{\pi m^2}\sqrt{\frac{\beta E_s}{2\pi}}\,n_s\,f(q,\omega)\,g^{\alpha\alpha}(Q = \frac{\pi q}{2m}) \tag{2.12}$$

with the different factors originating from

density $\qquad\qquad\qquad\qquad n_s$

velocity distribution $\qquad\quad f(q,\omega) = \frac{1}{cq}e^{-\frac{1}{2}\beta E_s\left(\frac{\omega}{cq}\right)^2}$

$$g^{xx}(Q) = \left(\frac{Q}{sinhQ}\right)^2$$

spatial distribution $\qquad\quad g^{yy}(Q) = \left(\frac{Q}{coshQ}\right)^2$

$$g^{zz}(Q) = \frac{\pi^2}{8}\frac{J}{D}\frac{1}{cosh^2Q}\quad.$$

The most important result here is the central peak found in the longitudinal structure factor. This is a characteristic feature of solitons . In ferromagnets the intensity of this peak increases proportional to the kink density. This is caused by the fact that in a ferromagnetic chain the soliton connects mathematically different but physically equivalent ground states (see Fig. 9), generating correlations this way.

For a comparison with experiments more extensive calculations are necessary. So out-of-plane fluctuations and 2-magnon processes have to be considered, the latter because they also give rise to a central scattering peak, which must be compared with the one caused by kinks.

2.3 Structure Factor in Easy-Plane Antiferromagnets

For an antiferromagnet one can also calculate the structure factor, but some more approximations are necessary [5]. The transverse spin-spin correlation function is given by:

$$< S^y(x,t)S^y(0,0) >= (-1)^n <\sin\phi_n(t)\sin\phi_0(0)> \tag{2.14}$$

with

$$\sin\phi(x,t) = \pm\tanh m\gamma(x - ut - x_0) \tag{2.15}$$

The behaviour of the transverse sublattice magnetization N_y for a sequence of Kinks is shown in Fig. 10. To go on with the calculation, one only considers the asymptotic behaviour of $\sin\phi$ by introducing a domain variable $\sigma(x,t)$ [8]:

$$\sin\phi(x,t) \rightarrow \sigma(x,t) = \pm 1 \tag{2.16}$$

Let now N_1 be the number of domain walls between 0 and x at $t = 0$ and N_2 be the number of domain walls passing x in the time betwen 0 and t. Then one gets:

$$\sigma(x,t)\sigma(0,0) = (-1)^{N_1+N_2} \tag{2.17}$$

In the simplest approximation we assume independent Poisson distributions for N_1 and N_2 ; the correlation function then reads:

$$<\sigma(x,t)\sigma(0,0) >= e^{-2(\bar{N}_1+\bar{N}_2)}$$
$$\bar{N}_1 = n_s x \qquad \bar{N}_2 = \frac{1}{\sqrt{\pi}}u_{th}t \tag{2.18}$$

After Fourier-transforming we obtain:

$$S^{yy} = \frac{\Gamma_D}{\omega^2 + \Gamma_D^2}\frac{\Gamma_S}{q^2 + \Gamma_s^2}$$
$$\Gamma_S = 4n_s \qquad \Gamma_D = \frac{4n_s u_{th}}{\sqrt{\pi}} = \frac{4n_s c}{\sqrt{4\pi\beta m}} \tag{2.19}$$

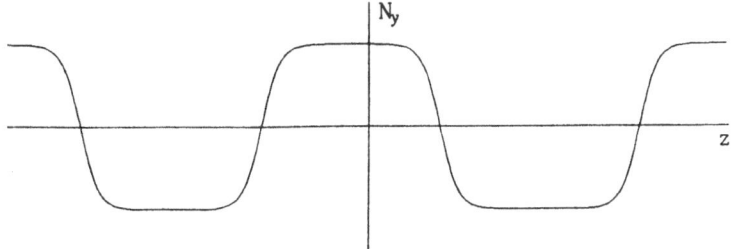

Fig. 10. Sequence of kinks in an antiferromagnetic chain

This result has to be compared with that for the longitudinal structure factor in the ferromagnet, because in the ferromagnetic ground state the spins point parallel to the field, but in an antiferromagnet they are arranged perpendicular to the field. Looking at eq.(2.19) again one finds a central peak. For vanishing soliton density this is the Bragg peak related to LRO induced by the magnetic field. At finite soliton density this Bragg peak broadens, and its widths $\Gamma_D, \Gamma_S \sim n_s$ are the signature of antiferromagnetic solitons.

3. Instabilities

3.1 The $CsNiF_3$ Soliton and Out–of–plane Fluctuations

To study the stability of the soliton against fluctuations out of the easy-plane it is convenient to use dimensionless variables for the energy functional. Then one measures time in units of $(JSm^2)^{-1}$ and space in units of m^{-1}. The easy plane is $\Theta = \frac{\pi}{2}$ so one replaces $\Theta' := \frac{\pi}{2} - \Theta$ but in the following we don't write the prime. The energy functional for the easy-plane ferromagnet in an external field minus ground state energy reads

$$E\{\Phi, \Theta\} \sim \int dz \left\{ \tfrac{1}{2} \left(\frac{\partial \Theta}{\partial z} \right)^2 + \tfrac{1}{2} \cos^2 \Theta \left(\frac{\partial \Phi}{\partial z} \right)^2 + \tfrac{1}{2} \lambda \sin^2 \Theta + (1 - \cos \Theta \cos \Phi) \right\} . \quad (3.1)$$

The only dimensionless parameter is then $\lambda = \frac{2DS}{\mu B}$.

To make an energetic stability analysis one expands around the soliton solution

$$\Phi = \Phi_{Sol} + \alpha$$
$$\Theta = \qquad \beta . \qquad (3.2)$$

Then one can write the energy in the following form

$$E = E_{Sol} + \int dz \left\{ \alpha L_1 \alpha + \beta \left(L_2 + \lambda \right) \beta \right\}, \qquad (3.3)$$

where the operators L_1, L_2 and their eigenvalues are

$$L_1 = -\frac{d^2}{dz^2} + 1 - 2 \operatorname{sech}^2 z \quad ; \quad \text{eigenvalues } l_1 = 0, k^2 + 1$$

$$L_2 = -\frac{d^2}{dz^2} + 1 - 6 \operatorname{sech}^2 z \quad ; \quad \text{eigenvalues } l_2 = -3, 0, k^2 + 1 . \qquad (3.4)$$

The solution of $(\alpha_0 L_1 \alpha_0) = 0$ is $\alpha_0 = 2 \operatorname{sech} z$ and represents a Goldstone mode. The deviation in β is stable only if $(\beta (L_2 + \lambda) \beta) > 0$, i.e. for $\lambda > 3$ [11,12,13].

To study dynamical instabilities one rewrites the equations of motion in these units

$$\frac{\partial \Theta}{\partial t} = \cos \Theta \frac{\partial^2 \Phi}{\partial z^2} - 2 \sin \theta \frac{\partial \Phi}{\partial z} \frac{\partial \Theta}{\partial z} - \sin \phi$$

$$\frac{\partial \Phi}{\partial t} = -\frac{1}{\cos \Theta} \frac{\partial^2 \Theta}{\partial z^2} + \lambda \sin \Theta - \sin \Theta \left(\frac{\partial \Phi}{\partial z} \right)^2 + \tan \Theta \cos \Phi , \qquad (3.5)$$

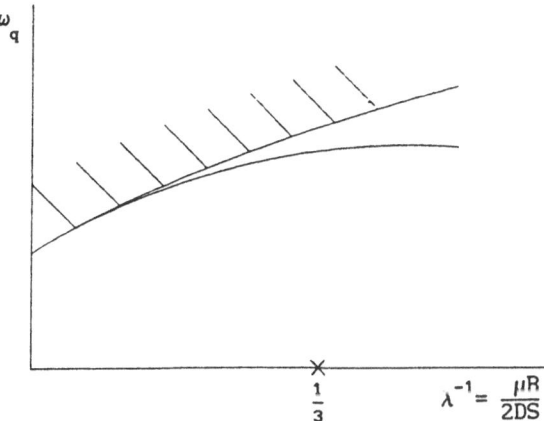

and makes the ansatz

$$\Phi(z,t) = \Phi_{\text{Sol}} + \phi(z,t)$$
$$\Theta(z,t) = \qquad \theta(z,t) \tag{3.6}$$

with time-dependent deviations ϕ and θ from the soliton solution. Then one obtains the coupled differential equations

$$-\frac{\partial \theta}{\partial t} = L_1 \phi$$
$$\frac{\partial \phi}{\partial t} = (L_2 + \lambda)\theta \tag{3.7}$$

with the operators L_1, L_2 of Eq.(3.4). To solve these equations one expands ϕ and θ in the eigenfunctions of L_1 and L_2. For the zero-frequency mode there are two independent solutions, because there are two coupled differential equations of first order.

$$\phi(z,t) = \sum_q c_q e^{-i\omega_q t}\phi_q(z) + (a_0 + b_0 t)\phi_0(z)$$
$$\theta(z,t) = \sum_q c_q e^{-i\omega_q t}\theta_q(z) + \qquad b_0\,\theta_0(z). \tag{3.8}$$

In Fig. 11 one can see frequencies of the bound states and the range of the continuum states. The translational mode coincides with the axis $\omega_q = 0$. There is no soft mode but instability due to translation. The energy for one fluctuation is determined by c_q, a_0, b_0 and reads

$$E = E_{\text{Sol}} + \text{const.}\Big\{b_0^2 \int dz\,\phi_0(z)\theta_0(z) + \sum_q |c_q|^2 \omega_q \Big\}. \tag{3.9}$$

The coefficient multiplying b_0^2 is proportional to $\frac{1}{\lambda - 3}$ and is related to the beginning of a spontaneous movement of the static soliton for $\lambda < 3$ [13].

3.2 Soliton Instabilities in the Discrete Chain

In the discrete model

$$E = JS^2 \sum_n \{1 - \sin\Theta_{n+1} \sin\Theta_n \cos(\Phi_{n+1} - \Phi_n) - \cos\Theta_{n+1} \cos\Theta_n$$

$$+ \tfrac{D}{J}\cos^2\Theta_n + \tfrac{\mu B}{JS}(1 - \sin\Theta_n \cos\Phi_n)\}. \qquad (3.10)$$

one can look for stationary states and their instabilitites too. By minimizing the energy $\frac{\delta E}{\delta \Phi_n} = \frac{\delta E}{\delta \Theta_n} = 0$ one can investigate two different cases.

For $\Theta \equiv \frac{\pi}{2}$ (i.e. discrete Sine-Gordon problem), any spin configuration is confined to the easy-plane and is necessarily a static structure described by

$$\sin(\Phi_{n+1} - \Phi_n) - \sin(\Phi_n - \Phi_{n-1}) = m^2 \sin\Phi_n . \qquad (3.11)$$

If one introduces

$$w_n := \Phi_n - \Phi_{n-1}$$

one obtains the mapping equations

$$\sin w_{n+1} = \sin w_n + m^2 \sin\Phi_n$$

$$\Phi_{n+1} = \Phi_n + w_{n+1} \qquad (3.12)$$

which one can iterate numerically. The map is similar to the standard map, which is studied by P. Bak [14]. The separatrix corresponds to the soliton solution. The soliton solution can have two different positions in relation two the underlaying lattice, central bond (CB) and central spin (CS) position. By numerical analysis one obtains two kinds of CB-solitons with different slope angles between $n = -1$ and $n = +1$. These angles depend on the parameter m and are equivalent for $m = m_c$ [15]. There are no soliton solutions in the discrete lattice for $m > m_c$. But as long as $m < m_c$ the solitons stay stable also for a whole range of anisotropies (see Fig. 12). The translational mode is even and has frequencies ω^2 which are slightly above zero because of small energy difference between the CS and CB structures. For $m = m_0$

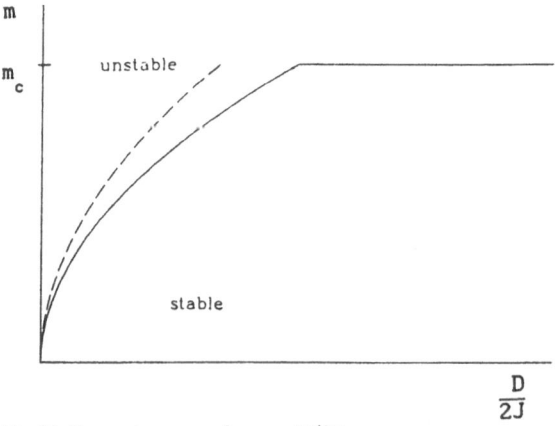

Fig. 12. Parameter space of m vs. $D/2J$

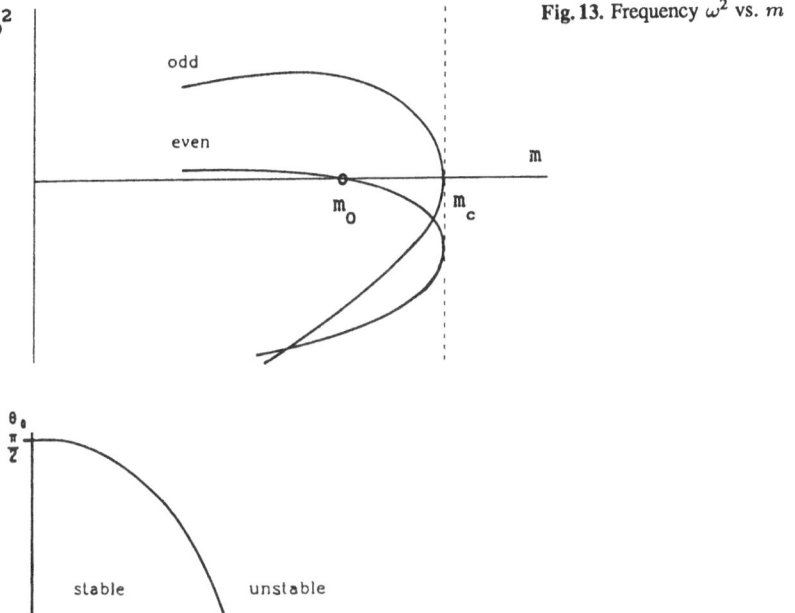

Fig. 13. Frequency ω^2 vs. m

Fig. 14. Parameter space Θ'_0 vs. m

the mode becomes unstable because of out–of–plane fluctuations. The odd solution with higher frequency becomes unstable at $m = m_c$ (see Fig. 13).

For $\Theta_n \neq \frac{\pi}{2}$ there exist solutions $\Phi(n-ut), \Theta(n-ut)$. Numerical analysis shows that the stability area depends on the out-of-plane deviation $\Theta'_0 = \frac{\pi}{2} - \Theta_n$ with which one starts the iteration. This is shown in Fig. 14. The value of m where the soliton becomes unstable decreases with increasing Θ'_0. When the soliton becomes unstable the width goes to zero, and the transition from $\Phi = 0$ to $\Phi = 2\pi$ takes place within one lattice spacing (see Fig. 14 in [15]).

4. Driven Sine–Gordon Chain with Damping

The driven Sine–Gordon chain with damping with the equation of motion being

$$\frac{\partial^2 \Phi}{\partial t^2} - \frac{\partial^2 \Phi}{\partial x^2} + \sin \Phi = \Gamma \sin(\omega_d t) - \varepsilon \frac{\partial \Phi}{\partial t} , \tag{4.1}$$

has been investigated by Bishop et al. [16]. Here Γ stands for the amplitude of the driving force and ε is the damping strength. If the right hand side were zero

169

one would obtain the unperturbed SG equation. The authors have chosen periodic boundary conditions and the variables were chosen as $\omega_d \equiv 0.6$ and $\varepsilon \equiv 0.2$. Γ was varied from zero to 2. Among the many results obtained we only mention the following:

- flat initial conditions

$$\Phi(x,0) = \dot{\Phi}(x,0) = 0$$

  For $\Gamma \lesssim 0.61585$ one only sees simple periodic motion. But above the threshhold value there is a large variety of phenomena to be seen. Chaotic as well as periodic motion or even *running periodic* motion where the power spetrum has a strong component at zero frequency.

- kink initial conditions

$$\Phi(L,t) - \Phi(0,t) = 2\pi$$

  In this case the threshhold value is $\Gamma \approx 0.59$. For lower values of the driving force the system exhibits small-amplitude oscillations. There are two more transitions at $\Gamma \approx 0.64$ from periodic kink oscillations to quasi-periodic motion, and at $\Gamma \approx 0.75$ where the system finally becomes chaotic. One sees chaotic motion on a small time scale but a rather simple periodic one on a larger time scale. The concluding remarks of the authors state that *spatial patterns tend to inhibit time chaos* which can also been seen in other systems.

References

1. H.J. Mikeska, M. Steiner, Advances in Physics **40**, 191 (1991)
2. C. Etrich, H.J. Mikeska: J.Phys. C**21**, 1583 (1988)
3. H.J. Mikeska: J.Phys. C**11**, L29 (1978)
4. M.G. Pini, A. Rettori: Phys.Rev. B**22**, 477 (1980)
5. H.J. Mikeska: J.Phys. C**13**, 2913 (1980)
6. K. Maki: J.Low Temp. Physics **41**, 327 (1980)
7. J. Villain: Physica **79B**, 1 (1975)
8. J.A. Krumhansl, J.R. Schrieffer: Phys.Rev. B**11**, 3535 (1975)
9. J.F. Currie, J.A. Krumhansl, A.R. Bishop, S.E. Trullinger: Phys.Rev. B**22**, 477 (1980)
10. M.D. Johnson, N.N. Chen, M. Fowler: Phys.Rev. B**34**, 7851 (1986)
11. E. Magyari, H. Thomas: Phys.Rev. B**25**, 531 (1982)
12. P. Kumar: Phys.Rev. B**25**, 483 (1982)
13. H.J. Mikeska, K. Osano: Z.Phys. B**52**, 111 (1983)
14. P. Bak: Phys.Rev.Lett. **46**, 791 (1981)
15. C. Etrich, H.J. Mikeska, E. Magyari, H. Thomas, R. Weber: Z.Phys. B**62**, 97 (1985)
16. A.R. Bishop, K. Fesser, P.S. Lomdahl, S.E. Trullinger: Physica **7D**, 259 (1983)

Solitary Nonlinear Excitations in Spin Systems: Experiment

H. Benner

Institut für Festkörperphysik, Technische Hochschule Darmstadt, W-6100 Darmstadt, Fed. Rep. of Germany

The soliton concept developed in the preceding chapter will now be discussed from an experimental point of view. Various techniques have been applied to probe the soliton dynamics of ferro- and antiferromagnetic sine-Gordon chains. We report on inelastic neutron scattering, nuclear magnetic resonance and electron spin resonance experiments in $CsNiF_3$ and TMMC. Particular interest was aimed at the instability of solitons in high magnetic fields.

1. Introduction

The mapping of a classical Heisenberg chain with a strong planar anisotropy and some additional symmetry-breaking anisotropy (resulting e.g. from a transverse magnetic field) to a sine-Gordon (sG) Hamiltonian, as proposed by *Mikeska* [1],

$$\mathcal{H} = \epsilon \int dz \left\{ \tfrac{1}{2} \left(\tfrac{\partial\Phi}{\partial z}\right)^2 + \tfrac{1}{2\,c^2}\left(\tfrac{\partial\Phi}{\partial t}\right)^2 + m^2(1 - \cos\Phi) \right\} \tag{1}$$

has turned out to be a successful concept for describing the low-temperature properties of both ferromagnetic (fm) and antiferromagnetic (afm) chains with different types of anisotropy [2-4]. According to this mapping, the field $\Phi(z,t)$ is related to the azimuthal angle of the spins with respect to their ground-state orientation, ϵ defines an energy scale set by the exchange interaction, and c and m denote some typical velocity and "mass" corresponding to the special type of anisotropy. The corresponding equation of motion

$$\frac{\partial^2\Phi}{\partial z^2} - \frac{1}{c^2}\frac{\partial^2\Phi}{\partial t^2} = m^2 \sin\Phi \tag{2}$$

is one of the few nonlinear and soliton-bearing wave equations which can be completely integrated. Its mathematical solutions - including magnons, breathers and solitons - and the corresponding thermodynamics are well known from literature [4-6].

In this chapter I will discuss the experimental aspects of probing solitons in real magnetic systems of sine-Gordon type. A soliton on a spin chain represents a microscopic structure which cannot be directly observed - like a solitary water wave - but can only be probed by its influence on macroscopic properties. Magnetic solitons have a well-defined excitation energy, so one might speculate about the possibility to excite them from outside - like magnons - by resonant microwave absorption or inelastic neutron scattering. Such attempts, however, fail for the lack of an appropriate *local* excitation mechanism. On the other hand, to excite them externally by microwave pulses or field gradients as proposed in the literature [7, 8] would require to apply magnetic fields of some Teslas strength on a length scale of a few nanometers within a time scale of nanoseconds - which so far seems impossible for technical reasons.

Hence, there remains only the chance to probe solitons which have been thermally excited. To separate their contribution from coexisting thermal excitations, an experimental proof firstly requires that the soliton excitations are dominating within a certain range of field and temperature, and secondly that they differ from others by characteristic properties. In this sense, measuring techniques probing the dynamic properties are generally more instructive than static ones.

1.1 Topology of Ferro- and Antiferromagnetic Solitons

Considering a fm chain with easy-plane anisotropy in a transverse magnetic field at sufficiently low temperature, there exists one ground-state where the spins are aligned in $-H$-direction. A soliton represents a 2π-twist of the ordered spin chain which can freely move along the chain without dispersing. In an afm chain the ground-state corresponds to a spin-flop configuration, i.e. the two sublattices, apart from a slight canting, are aligned antiparallel to each other and perpendicular to H. This way, there exist two degenerate, but topologically inequivalent ground-states which differ by the sign of spin orientation. Thus, an afm soliton represents a π-twist, reversing both sublattices when passing along the chain (Fig. 1). The difference in topology results in different spatial and temporal behaviour of the spin components and will largely affect the experimental evidence. While in the fm case spin fluctuations only

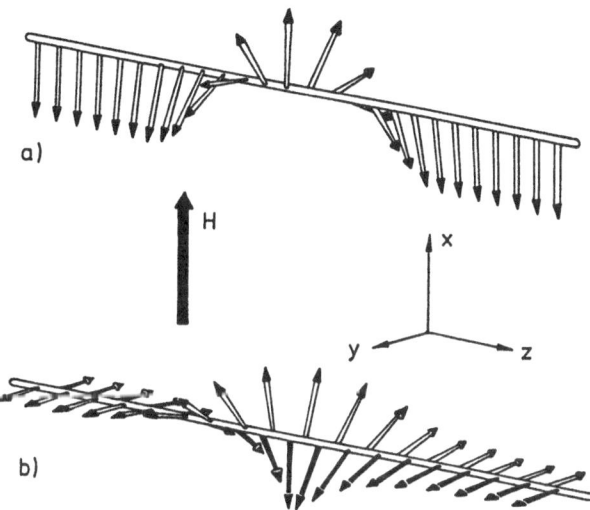

Fig. 1. Model of a sine-Gordon soliton in a classical ferromagnetic (a) and antiferromagnetic (b) spin chain.

originate from the 2π-soliton itself, it is the flipping of all spins *between* two neighbouring π-solitons (the so-called "flipping mode") which yields the essential contribution in the afm case. In either case, spin fluctuations in chain direction are largely suppressed by the strong planar anisotropy.

1.2   Experimental Methods

The dynamic properties of an electron spin system are usually discussed in terms of the dynamic structure factor $S^{\alpha\alpha}(q,\omega)$, $\alpha = x, y, z$, which is the spatial and temporal Fourier transform of the respective spin correlation function. A phenomenological derivation treating the solitons like a one-dimensional (1D) gas of non-interacting particles has been given in the preceding chapter. Let us, for example, consider the soliton contribution in a fm spin chain. The component parallel to the magnetic field H reads [1]

$$
\begin{aligned}
S^{xx}(q,\omega) &= \frac{1}{4\pi^2} \int dz \int dt \; e^{i(q_z z - \omega t)} \left\langle S^x(z,t)\, S^x(0,0) \right\rangle \\[2mm]
&= \frac{128\, S^2 \beta}{c\, q_z\, \pi^2} e^{-8\beta m} e^{-4\beta m \omega^2 / c^2 q_z^2} \left[M^x(q_z) \right]^2 \quad .
\end{aligned} \tag{3}
$$

Here, the characteristic parameters of (1), the soliton mass m and the characteristic velocity c, are defined by

$$m = \sqrt{g\mu_B H / 2JS} \tag{4}$$

$$c = 2S\sqrt{JA} . \tag{5}$$

J is the exchange integral between two neighbouring spins, A is the easy-plane anisotropy, and β denotes the inverse thermal energy normalized to $\epsilon = 2JS^2$. The essential temperature and field dependencies of $S^{xx}(q, \omega)$ result from the Boltzmann factor of the soliton density $n_s \sim \exp(-8\beta m) = \exp(-E_s/k_B T)$. Here, $E_s = 8m\epsilon$ denotes the activation energy of a fm soliton at rest which, following (4), is proportional to $\sqrt{H}$. The factor $M^x(q_z)$ results from the geometrical shape of the soliton, and the frequency dependent term of (3) describes its dynamic behaviour. According to the shape of a fm soliton, the contribution $S^{xx}(q, \omega)$ of the spin component parallel to H by far exceeds the contributions $S^{yy}(q, \omega)$ and $S^{zz}(q, \omega)$. So, all experiments on fm chains mainly probe the soliton-induced parallel structure factor which represents a quasi-elastic contribution close to the Bragg peak.

The first and most common technique to study magnetic solitons was inelastic neutron scattering [9,10]. The magnetic contribution to the differential cross-section is directly given by the dynamic structure factor [11]:

$$\frac{d^2\sigma}{d\omega\, d\Omega}\Big|_{magn.} \sim S^\perp(q, \omega) \tag{6}$$

Note that the magnetic scattering of unpolarized neutrons probes the spin components *perpendicular* to q, that means, if q is in $\hat{z}$-direction, then $S^\perp(q, \omega) = S^{xx}(q_z, \omega) + S^{yy}(q_z, \omega)$. Using polarized neutrons, all components of $S^{\alpha\alpha}(q, \omega)$ can in principle be measured separately. Complementary information can be obtained by nuclear magnetic resonance (NMR) [12-15] which, by means of the spin-lattice relaxation rate, only probes the dynamic structure factor integrated over q

$$\frac{1}{T_1} \sim \sum_\alpha \int dq \, B_q^\alpha \, S^{\alpha\alpha}(q, \omega_N) \ , \quad \alpha = x, y, z, \tag{7}$$

but offers the advantage of very high energy resolution (typically six orders of magnitude higher!). By Raman scattering [16] and by electron spin resonance

(ESR) [15,17] the influence of solitons can indirectly be detected from the soliton-induced broadening of the magnon linewidth, which in the limit $q \to 0$ is proportional to their density $n_s = 4\,m\sqrt{\beta\,m/\pi}\,\exp(-8\beta m)$ and average velocity $\bar{v}_s = c/\sqrt{4\pi\beta\,m}$:

$$\Gamma_{q=0} = \Gamma_{intr} + \Gamma_s ; \quad \Gamma_s = 4\,n_s\,\bar{v}_s = \tfrac{8}{\pi}\,m\,c\,e^{-8\beta m} . \tag{8}$$

The influence of solitons on static properties, such as specific heat [18-20], magnetization [21] and ordering temperature [22] has also been studied, but turns out to be less significant, owing to the competing contributions of other thermal excitations.

2. Ferromagnetic Solitons in CsNiF₃

Prototype of a fm sG chain is the well-known quasi-one-dimensional magnet CsNiF$_3$ [9]. Its crystal structure is of the hexagonal P6$_3$/mmc space group. The magnetic Ni^{2+} ions are ferromagnetically coupled by superexchange via the surrounding F$^-$ octahedra and form chains along the crystallographic c-axis which are separated by the large Cs$^+$ ions (Fig. 2). The magnetic behavior of the Ni^{2+} ions can be characterized by $S = 1$, $g = 2.26$, a single-ion anisotropy of easy-plane type $A/k_B = 5\,K$ (for classical spins) and a Heisenberg exchange integral $J/k_B = 11.8\,K$ [1] [15].

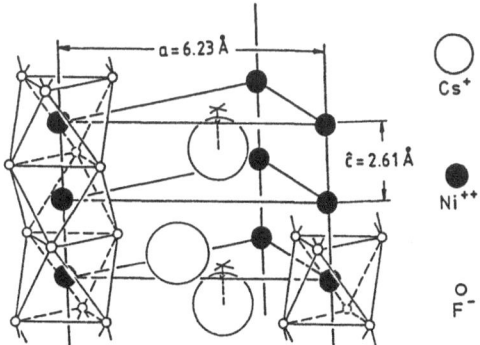

Fig. 2. Crystal structure of CsNiF$_3$

---

[1]Note that in our notation the exchange integral J differs from that of the preceding chapter by factor 2 and, in the afm case, by the sign.

2.1   Inelastic Neutron Scattering

First experimental evidence of magnetic solitons was obtained by *Kjems* and *Steiner* [9] from inelastic neutron scattering. Following (6), the differential cross-section for unpolarized neutrons consists of a magnetic contribution which is directly given by the dynamic structure factor of spin components perpendicular to q, and of a nuclear contribution which mainly results from scatterings at the Cs nuclei giving rise to a strong incoherent background. Since the latter is independent of temperature, it can be separated by a reference measurement at low temperature. The experiment was performed on a cold-neutron source triple-axis spectrometer. The transferred momentum $q = k_i - k_f$ (k_i and k_f being the initial and the final momentum of the scattered neutrons) remained fixed and was chosen to be $(0, 0, 1.9)$ r.l.u., that means, parallel to the $\hat{z}$-axis, but slightly outside a magnetic Bragg *plane*. (Note that the nuclear interaction, as usual, probes a 3D lattice of scattering nuclei, whereas the magnetic interaction probes a quasi-1D spin lattice. Hence, the magnetic Bragg peaks form planes perpendicular to the chain axis!) Having subtracted the dominating incoherent background, the remaining magnetic contribution shows a three-peak spectrum consisting of two magnon lines at $\pm 0.5\,\mathrm{meV}$ and a quasi-elastic contribution supposed to result from the solitons (Fig. 3). The quasi-elastic contribution could be separated by fitting the spectrum with a Gaussian line centered at $\omega = 0$, according to (3), and of two Lorentzians for the magnon peaks.

For verifying the predictions of the sG model, instead of the dynamic structure factor the authors used the integrated intensity

$$I_q = \int_{-\infty}^{+\infty} d\omega\ S^{xx}(q_z, \omega) = \frac{S^2 n_s}{m^2}\ \sqrt{\pi}\ [\,M^x(q_z)\,]^2 \tag{9}$$

in order to improve the accuracy of the analyzed data. For constant q_z one finds $I_q \sim T^{-1/2}\, H^{-1/4}\, \exp(-\alpha\sqrt{H}/T)$, where $E_s = \alpha\sqrt{H} \cdot k_B$ denotes the soliton activation energy. The dependence of the integrated intensity on magnetic field and temperature is shown in Fig. 4. The data follow the expected universal dependence on $\sqrt{H}/T$, the experimental value of α, however, was found to be about 30 % smaller than expected from the classical sG model: $\alpha_{th} = 8\,S\sqrt{2\,J S g\,\mu_B}\,/\,k_B \approx 15.2\,\mathrm{K}/\sqrt{\mathrm{kOe}}$.

Hence, very soon a controversial discussion arose whether the simple sG model with its drastic and partly inconsistent approximations could actually

176

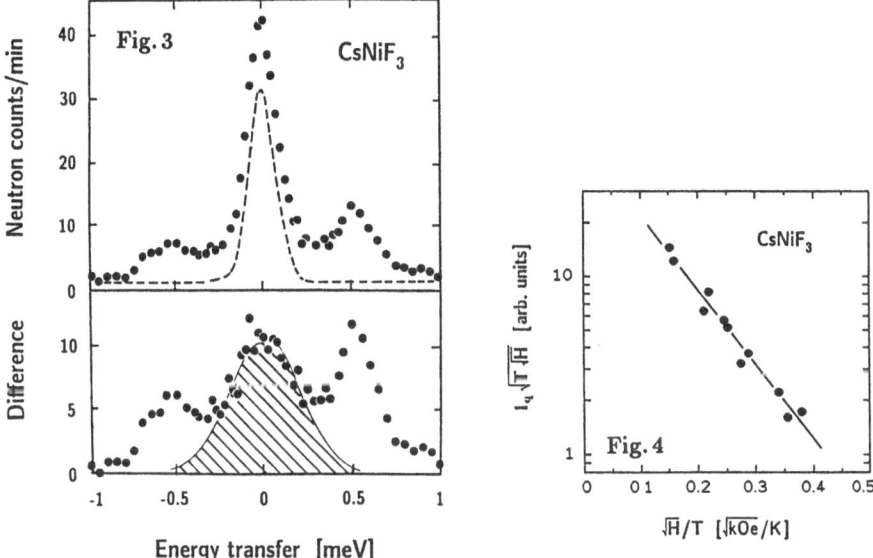

Fig. 3. Inelastic neutron scattering spectrum, observed in CsNiF$_3$ at $q = (0, 0, 1.9)$, $T = 9.3\,K$ and $H = 5\,kOe$. Dashed line represents the incoherent non-magnetic background obtained at high field and low temperature (data from [9]). b) Magnetic scattering contribution, evaluated as the difference of both upper spectra. The central peak (hatched area) is supposed to result from solitons.

Fig. 4. Universal field and temperature dependence of the integrated intensity I$_q$ (hatched area in Fig. 3). Experimental slope $\alpha_{exp} \approx 10\,K/\sqrt{kOe}$ (full line) is about 30 % smaller than expected from (9).

describe the properties of a real spin system. This controversy, however, turned out to become very fruitful, since it stimulated a number of theoretical papers [23-31] based on more realistic assumptions (e.g. finite anisotropy, discrete spin lattices, quantum properties or interactions with other thermal excitations; see also preceding chapter). Qualitatively, these extended models showed essentially the same behaviour as the simple sG model, and could be mapped to it by renormalizing either the soliton energy or the amplitude of the spin correlation functions. In such extended sense the sG model will now be used for the further interpretation of experimental results, and a detailed discussion of α_{exp} will be given below in connexion with our NMR data. On the other hand, subsequent and more detailed neutron scattering experiments by *Steiner et al.* [9], including investigations with polarized neutrons [32], have shown that only part of the observed central peak in $S^{xx}(q,\omega)$ could be attributed to solitons because of competing two-magnon contributions [33] whereas a central peak in $S^{yy}(q,\omega)$ should be exclusively due to solitons.

2.2 NMR Experiments

Nuclear magnetic resonance allows to study both static and dynamic properties resulting from soliton excitations. It is convenient to choose the ^{133}Cs nuclei ($I = 7/2$) as nuclear probes since they are located on highly symmetric lattice sites and are only weakly coupled to the magnetic chains. Hyperfine interactions give rise to a shift of resonance field which is proportional to the Ni^{2+} magnetization $M^z(H,T)$. From a careful analysis of the observed field and temperature dependencies [15], we found that $M^z(H, T)$ is dominated by two-magnon contributions and, therefore, less appropriate for probing the influence of solitons.

Their influence, however, shows up more distinctly in the relaxation time. Via hyperfine coupling with the Ni^{2+} electronic spins, the passing solitons cause incoherent fluctuations of the local field seen by each nucleus and force its relaxation the more effectively the larger their number. The spin-lattice relaxation rate $1/T_1$ is related to the dynamic structure factor by (7), where ω_N denotes the nuclear Larmor frequency and the $B_q{}^\alpha$ are coupling coefficients originating from both transferred and dipolar hyperfine couplings. When discussing the soliton dynamics of a fm chain, $S^{\alpha\alpha}(q,\omega)$ does not depend on q_x and q_y, and we may also neglect the weak q_z-dependence of $B_q{}^x$ and need only consider the spin component parallel to H. The resulting relaxation rate

$$\frac{1}{T_1} \sim \int dq_z \; S^{xx}(q_z, \omega_N) \; \sim \; \frac{1}{T} e^{-\alpha \sqrt{H}/T} \tag{10}$$

is proportional to the soliton density and follows an Arrhenius law where the activation energy E_s is proportional to $\sqrt{H}$.

Experimentally, the following result was obtained: For $3\,\mathrm{K} \leq T \leq 20\,\mathrm{K}$ and $3\mathrm{kOe} \leq H \leq 55\mathrm{kOe}$ a logarithmic plot of T/T_1 (Fig. 5) clearly shows the expected universal dependence on $\sqrt{H}/T$. The bending for $\sqrt{H}/T < 0.25$ (H in kOe, T in K), i.e. for low field and high temperature, can be understood by the soliton density becoming too high to satisfy the free soliton gas approximation. Here, the average distance between the solitons becomes smaller than their width. In contrast to that, for $\sqrt{H}/T > 0.6$, the soliton density becomes so low that excitations of lower energy are dominating [14]. In the range $0.25 \leq \sqrt{H}/T \leq 0.6$ which was identified to be the soliton regime, our data follow a straight line with a slope α representing the soliton energy. We obtained $\alpha_{\mathrm{exp}} = 10.5\,\mathrm{K}/\sqrt{\mathrm{kOe}}$, which is about 30 % below the theoretical

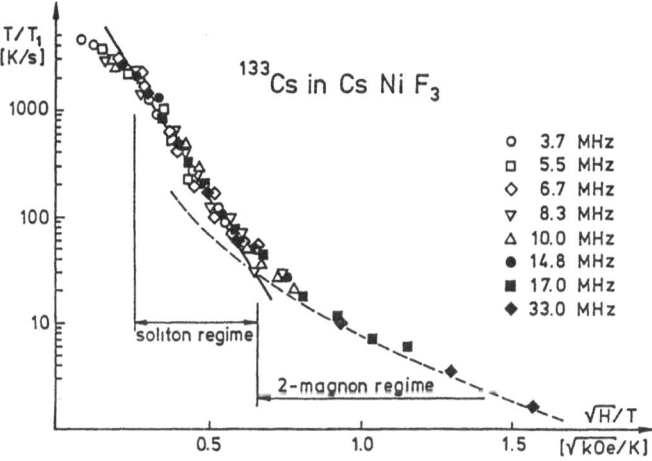

Fig. 5. Universal plot of the observed nuclear relaxation rate T/T_1 vs. H/T. For the soliton regime, the experimental slope (without any correction) amounts to $\alpha_{exp} \approx 11\,\text{K}/\sqrt{\text{kOe}}$. Dashed line: two-magnon contributions.

value expected for the sG model. A similar reduction was found in inelastic neutron scattering [9] and ESR experiments. [15].

For presenting a more quantitative discussion, one first had to separate the contributions of other thermal excitations, such as two-magnon contributions, which could be evaluated from the asymptotic behaviour at large values of $\sqrt{H}/T$ [15]. Attributing the remaining rate to the soliton contribution we found $\alpha_{exp} = 13\,\text{K}/\sqrt{\text{kOe}}$. Quantum corrections [23, 24, 30] yield a decrease of the classical sG energy of about 10%. A renormalization of the soliton mass due to out-of-plane fluctuations [24] could not account for the remaining deviation because of its different dependence on H and T, but a possible explanation results from soliton-magnon interference effects [25]. They cause a reduction of the soliton structure factor which is universal in $\sqrt{H}/T$ and increases for smaller values of this parameter. Thus, in total, the theoretical slope of T/T_1 is expected to become flatter. Although no adjustable parameters were used, the theoretical curve agreed quite well with our data [15].

A characteristic signature of solitons in real chains should be the occurrence of topological instabilities in high magnetic fields. For finite anisotropy A the spins in a moving soliton are tipped out of the easy-plane, and the tipping angle Θ increases with the soliton velocity [1]. Thus, with increasing H the gain of magnetic energy for the spins in the center of the fastest moving solitons will exceed the loss of anisotropy energy when declining out of the easy plane. As predicted by *Magyari* and *Thomas* [26], *Kumar* [27] and *Mikeska et al.* [29] there exists a critical field $H_c = 2 A S / 3 g \mu_B$, above which even a soliton at rest could decrease its energy by starting to move, and should become topologically unstable. Formally, this instability can be included in the soliton structure factor (3) by a renormalization of its amplitude and of the kinetic energy [29], which essentially means by rescaling the characteristic velocity c according to $c_H \approx c \sqrt{1 - H/H_c}$. Theoretically, for CsNiF$_3$ the critical transverse field H_c should be about 18 kOe. Surprisingly, our T_1 measurements did not show any marked deviation from sG behaviour even at the highest fields of more than 50 kOe. Neutron scattering experiments did not indicate any drastic instability behaviour either [32].

As a further check for soliton instabilities it was suggested [34] to support the tipping of the spins by applying an additional magnetic field H^z in chain direction. Then, already for rather low values of H^z (≈ 5 kOe) all slowly moving solitons (and antisolitons), which are the most efficient ones for T_1, were predicted to become unstable and to collapse, leading to a drastic decrease of the soliton density and, therefore, of the relaxation rate, too.

Experimentally, the additional longitudinal field was applied by rotating the sample in the magnet as shown in Fig. 6 f. This way, both the in-plane component $H^x = H \sin \vartheta$ and the out-of-plane component $H^z = H \cos \vartheta$ of the magnetic field were varied simultaneously, and the critical value of H^x, where a sudden decrease of $1/T_1$ should occur, was mapped to a critical angle ϑ_c. Figures 6 a-e show our experimental results for various fields and temperatures. In neither case a dramatic discontinuity occurs, and $1/T_1$ increases smoothly. Instead, the observed angular dependence of $1/T_1$ could be well simulated by means of the following assumptions: (i) two-magnon and soliton contributions were calculated within the sG model by taking into account only the in-plane component $H^x(\vartheta)$; (ii) the angular dependence of the hyperfine coupling coefficients $B_0^x(\vartheta)$ was determined by means of rotation matrices and included in the calculation of $1/T_1$. When comparing this model with our data we have distinguished between magnon contributions (dashed line) and the combined effect of magnons and solitons (solid line). In Fig. 6 a, $1/T_1$ is dominated by soliton contributions ($\sqrt{H}/T = 0.42$) and the field is far below H_c, so that the solitons should remain stable for all angles. In contrast to that, in Fig. 6 e, $\sqrt{H}/T$ takes the rather large value of 1. Then, nearly no soliton contribution remains which prevents any occurring instability from being

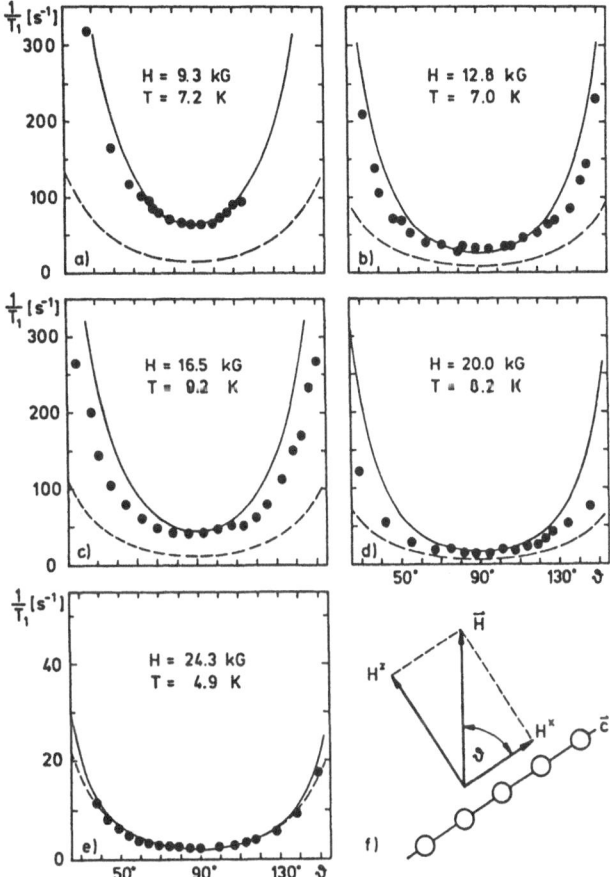

Fig. 6. Angular dependence of the nuclear relaxation rate for various fields and temperatures compared with theory; dashed lines: magnon contributions only, solid lines: combined magnon and soliton contributions.

detected. In between (Figs. 6 b-d), the data indeed tend to deviate from the theoretical angular dependence towards smaller relaxation rates, but there occurs no discontinuity. This is in agreement with more recent work predicting a continuous deformation of the soliton shape [35] and, accordingly, only continuous changes of n_s and $S^{xx}(q_z, \omega)$ [36].

2.3 ESR Experiments

Direct observation of solitons by ESR is not possible since the solitons mainly contribute for non-zero q at $\omega = 0$, see (3), while in ESR experiments the *uni-*

form mode ($q=0$) is probed at microwave frequencies ($\omega/2\pi = 10\text{-}100\,\text{GHz}$), which are far beyond the width of the quasi-elastic soliton peak. Note that within the sG limit the frequency of the uniform mode $\omega_0 = mc = \sqrt{2ASg\mu_B H}$ is proportional to $\sqrt{H}$.

Indirect information on solitons, on the other hand, can be obtained from observing the $q=0$ magnon, owing to interference effects between solitons and magnons, as already indicated by our NMR results. Decreasing anisotropy and 3D ordering effects [37] limit the regime where one might probe soliton properties to temperatures less than $10\,\text{K}$ and to magnetic fields larger than $2\,\text{kOe}$ which correspond to magnon gaps of $\omega/2\pi \geq 40\,\text{GHz}$. For this reason we confine our discussion to V-band microwave frequency ($70\,\text{GHz}$). The corresponding field is far below the critical field H_c, therefore, we cannot study any soliton instability with our ESR equipment.

Owing to soliton-magnon interference the magnon structure factor is affected in the following way: Solitons passing a magnon give rise to uncorrelated phase shifts which result in a broadening of the magnon linewidth. This broadening is proportional to the number of passing solitons. Especially for the uniform ($q=0$) mode this additional linewidth contribution is given by [25]

$$\Gamma_s(H,T) = 4\,n_s\,\bar{v}_s = \tfrac{8}{\pi}\,mc\,e^{-E_s/k_B T} \tag{11}$$

which again depends strongly on magnetic field and temperature. Since conventional ESR spectrometers do not allow a variation of frequency, our analysis was restricted to the temperature dependence of the linewidth. Converting the frequency width to magnetic field units one obtains a soliton-induced broadening of

$$\delta H = \tfrac{16}{\pi}\,H\,e^{-\alpha\sqrt{H}/T} \tag{12}$$

Due to the intrinsic field dependence of δH, the ESR lineshape - assumed to be a Lorentzian in frequency - becomes more and more asymmetric in field with increasing linewidth.

As expected from theory, the observed peak-to-peak linewidth δH_{pp} of the derivative ESR signal showed a strong monotonous increase with temperature (see insert in Fig. 7). In order to separate the soliton-induced broadening, one had to subtract the intrinsic linewidth obtained by extrapolation to zero tempe-

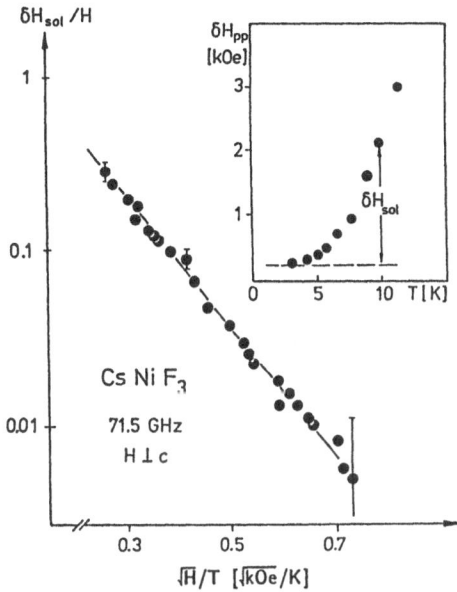

Fig. 7. Logarithmic plot of the soliton-induced broadening of the ESR linewidth vs. H/T.

rature. The observed residual width $\delta H_{pp}(T \to 0)$ was sensitively dependent on the shape of each sample and could partly be attributed to the inhomogeneity of the internal fields. A logarithmic plot of $\{\delta H_{pp}(T) - \delta H_{pp}(T \to 0)\}/H$ as shown in Fig. 7 yielded experimental evidence for the expected exponential dependence on $\sqrt{H}/T$ with a slope of $\alpha_{exp} \approx 10 \, K/\sqrt{kOe}$ resulting from the soliton density. Such naive evaluation, however, does not consider the change of lineshape with increasing linewidth, which systematically decreases the ratio between the experimental peak-to-peak linewidth δH_{pp} and the HWHM linewidth described by (12). Simulating the shape deformation, best agreement with our data was obtained for $\alpha_{exp} = (12 \pm 1) \, K/\sqrt{kOe}$, which is consistent with NMR and neutron data.

3. Antiferromagnetic Solitons in TMMC

A nearly ideal example of the afm sG chain is represented by the compound $(CH_3)_4NMnCl_3$ (TMMC) [10]. Its crystal structure is isomorphic to $CsNiF_3$ (Fig. 2), the magnetic Mn^{2+} ions ($S = 5/2$, $g = 2.01$), however, couple antiferromagnetically in c-direction ($J/k_B = -6.8 \, K$) and, in contrast to $CsNiF_3$, the planar anisotropy is of dipolar origin and only amounts to $0.2 \, K$.

Mapping the afm spin chains to a sG Hamiltonian, the soliton mass and velocity have the meaning

$$m = g\mu_B H/4|J|S \tag{13}$$

$$c = 4|J|S\sqrt{1 - A/4|J|} \quad , \tag{14}$$

and the energy scale is given by $\epsilon = |J|S^2/2$. As discussed above, the dominating contribution to the dynamic structure factor no longer results from spin fluctuations inside the solitons but from the flipping of the afm sublattices when a π-soliton is passing. Therefore, the staggered mode $S^{yy}(q \approx q_0, \omega)$, which may be considered as a broadened Bragg peak with a frequency width given by the sublattice flipping rate Γ [10], determines most of the experimental properties.

The soliton dynamics of TMMC has been widely studied by inelastic neutron scattering [10], NMR [12,13], antiferromagnetic resonance [17] and specific heat [18]. Rather than boring the reader by presenting analogous investigations in detail, I confine myself to NMR experiments and point out possibilities how to manipulate the π-solitons by the experimental conditions.

3.1   Ballistic and Diffusive Solitons in Pure and Doped Samples

Probing the soliton dynamics of afm sG chains by NMR, the relevant contribution to the relaxation rate is related to the transverse structure factor according to

$$\frac{1}{T_1} \sim B_{q_0}^y \int_0^{q_0} dq_z \, S^{yy}(q_z, \omega_N) \quad . \tag{15}$$

Following *Mikeska* [3], the respective spin correlation functions decay exponentially like $\exp\{-2\bar{N}(t)\}$, where $\bar{N}(t)$ denotes the mean number of uncorrelated flips caused by passing solitons at some reference site. Assuming a 1D gas of non-interacting solitons, the average number of flips will be proportional to time, to the soliton density n_s and to their mean velocity $\bar{v}_s$:

$$2\bar{N}(t) = 4 n_s \bar{v}_s t \equiv \Gamma t \quad . \tag{16}$$

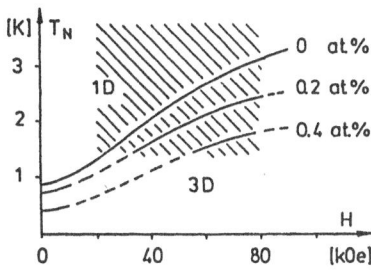

Fig. 8. Effect of Cu^{2+} doping on the 3D ordering temperature in TMMC (hatched area: H and T regimes available by our equipment).

The Fourier transform of $\exp\{-2\bar{N}(t)\}$ yields a Lorentzian with a width $\Gamma = \alpha H \exp(-\alpha H/T)$. Again, the parameter α represents a measure for the soliton energy $E_s = 8m\epsilon \equiv \alpha H \cdot k_B$, which in the afm case is directly proportional to H rather than to $\sqrt{H}$. The value of α can be determined from experiment. Describing TMMC by the classical sG model one expects $\alpha_{th} = 0.34\,\text{K/kOe}$ whereas quantum corrections [23] and the tilting of the spins out of the easy plane [38] lead to a reduction of 30%. For the limiting cases of very low and very high ratios Γ/ω_N the asymptotic behaviour of the relaxation time is given by [13]

$$T_1 \sim He^{-\alpha H/T}, \quad \Gamma \gg \omega_N \tag{17a}$$

$$T_1 \sim He^{+\alpha H/T}, \quad \Gamma \ll \omega_N \tag{17b}$$

and can easily be checked using the same universal plot. Because of the strong exponential dependence, the crossover regime between the two cases will be rather small, showing a well-defined minimum at $\Gamma \approx \omega_N$.

The first NMR experiments probing the soliton dynamics in TMMC were performed on very pure and carefully grown single crystals [12] to ensure that the soliton motion was not affected by any scattering at impurities or defects. In such samples only the low frequency limit (17a) could be observed, since the available soliton regime was restricted to small values of H/T due to the occurence of 3D order in higher magnetic fields. The obtained value $\alpha_{exp} = 0.26\,\text{K/kOe}$ was in accordance with neutron scattering data [10] and could also be verified by AFMR experiments [17].

There exists, however, a possibility to suppress the 3D order by doping the chains with impurities. As shown in Fig. 8, doping concentrations of less than one per cent will drastically reduce the ordering temperature. The

185

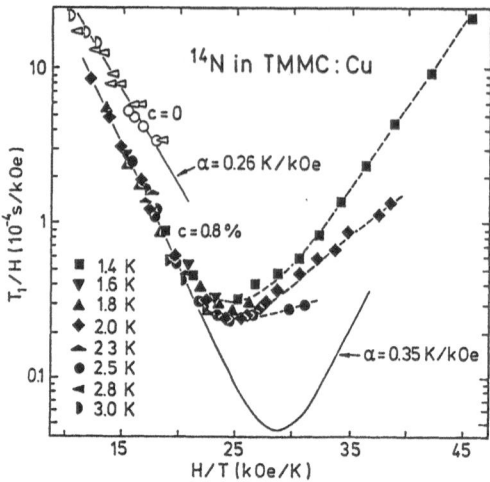

Fig. 9. Experimental data of T_1/H for pure ($c=0$) and Cu^{2+} doped ($c=0.8$ at.%) TMMC samples, compared with the universal behaviour in H/T predicted for ballistic solitons (full lines).

question remained how would the solitons be affected by restricting their mobility on the chains. To clarify this point, further NMR measurements were performed both on pure single crystals and on doped ones containing 0.2...0.8 at.% of Cu^{2+} or Cd^{2+}. The relaxation time T_1 of ^{14}N nuclei, which are symmetrically located between the chains, was recorded as a function of temperature and of the transverse magnetic field within the ranges $20\,kOe \leq H \leq 80\,kOe$ and $1.4\,K \leq T \leq 3\,K$. All data presented in Fig. 9 were obtained in the 1D regime. This condition could be checked by observing the ^{14}N spectrum which shows a characteristic splitting of lines at the 3D ordering transition.

While the earlier results [12] could be reproduced for the pure samples, the above model failed for the doped ones. Neither the expected universal H and T dependence of (17), nor the minimum position was described correctly, and the slope of the low-frequency limit would have implied a non-realistic increase of the soliton energy of 50 %. Obviously, a change of the dynamic behaviour had occurred.

To explain our data we assumed that, due to scattering of the solitons at the impurities, their motion along the chain can be described by a 1D random walk and becomes diffusive. The number of sublattice flips within a time t then reads

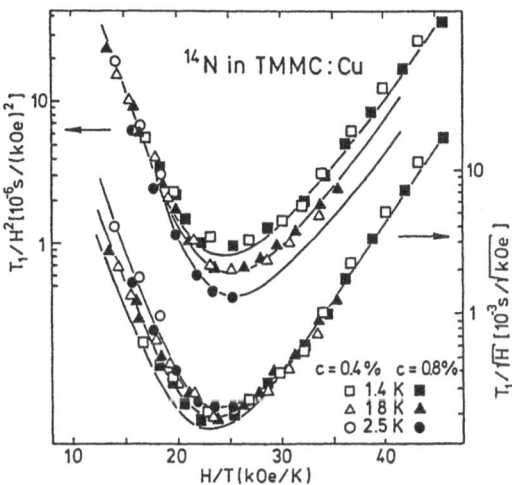

Fig. 10. Experimental data of T_1/H^2 (left-hand scale) and $T_1/\sqrt{H}$ (right-hand scale) for Cu^{2+} doped TMMC samples ($c = 0.4$ and 0.8 at. %), compared with the universal behaviour in H/T predicted for diffusive solitons (full lines: $T = 1.4$, 1.8 and 2.5 K).

$$\bar{N}(t) = 2\, n_s \sqrt{Dt} \tag{18}$$

and the respective spin correlation function should decay according to $\exp(-\sqrt{t})$ instead of $\exp(-t)$. This leads to the following asymptotic expressions for T_1 [13]:

$$T_1 \sim \Delta \qquad , \quad \Delta \gg \omega_N \tag{19a}$$

$$T_1 \sim \sqrt{\omega_N/\Delta} \quad , \quad \Delta \ll \omega_N \tag{19b}$$

The flipping rate, which now contains the square of the soliton density, is given by

$$\Delta = \alpha^2 H^2 \tau_D e^{-2\alpha H/T} \,. \tag{20}$$

This rate depends strongly on the soliton energy and on some characteristic time τ_D which is related to the diffusion constant and to the thermal velocity of the solitons by $\tau_D = \bar{v}_s{}^2 D$. Thus, the diffusive model predicts universal dependence on H/T for $T_1/H^2 \sim \exp(-2\alpha H/T)$ in the low-frequency limit and for $T_1/\sqrt{H} \sim \exp(+\alpha H/T)$ in the high-frequency limit. In contrast to the model of ballistic solitons, the exponents at both limits differ by a factor of 2.

The asymptotic behaviour in both limits is quite well illustrated in Fig. 10 for two of the Cu^{2+}-doped samples, confirming that the diffusive soliton model yields an adequate description [13]. The evaluation of both asymptotic slopes ($\alpha_{exp} = 0.25\,\text{K/kOe}$) showed that in spite of drastic changes of the dynamics the soliton activation energy was hardly affected. Surprisingly, for the impurity concentrations investigated (0.2, 0.4, and 0.8 at.% Cu), the diffusion parameter τ_D was nearly independent of concentration, amounting to about $5 \cdot 10^{-11}$s. It depends, however, very sensitively on the special kind of dopants. For manganese chains containing the same concentrations of non-magnetic Cd^{2+} impurities, τ_D was found to be decreased by a factor of 5 [39].

3.2 Longitudinal Solitons

Analysis of the stability of afm solitons including the out-of-plane spin components [40,41] has shown that for very high fields the transverse magnetic field H and the anisotropy A are interchanging their roles: The magnetic field then acts as the dominating anisotropy forcing the spins into the yz-plane, whereas the dipolar anisotropy yields the symmetry-breaking field. The crossover field H_A is defined by the condition that the magnetic energy of the spin-flop state becomes as large as the anisotropy energy:

$$H_A = 4S\sqrt{A\,|J|}\,/\,g\mu_B \ . \tag{21}$$

For TMMC, H_A amounts to about 90 kOe. A new *longitudinal* type of sG soliton was predicted which is represented by a π-turn in the yz-plane instead of the xy-plane, whereas the ground-state orientation of the sublattices outside the soliton remains unaffected. The activation energy of these longitudinal solitons does not depend on the external magnetic field but only on exchange coupling and dipolar anisotropy; $E_A = \alpha H_A \sim \sqrt{A\,J}$. While the common transverse type of sG-like soliton becomes unstable for $H > H_A$, the longitudinal type remains stable for arbitrary fields (although in low fields its density remains rather small because of the higher excitation energy).

In contrast to the fm case, the instability of the transverse afm solitons could be well observed by NMR. Though T_1 is not directly sensitive to a distortion of the soliton shape, the continuous changeover to a field-independent activation energy shows up in the sublattice flipping rate according to

$$\Gamma = 4\,n_s\,\bar{v}_s \approx \alpha H_A\,e^{-\alpha H_A/T} \tag{22}$$

and

$$\Delta = \Gamma^2 \tau_D \approx \alpha^2 H_A^2 \tau_D e^{-2\alpha H_A / T} \tag{23}$$

for the ballistic and for the diffusive soliton model, respectively. T_1 measurements on pure TMMC in very high magnetic fields ($120\,\text{kOe} \le H \le 170\,\text{kOe}$) have clearly proven that the flipping rate is independent of H [37] as expected for longitudinal solitons. The observed activation energy of $(22 \pm 2)\,\text{K}$ agrees well with the predicted value $\alpha\,H_A$.

3.3 Soliton Pairs below T_N

Until recently, magnetic sG solitons were considered as a specific property of the 1D phase. For afm chains, however, it was shown [42] that solitons should also occur in the 3D ordered phase below T_N, and that the phase transition could be understood to be induced by a pairing of solitons. This means that π-solitons on the same chain, moving independently above T_N, are bound into pairs. The pairing is effected through weak interchain interactions lifting the degeneracy of the ground-state and giving rise to long-range afm order. The weak (dipolar) couplings between neighbouring chains tend to align facing spins parallel to each other. The occurrence of two separate π-solitons (or antisolitons) on one of the chains would reverse the afm sublattices on a certain part of the chain and align the facing spins of neighbouring chains in an unfavourable anti-parallel orientation. The corresponding increase of energy can be minimized if both solitons move towards each other up to a certain minimum distance forming a $+2\pi$ (-2π) twist of the chain, since for topological reasons they cannot annihilate. This mechanism resembles very much e.g. that of polaron formation in trans-polyacetylene chains [43].

Describing the interchain interactions by a staggered mean field as the simplest approach, *Hołyst* [42] was able to map the corresponding equation of motion to a double-sine-Gordon equation

$$\frac{\partial^2 \Phi}{\partial z^2} - \frac{1}{c^2} \frac{\partial^2 \Phi}{\partial t^2} = \frac{m^2}{2} \sin 2\Phi + \eta \sin \Phi \tag{24}$$

where c and m keep their usual meanings defined by $(13, 14)$, and η describes the mean interchain coupling $\eta = 6\,J'/\,|J|$. Analytic solutions of this equation - which, in fact, resemble two coupled π-kinks - and the corresponding

189

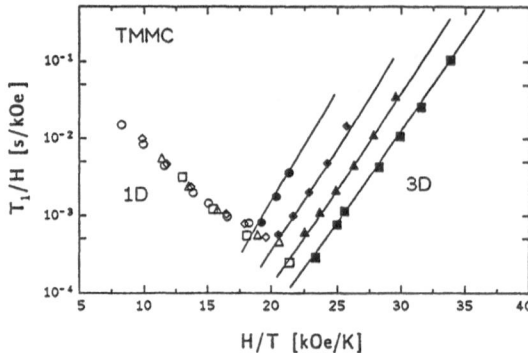

Fig. 11. Comparison of T_1/H data from the 1D and 3D regimes in pure TMMC. Open symbols: $T \geq T_N$, full symbols: $T < T_N$; $H = 34.7\,\text{kOe}$ (o), $41.1\,\text{kOe}$ ($\diamond$), $47.3\,\text{kOe}$ ($\triangle$) and $53.7\,\text{kOe}$ ($\square$). Full lines correspond to the model of soliton pairs described by eqs. (25, 26).

thermodynamics have been treated in the literature [42, 44]. The most important effect of π-soliton pairing is the increase of activation energy by roughly a factor of 2:

$$E_{2\pi} = 2 E_\pi \{ \sqrt{1+x} + x \operatorname{arsinh} \frac{1}{\sqrt{x}} \} \,, \qquad x = \frac{96 \, J' |J| S^2}{(g \mu_B H)^2} \qquad (25)$$

approaching $2 E_\pi = 2 \alpha H \cdot k_B$ in the limit $x \to 0$, i.e. in case of very weak interchain coupling. This doubling of activation energy results in a much lower soliton density, and gives rise to a spin-lattice relaxation rate of [45]

$$\frac{1}{T_1} \sim \frac{H^2}{T} \, e^{-2 E_\pi / k_B T} \,. \qquad (26)$$

The factor 2 in the soliton energy should not be confused with diffusive behaviour; here, we are dealing with *pure* samples, and the doubling of the exponent - corresponding to the 3D regime - occurs in *high* magnetic fields, whereas for the doped samples the situation was just reverse.

Experimental evidence for soliton pairs in pure TMMC has been obtained quite recently [46] by extending our field and temperature dependent T_1 measurements to the 3D regime, i.e. to $H > 40\,\text{kOe}$ and $T < 2\,\text{K}$ (cf. Fig. 8). Data from both the 1D and 3D regimes are compared in Fig. 11 using the same universal plot as above for the ballistic solitons. The 3D data, taken at four different fields, no longer remain universal - as expected from the different prefactors in (26) and from the dependence on J'/H^2 in (25).

190

However, the corresponding theoretical curves are in very good agreement with our data, proving, in fact, the expected doubling of activation energy: $E_{2\pi}/H \geq 0.6\,\mathrm{K/kOe}$. From the distances of the different curves, which are expected to scale with J'/H^2, the interchain coupling was evaluated to amount to $J'/k_B = (0.9 \pm 0.1)\,\mathrm{mK}$, in accordance with earlier results [47]. Correcting the experimental slope for the term in braces appearing in (25), we finally obtained $2\alpha_{\pi,\mathrm{exp}} \approx (0.57 \pm 0.03)\,\mathrm{K/kOe}$, which almost exactly agrees with our previous 1D result.

4. Conclusions

Summarizing, I have tried to show that the soliton concept in ferromagnetic and in antiferromagnetic chains, on the whole, is well supported by inelastic neutron scattering, NMR and ESR measurements. In CsNiF$_3$ the experimental evidence of solitons requires rather detailed analysis, since their contribution to the scattering cross-section, to the relaxation rate or to the FMR linewidth is competing with other thermal excitations of equivalent importance. NMR experiments aiming at the instability of solitons in high magnetic fields were less conclusive, since the expected out-of-plane instability is largely hidden by two-magnon contributions. In TMMC the situation appears more evident. Here, owing to the different topology of antiferromagnetic solitons, the soliton-induced flipping of sublattices yields the dominating contribution, and the observed behaviour could be well understood in terms of the sine-Gordon model. I have discussed some possibilities how to play with π-solitons by variation of the experimental conditions: Doping the chains with impurities changes the soliton dynamics from ballistic to diffusive, but does not affect their excitation energy. Increasing the magnetic field changes the plane of spin rotation from transverse (xy) to longitudinal (yz) but retains their sine-Gordon character. Decreasing the temperature below T_N, finally, results in a pairing of solitons, which even lets them survive in the presence of 3D long-range order. More spectacular ways of manipulation, e.g. by strong external HF fields, aiming at the non-thermal excitation of solitons [7,8] or at the excitation of competing chaotic structures [48] could not be realized so far in the presently known model systems, mainly for technical reasons. They remain a challenge for future investigations.

References

[1]   H. J. Mikeska: J. Phys. C 11, L 29 (1978)

[2]   K. Maki: J. Low Temp. Phys. 41, 327 (1980)

[3]   H. J. Mikeska: J. Phys. C 13, 2913 (1980)

[4]   K. M. Leung, D. W. Hone, D. L. Mills, P. S. Riseborough, S. E. Trullinger:
      Phys. Rev. B 21, 4017 (1980)

[5]   J. Rubinstein: J. Math. Phys. 11, 258 (1970)

[6]   J. F. Currie, J. A. Krumhansl, A. R. Bishop, S. E. Trullinger:
      Phys. Rev. B 22, 477 (1980)

[7]   D. Hackenbracht, H. G. Schuster: Z. Phys. B 42, 367 (1981)

[8]   M. M. Bogdan, A. M. Kosevich, I. V. Manzhos: Fiz. Nizk. Temp. 11, 991 (1985)
      [English transl.: Sov. J. Low Temp. Phys. 11, 547 (1985)]

[9]   J. K. Kjems, M. Steiner: Phys. Rev. Lett. 41, 1137 (1978);
      M. Steiner, K. Kakurai, J. K. Kjems: Z. Phys. B 53, 117 (1983)

[10] L. P. Regnault, J. P. Boucher, J. Rossat-Mignod, J. P. Renard, J. Bouillot,
      W. G. Stirling: J. Phys. C 15, 1261 (1982)

[11] see e.g. S. W. Lovesey: Theory of neutron scattering from condensed matter, Vol. 2
      (Clarendon, Oxford 1984) pp. 1-11

[12] J. P. Boucher, J. P. Renard: Phys. Rev. Lett. 45, 486 (1980)

[13] J. P. Boucher, H. Benner, F. Devreux, L. P. Regnault, J. Rossat-Mignod, C. Dupas,
      J. P. Renard, J. Bouillot, W. G. Stirling: Phys. Rev. Lett. 48, 431 (1982)

[14] T. Goto, Y. Yamaguchi: J. Phys. Soc. Jpn. 50, 2133 (1981)

[15] H. Benner, H. Seitz, J. Wiese, J. P. Boucher: J. Magn. Magn. Mater. 45, 354 (1984) ;
      H. Seitz, H. Benner: Z. Phys. B 66, 485 (1987)

[16] J. Cibert, Y. Merle d'Aubigné: Phys. Rev. Lett. 46, 1428 (1981)

[17] H. Benner, J. Wiese, R. Geick, H. Sauer: Europhys. Lett. 3, 1135 (1987)

[18] F. Borsa: Phys. Rev. B 25, 3430 (1982)

[19] I. Harada, K. Sasaki, H. Shiba: J. Phys. Soc. Jpn. 51, 3069 (1982)

[20] A. P. Ramirez, W. P. Wolf: Phys. Rev. Lett. 49, 227 (1982)

[21] J. Cibert, Y. Merle d'Aubigné: J. Magn. Magn. Mater. 31-34, 1135 (1983)

[22] J. P. Boucher: Solid State Commun. 33, 1025 (1980)

[23] K. Maki: Phys. Rev. B 24, 3991 (1981)

[24] H. J. Mikeska: J. Appl. Phys. 52, 1950 (1981); Phys. Rev. B 26, 5213 (1982)

[25] E. Allroth, H. J. Mikeska: Z. Phys. B 43, 209 (1981)

[26] E. Magyari, H. Thomas: Phys. Rev. B 25, 531 (1982)

[27] P. Kumar: Phys. Rev. B 25, 483 (1982); Physica 5D, 359 (1982)

[28] V. K. Samalam, P. Kumar: Phys. Rev. B 26, 5146 (1982)

[29] H. J. Mikeska, K. Osano: Z. Phys. B 52, 111 (1983);
C. Etrich, H. J. Mikeska: J. Phys. C 16, 4889 (1983)

[30] P. S. Riseborough: Z. Phys. B 57, 289 (1984)

[31] K. Sasaki: Phys. Rev. B 33, 7743 (1986)

[32] K. Kakurai, R. Pynn, B. Dorner, M. Steiner: J. Phys. C 17, L 123 (1984);
K. Kakurai, M. Steiner, R. Pynn, B. Dorner: J. Magn. Magn. Mater. 54-57, 835 (1986)

[33] G. Reiter: Phys. Rev. Lett. 46, 202 (1981); Erratum: Phys. Rev. Lett. 46, 518 (1981)

[34] D. Hackenbracht, H. G. Schuster: Phys. Rev. B 27, 6916 (1983)

[35] R. Liebmann, M. Schöbinger, D. Hackenbracht: J. Phys. C 16, L 633 (1983)

[36] C. Etrich, H. J. Mikeska: J. Phys. C 21, 1583 (1988)

[37] J. P. Boucher, L. P. Regnault, H. Benner, in: Nonlinearity in Condensed Matter,
Springer Ser. Solid-State Sci., Vol. 69 (Springer, Berlin, Heidelberg 1987) pp. 24-36

[38] M. E. Gouvea, A. Pires: Phys. Rev. B 34, 306 (1986)

[39] H. Benner, J. P. Boucher, J. P. Renard, H. Seitz: J. Magn. Magn. Mater. 31-34, 1233 (1983)

[40] I. Harada, K. Sasaki, H. Shiba: Solid State Commun. 40, 29 (1981)

[41] N. Flüggen, H. J. Mikeska: Solid State Commun. 48, 293 (1983)

[42] J. A. Hołyst, A. Sukiennicki: J. Phys. C 18, 2411 (1985);
J. A. Hołyst: Z. Phys. B 74, 341 (1989)

[43] S. Roth, H. Bleier: Adv. Phys. 36, 385 (1987)

[44] K. M. Leung: Phys. Rev. B 26, 226 (1982); Phys. Rev. B 27, 2877 (1983);
C. A. Condat, R. A. Guyer, M. D. Miller: Phys. Rev. B 27, 474 (1983)

[45] J. A. Hołyst, H. Benner: Solid State Commun. 72, 385 (1989)

[46] H. Benner, J. A. Hołyst, J. Löw: Europhys. Lett. 14, 383 (1991)

[47] R. Dingle, M. E. Lines, S. L. Holt: Phys. Rev. B 187, 643 (1969);
M. Steiner, J. Villain, C. G. Windsor: Adv. Phys., 25, 87 (1976)

[48] A. R. Bishop, K. Fesser, P. S. Lomdahl, S. E. Trullinger: Physica 7 D, 259 (1983)

Ballistic Electrons and Hamiltonian Chaos in Semiconductor Microstructures

T. Geisel

Institut für Theoretische Physik und Sonderforschungsbereich
Nichtlineare Dynamik, J.W. Goethe Universität Frankfurt,
W-6000 Frankfurt 11, Fed. Rep. of Germany

1. Introduction

Semiconductor physics and technology have reached a stage, where the artificial creation of diverse microstructures can be realized. High-purity AlGaAs/GaAs heterojunctions permit a ballistic motion of charge carriers in two dimensions with elastic mean free paths of the order of 10 μm. In addition, lateral structures such as quantum dots, quantum wires, or lateral surface superlattices are imposed in search of novel electronic properties for future devices. E.g. the lateral surface superlattices (LSSL) serve to break the lateral free-particle behaviour of the electron and to produce minigaps in the band structure [1].

These lateral structures represent a two-dimensional potential for the charge carriers. I will demonstrate in this lecture that the ballistic motion of charge carriers typically exhibits chaotic behaviour. This phenomenon will dominate the dynamical properties at low temperatures. At present the spatial scales of the structures are still larger ($\geq$ a factor of 10 for LSSLs) than the Fermi wavelength. The particle dynamics can thus be described by classical approximations, where the chaotic behaviour shows up. On the other hand, as the spatial scales are reduced further, these systems will also become interesting objects and a testing field for studies in quantum chaos.

It is the purpose of this lecture to outline the relevance of Hamiltonian chaos, KAM-theory, etc. and to illustrate the subtle phenomena that are to be expected in these systems. To be specific we shall focus on lateral surface superlattices. We have shown some time ago that the chaotic particle dynamics is an anomalous diffusion process associated with $1/f$-noise in the current fluctuations [2]. The explanation of this phenomenon is based on the existence of a self-similar hierarchy of cantori (i.e. broken KAM-tori with a cantor set structure) in phase space and thus touches some of the finest recent achievments in the theory of nonintegrable Hamiltonian systems. More recently we have shown that even in the case of an integrable superlattice potential the addition of a magnetic field gives rise to similar phenomena [3]. The latter problem is also of interest as the classical limit of the problem of a Bloch electron in an incommensurate magnetic field studied by Hofstadter and others [4]. In order to circumvent the unaccessibly high magnetic fields required to observe his self-similar electronic band structure, Hofstader already suggested to use artificial 2d superlattices with much larger lattice spacing than in natural crystals. Thereby, however, the electron dynamics also approches the classical limit. Meanwhile such superlattices are available for experiments. Lateral superlattices with 1d modulations [5,6] and 2d modulations [7] have been realized with lattice parameters down to about 200 nm. This scale is still large enough to explain experimental observations on the basis of classical approximations for the dynamics of wave packets [8].

The lecture is organized as follows. The next section gives a short introduction to the phenomena of nonintegrable Hamiltonian systems. Section 3 reports numerical results on chaotic particle dynamics, anomalous diffusion, and $1/f$-noise in LSSLs. Section 4 gives an analytic description in terms of a statistical theory for a trapping mechanism of orbits in a self-similar hierarchy of cantori. Due to space limitations the paper focusses on the case of a plain LSSL in the absence of external fields. We treat the magnetic field case in another recent publication [3].

2. Chaos in Hamiltonian Systems

The later sections involve results of the theory of nonintegrable Hamiltonian systems. For this reason I first give a brief introduction into this field. The reader looking for more detailed reviews will find them in Ref. [9].

A Hamiltonian system with N degrees of freedom is described by N pairs (q_i, p_i) of canonically conjugate variables and a Hamiltonian $H(q_1, \ldots, q_N, p_1, \ldots, p_N)$. The q_i and p_i form the vectors $q = (q_1, \ldots, q_N)$ and $p = (p_1, \ldots, p_N)$ and span the 2N-dimensional phase space.

The canonical equations of motion are

$$\dot{p} = -\nabla_q H(q, p)$$
$$\dot{q} = \nabla_p H(q, p) .$$

(1)

The system is called integrable, if a canonical transformation exists which transforms (q, p) into new canonically conjugate variables Θ and I, such that the new Hamiltonian H' is independent of Θ, i.e. $H' = H'(I)$ and Θ is defined mod 2π. The Hamiltonian equations of motion become

$$\dot{I} = -\nabla_\Theta H'(I)$$
$$\dot{\Theta} = \nabla_I H'(I) = \omega = \text{ const. in time.}$$

(2)

and can be integrated to

$$I(t) = I(0) = \text{const.}$$
$$\Theta(t) = \omega(I)t + \Theta(0) .$$

(3)

I and Θ are called *action-angle variables*. The components I_i are constants or integrals of motion, the components $\omega_i = \partial H'/\partial I_i$ are called *frequencies* and their reciprocal value multiplied by 2π are *periods* T_i, i.e

$$T_i = \frac{2\pi}{\omega_i} .$$

(4)

195

It is obvious that (for $N > 1$)an orbit does not explore the whole energy surface in phase space, but a manifold M of lower dimension than the energy surface. M can be identified with an N-torus T^N (the cartesian product of N circles S^1 : $T^N = S^1 \times S^1 \times \ldots S^1$). These tori are called *invariant tori* as every initial condition on a torus generates an orbit that remains on the torus forever. The $\{I_i\}$ define the torus and the $\{\theta_i\}$ define the position on the torus.

An orbit is closed or periodic, if there is a time τ such that

$$\omega\tau = 2\pi N , \tag{5}$$

where N is a non-zero vector of integers n_i. In this case the frequencies ω_i are said to be commensurate. The ω_i are incommensurate and the motion is called quasiperiodic, if the condition (5) cannot be fulfilled. Then the orbit is not closed and it can be shown that it densely covers the torus.

For systems with more than one degree of freedom it is useful to introduce the *Poincaré surface of section*. The orbits generated by Hamilton's equations define a map in the section the *Poincaré return map*. In a system with two degrees of freedom, the invariant torus is the two dimensional torus ($S^1 \times S^1$). The surface of section may be taken e.g. at $\Theta_1 = 0$. The return map is defined by the intersection of the orbits in a fixed sense, e.g. $\dot{\Theta}_1 > 0$. The invariant tori of the flow provide invariant circles of the map. If the orbit is commensurate, only isolated points will appear on the corresponding invariant circle. If it is incommensurate, the circle will be densely filled by the iterations of the map. For integrable systems the return map has the simple form

$$
\begin{aligned}
I_2(t + T_1) &= I_2(t) \\
\Theta_2(t + T_1) &= \Theta_2(t) + \omega_2 T_1 = \Theta_2(t) + 2\pi\frac{\omega_2}{\omega_1} ,
\end{aligned}
\tag{6}
$$

where $T_1 = 2\pi/\omega_1$ denotes the time between successive returns to the surface of section $\Theta_1 = 0$. We may write this in a discrete form by defining $\Theta_n = \Theta_2(nT_1 + t_0)$ and $I_n = I_2(nT_1 + t_0)$, where $n = 0, 1, 2, \ldots$:

$$
\begin{aligned}
I_{n+1} &= I_n \\
\Theta_{n+1} &= \Theta_n + 2\pi\frac{\omega_2}{\omega_1} = \Theta_n + 2\pi\alpha .
\end{aligned}
\tag{7}
$$

The map is characterized completely by the rotation number $\alpha := \omega_2/\omega_1$.

We now discuss what happens if an integrable system $H_0(I)$ is perturbed by a small nonintegrable term $\epsilon H_1(\Theta, I)$

$$H(\Theta, I) = H_0(I) + \epsilon H_1(\Theta, I) \tag{8}$$

with $H_1(\Theta, 0) = 0$. I and Θ still are canonical coordinates but no longer action-angle variables. The return map changes to

$$
\begin{aligned}
I_{n+1} &= I_n + \epsilon f(\Theta_n, I_n) , \\
\Theta_{n+1} &= \Theta_n + 2\pi\alpha + \epsilon g(\Theta_n, I_n) ,
\end{aligned}
\tag{9}
$$

with $g(\Theta_n, 0) = f(\Theta_n, 0) = 0$. The point $(0,0)$ is still an elliptic fixed point of the return map. Thus the surface of section can still be taken perpendicular to a closed orbit.

The main question is what will happen to the invariant tori. Are they destroyed by the nonintegrable perturbation ? The answer to this question is given by the celebrated *KAM-theorem* (after Kolmogorov, Arnold and Moser) [9]. There still will be invariant tori for small but finite values of the perturbation parameter ϵ, but with increasing ϵ the tori are destroyed successively. At which value of ϵ a torus is destroyed depends on its rotation number α. The more irrational α, the longer the torus will survive.

The meaning of more or less irrational values of α may be understood in terms of their continued-fraction expansion

$$\alpha = a_0 + \cfrac{1}{a_1 + \cfrac{1}{a_2 + \cfrac{1}{a_3 + \cdots}}} \quad , \quad a_0 \in \mathbb{Z}, \ a_i \in \mathbb{N}, \tag{10}$$

whose rational approximants are

$$\alpha_n = \frac{r_n}{s_n} = a_0 + \cfrac{1}{a_1 + \cfrac{1}{a_2 + \cfrac{1}{a_3 + \cfrac{1}{\ddots \atop a_{n-1} + \cfrac{1}{a_n}}}}}. \tag{11}$$

One may say that a number is the more irrational the slower its continued fraction converges. In this sense the most irrational number (α_M) is the number which results by taking all $a_i = 1$

$$\alpha_M = 1 + \cfrac{1}{1 + \cfrac{1}{1 + \cfrac{1}{1 + \cdots}}} = \frac{\sqrt{5} + 1}{2} \quad , \tag{12}$$

which is known as the *golden mean*. The torus whose rotation number is $\alpha = \alpha_M$ will be the last to be destroyed. More exactly the KAM-theorem states that those tori are preserved whose winding numbers α satisfy the relation

$$|\alpha - \frac{r}{s}| > \frac{k(\epsilon)}{s^{2.5}} \quad , \tag{13}$$

for all integers r and s. Here k is a number, independent of r and s, which tends to zero with ϵ.

197

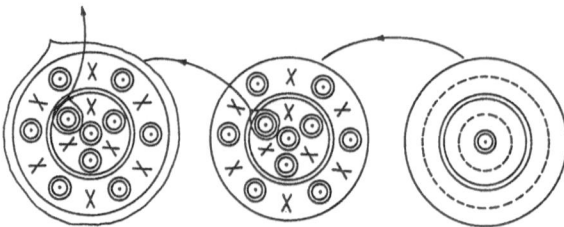

Fig. 1. In the Poincaré surface of section the unperturbed rational tori (dashed lines) give rise to elliptic (dots) and hyperbolic fixed points (crosses). An enlargement of the vicinity of an elliptic fixed point shows a self-similar structure, which repeats hierarchically down to smallest scales. Solid lines denote irrational tori.

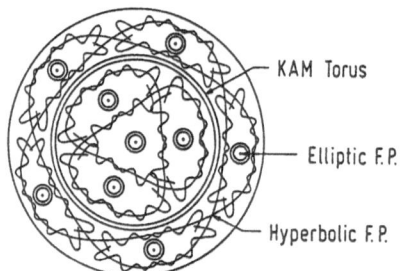

KAM Torus

Elliptic F.P.

Hyperbolic F.P.

Fig. 2. Complicated dynamics arises from intersections of the stable and unstable manifolds of the hyperbolic fixed points.

The rational tori on the other hand are more sensitive to small perturbations, which lead to resonances between degrees of freedom. The *Poincaré-Birkhoff theorem* asserts that a rational torus of period s gives rise to $2ks$ fixed points ($k = 0, 1, 2, \dots$). Among these, ks are elliptic fixed points and ks are hyperbolic fixed points. They are arranged in an alternating sequence along the unperturbed rational torus as shown in Fig. 1. Every elliptic fixed point in turn is surrounded itself by irrational tori and rational tori perturbed to elliptic and hyperbolic fixed points. This scheme repeats down to arbitrarily small scales, giving rise to a self-similar hierarchy of fixed points and tori.

Chaotic motion shows up in the vicinity of the hyperbolic fixed points where the orbits are sensitive to the perturbation and to their initial conditions. The topological dynamics is organized by homoclinic tangles [9] as illustrated in Fig. 2. These are generated by intersections of the stable and unstable manifolds of the hyperbolic fixed points. They must intersect each other an infinity of times along a single manifold without intersecting themselves. It was realized by Poincaré that the orbits in the vicinity of the homoclinic tangle must show complicated dynamics.

The KAM - tori are invariant manifolds, i.e. an orbit on such a manifold will remain on it forever. Chaotic orbits thus cannot penetrate KAM-tori and the latter act like dynamical barriers. This is particularly significant in the case of two degrees of freedom, where two-dimensional tori divide the three-dimensional energy surface into pieces that cannot be reached on a single orbit. As KAM- tori disappear with increasing perturbation, they turn into invariant cantor sets, partial barriers which can be penetrated and which have been christened *cantori*. This happens in all levels of the self-similar hierarchy of KAM-tori illustrated in Fig. 1. One thus finds a hierarchy of partial barriers nested within each other.

3. Nonlinear Dynamics of a Particle in a Lateral Surface Superlattice

The behaviour of electrons in semiconductor heterojunctions is described in principle by wavepackets of Bloch waves. As a lateral superlattice potential of sufficiently large lattice spacing is applied ($a \gg \lambda_F$), the motion of wavepackets in this potential can be approximated on the basis of (quasi-) classical dynamics. In the following we therefore investigate the dynamics of a classical particle moving conservatively in a two-dimensional superlattice potential [2]. The motion is chaotic, quasiperiodic, or periodic, depending on the initial condition in phase space. The main result is the observation and analysis of a new mechanism for $1/f$-noise as a generic phenomenon, which is closely related to the generic structure of phase space of nonintegrable Hamiltonian systems. The power spectral density $S(\omega)$ of velocity fluctuations was found to diverge like $\omega^{-\alpha}$ (with $0.7 \leq \alpha \leq 1.1$) for diffusive chaotic motions. This $1/f$-noise is associated with a divergence of the mean-square displacement of the particle like $t^{1+\alpha}$ for $\alpha < 1$ and t^2 for $\alpha \geq 1$, i.e. the diffusion process is anomalous. Although the model is similar to the Sinai billiard or periodic Lorentz gas, the origin of the observed $1/f$-noise is not related to the long-time tails found there [10], and it is entirely different from the one in dissipative chaotic systems [11–13]. Instead it is caused by a trapping of orbits in a self-similar hierarchy of cantori in phase space and will be described in terms of a renewal process and a random walk on a hierarchical lattice. The latter is related to a Markov tree model [14] that was proposed to explain long-time tails in area-preserving maps [15]. I emphasize, however, that the Markov tree model alone does not exhibit $1/f$-noise, but on the contrary leads to a vanishing velocity power spectrum at $\omega = 0$, when applied to a map on a compact space. This is because trapping due to cantori in this space inhibits diffusion instead of causing accelerated anomalous diffusion.

The model we consider describes the conservative motion of a particle in an analytic two-dimensional potential $V(x,y)$. In our application, the mass of the particle represents the effective mass of the electron. At sufficiently low temperature one may neglect thermal fluctuations. The particle thus is not subject to external random forces as would be the case in models of Brownian motion [16]. It moves without friction according to Newton's laws and conserves the total energy E representing the Fermi energy E_F. The potential parameters of the lateral surface superlattices can be varied in a large range. Using the

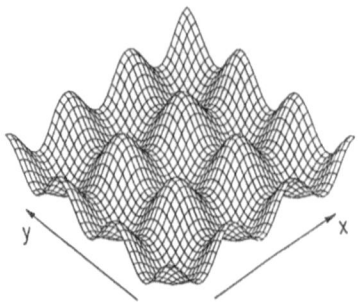

Fig. 3. Periodic Potential $V(x, y)$ with $A = 2.5$, $B = 1.5$, and $C = 0.5$.

technique e.g. of Ref. [6], one can obtain amplitudes of the potential between 0 and 1 eV. We start from a general 2-d Fourier expansion of a periodic potential

$$V(x, y) = V(\mathbf{r}) = \sum_{\mathbf{G}} V_{\mathbf{G}} \, e^{i\mathbf{G} \cdot \mathbf{r}} \qquad (14),$$

where $\mathbf{G}$ is a reciprocal lattice vector. We will consider the three lowest order terms

$$V(x, y) = A + B(\cos x + \cos y) + C \cos x \cos y \qquad (15)$$

where the third term is needed for the generic situation of a nonintegrable Hamiltonian. For numerical calculations we will mainly use the parameters $A = 2.5$, $B = 1.5$, and $C = 0.5$. This potential has the form of an egg-carton as is shown in Fig. 3. It has potential minima in the centers of the cells, potential maxima at the four corners, and saddle points at the midpoints of the edges. Without the coupling term ($C = 0$), the potential looks qualitatively the same. However, as this is the special integrable case, it does not exhibit diffusive and other chaotic motions.

With the above choice of parameters the potential minima are $V_{\min} = 0$, the maxima $V_{\max} = 6$, and the saddles have the potential $V_S = 2$. For total energies $E \leq V_S$ the particle is confined to a single cell. For energies $E > V_S$ the particle can make transitions to adjacent cells. Depending on the initial conditions we found that the particle is drifting or shows diffusive motion. It is worthwhile to emphasize that the diffusion arises deterministically, while in solid state physics one is used to consider random forces as the origin of diffusion [16]. With increasing energies $E > V_{\max}$ an integrable (free particle) situation is approached. We investigate the most interesting case $V_S < E < V_{\max}$ where the diffusive motion has a persistent character with long free paths, which are interrupted by episodes where the particle is trapped for a while in a cell. This is reminiscent of the low-friction limit of Brownian motion models [16], where a particle that is excited to a momentary energy above the potential barrier thermalizes so slowly to lower energies that it can perform a long free path.

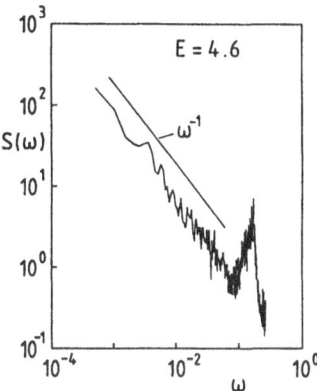

Fig. 4. Velocity power spectral density of diffusive motions. In a broad energy range above the saddle point energy V_S one finds $1/f$-noise $S(\omega) \sim \omega^{-\alpha}$ as exemplified for $E = 4.6$.

Assuming mass $= 1$, the potential (15) leads to the equations of motion

$$
\begin{aligned}
\frac{dv_x}{dt} &= (B + C \cos y) \sin x \\
\frac{dv_y}{dt} &= (B + C \cos x) \sin y
\end{aligned}
\tag{16}
$$

which were integrated numerically using the Adams method. As a criterion for the numerical accuracy we have verified the conservation of the total energy E. For a characterization of the diffusion process we have performed a power spectral analysis of the velocity, which has proven useful and efficient for diffusion in one-dimensional maps [13] and in the Sinai billiard [10]. The velocity power spectral density is defined as

$$
S(\omega) = \int_{-\infty}^{\infty} <\mathbf{v}(t) \cdot \mathbf{v}(0)> e^{-i\omega t} dt
\tag{17}
$$

and for symmetry reasons simplifies to

$$
S(\omega) = 2 \int_{-\infty}^{\infty} <v_x(t) v_x(0)> e^{-i\omega t} dt.
\tag{18}
$$

We have determined $S(\omega)$ for different energies E using Fast Fourier Transform, the Wiener-Khintchine theorem and the segment-averaging method [2]. Fig. 4 shows an example of $S(\omega)$ for diffusive motion at the energy $E = 4.0$ in a log-log plot where $S(\omega)$ behaves like ω^{-1}. We have carried out more detailed numerical studies which have shown that $S(\omega)$ diverges like $\omega^{-\alpha}$ ($0.7 \leq \alpha \leq 1.1$) in a range of energies E extending from the saddle-point energy $V_S = 2.0$ to $E = 4.6$. The values of the exponent α are collected in Table 1. Near the energy V_S an accurate spectral analysis becomes more and more time consuming. In the

Table 1. Variation of the noise exponent α with the particle energy E

E	2.5	3.0	4.0	4.25	4.35	4.6
$\alpha\,(\pm 0.1)$	0.8	0.7	0.75	1.0	1.1	1.0

energy range from $E = 5.0$ to $E = V_{\max} = 6.0$ (the potential maximum) $S(\omega)$ converges to a finite value for $\omega \to 0$. In the transition region $4.6 < E < 5.0$ the situation is ambiguous.

Considering that $S(\omega)$ is closely related to the mean-square displacement, and in particular that $S(\omega = 0) = 2D$ (where D is the diffusion coefficient), we can conclude that the diffusion is anomalously accelerated for $2.0 < E \leq 4.6$. For $5.0 \leq E \leq 6.0$ there is a normal diffusion process characterized by a finite diffusion coefficient. Intuitively one might expect that diffusion increases with increasing kinetic energy. What we observe, however, is the opposite; the strength of diffusion and the diffusion coefficient (where it exists) decrease with increasing energy.

In order to understand the origin of the observed anomalous diffusion and $1/f$-noise we have determined Poincaré surfaces of section at the boundaries of the cells at $x = 2\pi n$ (Fig. 5). Whenever the particle left a cell in $\pm x$-direction we have recorded its y-coordinate $(mod\,2\pi)$ and its v_y-coordinate. Every point in Fig. 5 thus represents a motion of the particle leaving the cell in the perpendicular $\pm x$-direction, and localized motions within a well cannot show up. As the distinction will be important, I will reserve the term *orbit* only for the discrete dynamics within the Poincaré surface of section and the term *trajectory* only for the particle motion in the perpendicular direction.

For diffusive motions one typically finds surfaces of section as shown in Fig. 5 for the energy $E = 4.0$. In the island there are periodic and quasiperiodic orbits, which are not shown. They pertain to unlimited free paths (drift motions), as

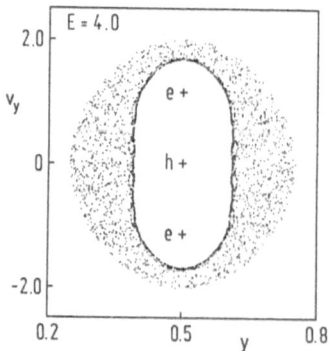

Fig. 5. Poincaré surface of section at the boundaries of the cells ($x = 2\pi n$) for energy $E = 4.0$. The points represent particle trajectories leaving the cell in the perpendicular direction. Position is measured in units of 2π. The letters e and h indicate positions of elliptic and hyperbolic fixed points.

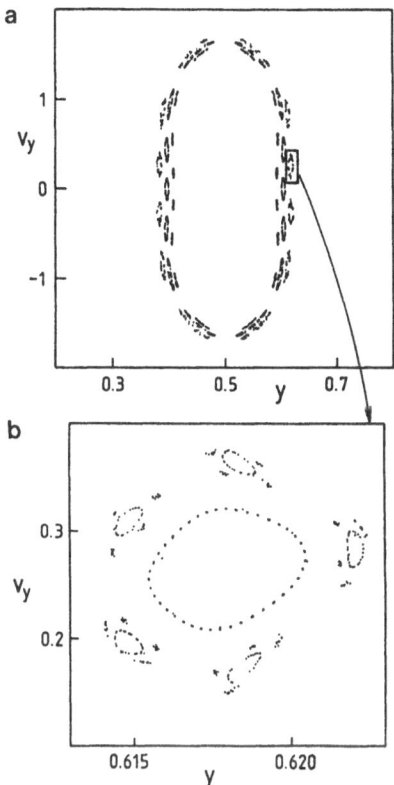

Fig. 6. a) Isolation of island chains near the boundary of the central island of Fig. 5.
b) Magnification of the box shown in Fig. 6a above, displaying the self-similar hierarchy of
daughter islands within the chaotic sea.

the particle consecutively crosses the edges of the cells. On the other hand,
orbits in the chaotic sea surrounding the island remain there only a *finite* time.
(Hereby we mean that the particle trajectory turns into a local motion within
a well, before it can reach the surface of section again.) When they reach its
outer boundary, the energy condition to cross the saddle is no longer fulfilled.
The free path of the particle thus persists only a finite time and gives rise to
diffusion.

Near the inner boundaries of the chaotic sea in Fig. 5 the orbits seem to
have a higher density, which we attribute to the finite observation time. This
fact also points to the origin of $1/f$-noise. The orbit can be seen to stick near
daughter islands surrounding the central islands in Fig. 5. To illustrate this in
more detail for $E = 4.0$ Fig. 6a shows three island chains indicated by rep-
resentative quasiperiodic orbits. They were isolated by selecting special initial
conditions. The magnification in Fig. 6b reveals three levels in a hierarchy of
daughter islands around daughter islands. We know that generically this hierarchy continues ad infinitum [17]. Every island in the chaotic sea is encircled
by cantori [18], partial barriers which the orbit can penetrate. The deeper the

orbit enters into the hierarchy of nested cantori, the longer it remains trapped before it can leave the chaotic sea (hereby I mean that the particle trajectory turns into a local motion within a well, before it can reach the Poincaré section again). This hierarchy of time scales is the origin of the observed $1/f$-noise.

When the energy E is increased the role of the nonintegrable perturbation increases. The permeability of the cantori increases and the islands of stability shrink or disappear. The fine structure of stability islands is thus destroyed. The trapping mechanism responsible for long free paths does not operate any more. This is why the diffusion of the particle is no longer anomalous for higher energies (e.g. $E = 5.5$). Anomalous diffusion may also disappear when the symmetry is changed such that a rectilinear infinite free path does not exist. In a study of a particle in a hexagonal periodic potential only normal diffusion was observed [19].

4. Transport Theory

In this section I give a statistical formulation of the mechanism outlined above. In a first step we describe the free paths by means of renewal theory. Let us recall that trapping of the orbit in the chaotic sea implies a free path of the particle trajectory. We treat successive free paths as statistically independent events and describe their duration T by a probability density $\Psi(T)$. This is justified because successive free paths are interrupted by localized chaotic motions, which have a randomizing effect. This behaviour is found also in intermittent chaotic systems [11–13,20] where similar assumptions are used.

One may think of using the distribution $\Psi(T)$ in a continuous-time random walk theory (CTRW) [21]. This would correctly describe the asymptotic divergence of the mean-square displacement $< [\mathbf{r}(t) - \mathbf{r}(0)]^2 >$ to leading order. It is not sufficiently accurate in some cases, however, to describe the power spectrum $S(\omega)$ for $\omega \to 0$, which is more sensitive to details of the motion. The following derivation of the velocity autocorrelation function

$$C(t) =< \mathbf{v}(t) \cdot \mathbf{v}(0) >= 2 < v_x(t)v_x(0) > \tag{19}$$

avoids this problem. We imagine the process along the time axis as a sequence of free paths with velocity $\pm v_0$ and neglect the (short) episodes of localized motion within a potential well. Picking an arbitrary time $t' = 0$ on the time axis, the probability $w_0(T)dT$ that the particle is in a free path of duration T is proportional to $\Psi(T)$ and also proportional to T

$$w_0(T)dT = \frac{T}{< T >}\Psi(T)dT. \tag{20}$$

The denominator ensures the correct normalization and is the average duration of a free path

$$< T >= \int_0^\infty T\Psi(T)dT. \tag{21}$$

The probability that this path persists until time t is the fraction $(T - t)/T$. The probability $w_{0,t}(T)dT$ that the particle is in a free path of duration T between time 0 and t is thus

$$w_{0,t}(T)dT = \frac{T - t}{<T>}\Psi(T)dT. \tag{22}$$

Using the average longitudinal velocity $v_0 = L/T$ mentioned above, we obtain the velocity autocorrelation function $C(t)$ from the integral of Eq. (22) over all possible durations

$$C(t) = \frac{v_0^2}{<T>}\int_t^\infty (T - t)\Psi(T)dT. \tag{23}$$

This equation only includes events consisting of a single path between times 0 and t. Sequences consisting of two or more free paths (independent events) between 0 and t have equal probability to end with a positive or negative velocity ($\pm v_0$) and thus do not contribute to Eq. (23). We have shown this in more detail for the Sinai-billiard previously [10]. Denoting the Laplace transforms of $C(t)$ and $\Psi(T)$ by $\tilde{C}(s)$ and $\tilde{\Psi}(s)$, Eq. (23) turns into

$$\tilde{C}(s) \propto \frac{1}{s} + \frac{1}{<T>}\frac{\tilde{\Psi}(s) - 1}{s^2}. \tag{24}$$

From Eq. (24) we obtain all statistical quantities of interest, e.g. the power spectral density Eq. (17) as

$$S(\omega) = \lim_{\varepsilon \to 0}[\tilde{C}(\varepsilon + i\omega) + \tilde{C}(\varepsilon - i\omega)]. \tag{25}$$

For the mean-square displacement of the particle

$$\sigma^2(t) = <[\mathbf{r}(t) - \mathbf{r}(0)]^2> \tag{26}$$

we use the identity

$$\sigma^2(t) = 2\int_0^t (t - \tau)C(\tau)d\tau \tag{27}$$

whose Laplace transform is $\widetilde{\sigma^2}(s) = 2s^{-2}\tilde{C}(s)$. Thus from Eq. (24)

$$\sigma^2(t) = 2\mathbf{L}^{-1}\{s^{-3} + <T>^{-1} s^{-4}[\tilde{\Psi}(s) - 1]\} \tag{28}$$

where $\mathbf{L}^{-1}$ denotes the inverse Laplace transform. Finally, the diffusion coefficient, where it exists is defined as $D = \lim \sigma^2(t)/2t$, and from Eq. (26) follows as

$$D = \int_0^\infty <\mathbf{v}(t) \cdot \mathbf{v}(0)> dt \tag{29}$$

or $D = (1/2)S(\omega = 0)$.

In a second step we would like to obtain $\Psi(T)$ analytically for the mechanism of cantorus trapping. Too little, however, is known with mathematical rigour about the transport across cantori. In fact, so far it has not even been shown

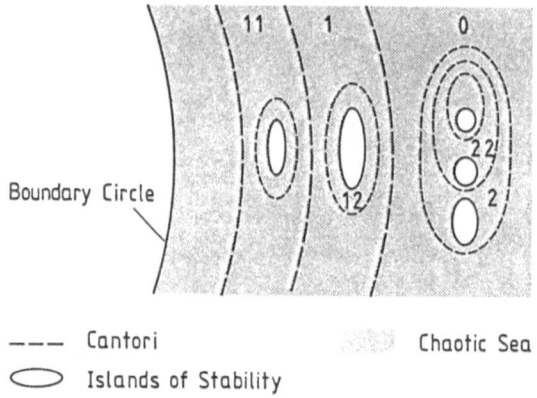

Boundary Circle

——— Cantori   :::: Chaotic Sea

⬭ Islands of Stability

Fig. 7. Schematic representation of the nested hierarchy of cantori. The numbers indicate Markov states.

that the chaotic motion is ergodic (in the abstract mathematical sense) and mixing for nonintegrable Hamiltonian systems satisfying the KAM-theorem, although this is believed by many authors on the basis of numerical results. In this situation we are forced to make a number of assumptions below, which reduce the dynamics of our model (Eqs. 16) to a simpler statistical model. But first we summarize what *is* known rigorously for analytic nonintegrable Hamiltonian systems with two degrees of freedom. Every island in the chaotic sea is encircled by a set of cantori belonging to a continuum of rotation numbers [18]. Those with sufficiently irrational rotation number (e.g. in terms of a continued fraction expansion) have a small flux across them and act as barriers [22]. The latter cantori form a sequence converging to the boundary circle of the island, as is illustrated schematically in Fig. 7. The boundary circle is believed to always be a critical KAM-torus, i.e. a KAM-torus at the transition to destruction. It was shown that critical tori exhibit scaling properties [23], which also imply scaling for the fluxes across the encircling low-flux cantori [22]. Embedded between these cantori are island chains (daughter islands) where the scheme repeats on a finer scale (Fig. 7). Thus it is confirmed that there are sequences of partial barriers nested hierarchically within each other, and that the hierarchical trapping mechanism must be present. The question is how the transport across cantori occurs in detail, and whether the dynamics might be complicated by additional phenomena.

The hierarchical nesting of cantori implies that the accessible regions between the cantori are organized on a tree. This is illustrated in Fig. 8 by the example of a binary tree. The numbered circles, i.e. the sites of the lattice correspond to the accessible regions, whereas the lines represent crossing of a cantorus. In order to enter the hierarchy from the outer region of the chaotic sea (stem of the tree), an orbit can cross one of several possible cantori in each step, i.e. move towards the center of an island, or cross a cantorus of a daughter island. One may assume for simplicity that successive crossings of low-flux cantori can be treated as a Markov process. The chaotic dynamics thus

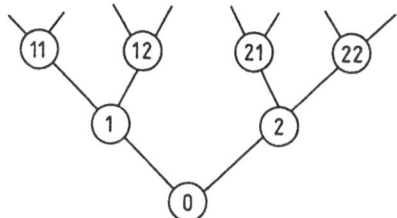

Fig. 8. In a Markov approximation, transitions across low-flux cantori correspond to a random walk on a hierarchical lattice as is illustrated here by a binary tree.

becomes statistically equivalent to a random walk on a hierarchical lattice or Markov tree [14], as will be described below. Let us assume a constant branching ratio m and number the accessible sites (regions) by a sequence of integers $\ell = \ell_1 \ell_2 \cdots \ell_n$ as illustrated in Figs. 7 and 8. The stem of the tree is the site $\ell = \ell_1 = 0$ and denotes the outer region of the chaotic sea. With each branching (cantorus crossing) an integer ℓ_{n+1} is added to the sequence, where $\ell_{n+1} = 1$ stands for crossing of a cantorus of the same island, and $\ell_{n+1} = 2, \cdots, m$ refers to daughter islands.

This random-walk problem is related to theories of diffusion in ultrametric spaces [24–26]. It should be pointed out, however, that in those cases one is mainly interested in the motion among the leaves of the tree, whereas here we have to consider hopping on the entire hierarchy of branches. Let us denote the transition rate from a site ℓ to a neighbouring site ℓ' by $w(\ell|\ell')$. These transition rates define the random walk. They are assumed to scale by fixed scaling coefficients ε_k and δ_k according to

$$w(1k|1) = \varepsilon_k w(1|0) \tag{30}$$

$$w(0|k) = \delta_k w(k|0) \tag{31}$$

and analogously for higher branches

$$w(\ell j k|\ell j) = \varepsilon_k w(\ell j|\ell) \tag{32}$$

$$w(\ell|\ell k) = \delta_k w(\ell k|\ell) \tag{33}$$

in order to account for the scaling near critical KAM-tori [22,23]. The set of transition rates is defined by fixing $w(k|0)$, which are assumed as $w(k|0) = \varepsilon_k w_0$, where w_0 is a constant.

Let $p_t(\ell|\ell')$ denote the conditional probability that the first transition to ℓ' occurs at time t, if at time 0 there was a transition to ℓ. The analogous probability for direct transitions (without intermediate transitions to other sites) is denoted by $d_t(\ell|\ell')$ and is easily determined from the transition rates. The conditional probability p_t must satisfy a set of coupled integral equations [14]

$$p_t(k|0) = d_t(k|0) + \sum_{j=1}^{m} \int_0^t d_\tau(k|kj)p_{t-\tau}(kj|0)d\tau. \tag{34}$$

207

This equation expresses the fact that the walker can jump from k to 0 either directly, or by jumping to a higher site kj at some intermediate time τ and descending to the stem 0 in the remaining time. The latter transitions fulfill a similar integral equation

$$p_t(kj|0) = \int_0^t p_\tau(kj|k)p_{t-\tau}(k|0)d\tau \tag{35}$$

i.e. in order to go from kj to 0, the walker must hop to the connecting site k at some intermediate time τ. The system of equations is solved by Laplace transform and a scaling ansatz

$$p_t(kj|k) = \varepsilon_j p_{\varepsilon_j t}(k|0). \tag{36}$$

Eq. (35) thus becomes

$$p_t(kj|0) = \varepsilon_j \int_0^t p_{\varepsilon_j \tau}(k|0)p_{t-\tau}(k|0)d\tau. \tag{37}$$

Using Eq. (37) in Eq. (34), one observes that the unknown functions $p_t(k|0)$ occur in products on the r.h.s. of Eq. (34). Laplace transformation leads to a nonlinear functional equation [14], which is expressed conveniently by introducing a scaling function $h(\tau)$ as

$$p_t(k|0) = w_0\varepsilon_k h(\alpha w_0 \varepsilon_k t) \tag{38}$$

where $\alpha = 1 + \sum_k \varepsilon_k \delta_k$. In terms of the Laplace transform $\tilde{h}(s)$, Eq. (34) becomes

$$\tilde{h}(s)\left[s + 1 - \alpha^{-2} \sum_{k=1}^m \varepsilon_k \delta_k \tilde{h}(s/\varepsilon_k)\right] = 1. \tag{39}$$

This equation is solved if $\tilde{h}(s)$ has the expansion

$$\tilde{h}(s) = f(s) + s^\mu g(s) \tag{40}$$

where $f(s)$ and $g(s)$ are analytic functions (that need not be determined) and the exponent μ must satisfy

$$\sum_{k=1}^m \delta_k \varepsilon_k^{1-\mu} = 1. \tag{41}$$

Eq. (40) refers only to cases $\mu \neq 1$. In the case $\mu = 1$, i.e. for

$$\sum_{k=1}^m \delta_k = 1 \tag{42}$$

one must account for logarithmic corrections. For this case we have found a solution of Eq. (39) as

$$\tilde{h}(s) = \alpha + as(\gamma + \ln s) + \cdots \tag{43}$$

where a is a constant and γ is Euler's constant. The inverse Laplace transforms of Eqs. (40,43) together with Eq. (38) yield the time-dependent probabilities.

From Eq. (38) we can now determine the distribution $\Psi(T)$ of the durations of free paths. Recall that a free path of the particle persists as long as the discrete orbit is trapped in the hierarchy of cantori. It ends, when the orbit reaches the outer region of the chaotic sea. In the simplified model of the random walker, the duration of the free path is then determined by the time interval between a transition from 0 to a trapped state k and the first transition back to 0. Since within this model the transitions are taken as independent events, the previous and following history of the walker is irrelevant, and the probability for those transitions is proportional to $\sum_k p_T(k|0)$. Therefore also the distribution of free times is

$$\Psi(T) \propto \sum_{k=1}^{m} p_T(k|0) \tag{44}$$

and with Eq. (38) becomes

$$\Psi(T) \propto \sum_{k=1}^{m} w_0 \varepsilon_k h(\alpha w_0 \varepsilon_k T). \tag{45}$$

For $\mu \neq 1$ its Laplace transform follows from Eq. (40)

$$\tilde{\Psi}(s) \propto \frac{1}{\alpha} \sum_{k=1}^{m} \left[f\left(\frac{s}{\alpha w_0 \varepsilon_k}\right) + s^{\mu} (\alpha w_0 \varepsilon_k)^{-\mu} g\left(\frac{s}{\alpha w_0 \varepsilon_k}\right) \right]. \tag{46}$$

For $\mu < 1$ the limiting behaviour for small s is further simplified to

$$\tilde{\Psi}(s) = 1 + cs^{\mu} + \cdots \tag{47}$$

where the normalization condition $\tilde{\Psi}(s = 0) = 1$ was used, and c is a constant. This is associated with an asymptotic algebraic decay of $\Psi(T)$

$$\Psi(T) \sim T^{-1-\mu}. \tag{48}$$

For $\mu = 1$, we obtain from Eqs. (43,45)

$$\tilde{\Psi}(s) = 1 + bs(\gamma + \ln s) + \cdots \tag{49}$$

with $b = (a/m\alpha^2 w_0) \sum_k \varepsilon_k^{-1}$. In this case the asymptotic time decay is $\Psi(T) \sim T^{-2}$ and may thus be included in Eq. (48).

With the above expressions for $\tilde{\Psi}(s)$ we have calculated the power spectral density from Eqs. (24,25) and the mean-square displacement from Eq. (28). The low frequency behaviour of $S(\omega)$ results as

$$S(\omega) \sim \omega^{\mu-2} \tag{50}$$

where for $\mu \leq 1$ we had to introduce a low-frequency cutoff as e.g. in Ref. [13]. The asymptotic divergence of the mean-square displacement results as

$$\sigma^2(t) \sim \begin{cases} t^{3-\mu} & \text{for } \mu \geq 1 \\ t^2 & \text{for } \mu \leq 1 \end{cases} \tag{51}$$

The mechanism treated in this section is thus able to qualitatively explain and recover the observed $1/f$-spectrum by Eq. (50). The actual value of μ, however, remains an open problem. Identifying $2 - \mu$ with the exponent α of Fig. 4 and Table 1 shows us that μ apparently fluctuates between 0.9 and 1.3. A pure $1/f$-decay is associated with $\mu = 1.0$. Within the model, μ is determined implicitly by Eq. (41) and thus depends on the coefficients δ_k and ε_k which express the scaling of the fluxes across cantori Eqs. (32,33). The coefficients δ_1 and ε_1 are determined by the scaling in the vicinity of a critical KAM-torus [23,22] and have the values $\delta_1 = 0.139$ and $\varepsilon_1 = 0.382$. It is less straightforward to obtain good estimates for the other coefficients. Using a renormalization scheme for island-around-island sequences, Meiss and Ott [14] have obtained estimates for the coefficients ε_2 and δ_2 of a binary tree model. These values would lead to an exponent $\mu = 1.96$, whereas from the numerical power spectra we deduce a value in the vicinity of $\mu = 1$. This discrepancy can be reduced to some extent by considering a higher than twofold branching. On the other hand it is clear that the Markov tree model can only represent a schematic description of the dynamics, and discrepancies must be expected. They may be due e.g. to the fixed branching ratio and to the fact that the Markov assumption might not be completely fulfilled.

5. Conclusion

In this lecture I have tried to demonstrate that the ballistic motion of charge carriers in lateral microstructures is typically expected to exhibit chaotic behaviour, to show the type of phenomena that can arise, and to present a theoretical framework for their description. The latter is based on concepts of the theory of nonintegrable Hamiltonian systems and on recent progress in the understanding of transport across dynamical barriers (cantori). To be specific I have focussed only on lateral surface superlattices (LSSLs) in the absence of external fields. I am convinced that this is only the beginning of a very fruitful field of research. We recently investigated the chaotic behaviour of particles in LSSLs in an applied magnetic field [3], but other problems and other mircostructures remain to be studied.

The main phenomenon which we have encountered in LSSLs is a chaotic diffusion process of the ballistic particle. It arises deterministically in the absence of random forces and impurities and thus leads to an unexpected reduction of the elastic mean-free-path in these samples (independent of impurities). Furthermore the diffusion process typically shows an anomalous nonlinear growth of the mean-square displacement and $1/f$-noise in the current fluctuations. It was demonstrated that this is caused by a physically new mechanism for $1/f$-noise based on transport in a self-similar hierarchy of cantori in phase space.

Note added:

Since the completion of this manuscript there has been a rapid growth of experimental and theoretical activities dealing with chaotic Hamiltonian dynamics in semiconductor microstructures. Strong experimental support for the relevance of chaotic trajectories stems from magnetoresistance measurements of antidot lattices [27] and its theoretical analysis [28], where the trapping mechanism in cantorus hierarchies outlined in section 3 and 4 is manifested by magnetoresistance peaks. The interested reader may find more recent material on chaotic Hamiltonian dynamics in LSSLs and its quantum aspects in Refs. [28–31].

References

1. H. Sakaki, K. Wagatsuma, J. Hamasaki, and S. Saito: Thin Solid Films **36**, 497 (1976); R. T. Bate: Bull. Am. Phys. Soc. **22**,407 (1977); A. C. Warren, D. A. Antoniadis, H. I. Smith, and J.Melngailis: IEEE Electron. Device Lett. **6**, 294 (1985)
2. T. Geisel, A. Zacherl, and G. Radons: Phys. Rev. Lett. **59**,2503 (1987); and Z. Phys. **B 71**, 117 (1988)
3. T. Geisel, J. Wagenhuber, P. Niebauer, G. Obermair: Phys. Rev. Lett. **64**, 1581 (1990)
4. D. R. Hofstadter: Phys. Rev. **B 14**, 2239 (1976)
5. R. R. Gerhardts, D. Weiss, K. v.Klitzing: Phys. Rev. Lett. **62**, 1173 (1989)
6. R. W. Winkler, J. P. Kotthaus, and K. Ploog: Phys. Rev. Lett. **62**, 1177 (1989)
7. D. K. Ferry, G. Bernstein, and Wen-Ping Liu: In *Physics and Technology of Submicron Structures*, edited by H. Heinrich, G. Bauer, F. Kuchar, Springer Series in Solid State Sciences Vol. 83 (Springer-Verlag, Berlin 1988), p.37; K.Ismail, W. Chu, A. Yen, D.A. Antoniadis, and H. I. Smith: Appl. Phys. Lett. **54**, 460 (1989); P. H. Beton, E. S. Alves, M. Henini, L. Eaves, P. C. Main, O. H. Hughes, G. A. Toombs, S. P. Beaumont, and C. D. W. Wilkinson (to be published); A. Lorke, J. P. Kotthaus, and K. Ploog (to be published);
8. C. W. J. Beenakker: Phys. Rev. Lett. **62**, 2020 (1989); C. W. J. Beenakker and H. van Houten: Phys. Rev. Lett. **63**, 1857 (1989)
9. see e.g., M. V. Berry: In *Topics in Nonlinear Dynamics*, edited by S. Jorna, AIP Conference Proceedings No. 46 (American Institute of Physics, New York, 1978), p.16, and D.F. Escande: Phys. Rep. **121**, 163 (1985)
10. A. Zacherl, T. Geisel, J. Nierwetberg, G. Radons: Phys. Lett. **114** A, 317 (1986); J. P. Bouchaud, P. Le Doussal: J. Stat. Phys. **41**, 225 (1985)
11. P. Manneville: J. Phys. (Paris) **41**, 1235 (1980); Y. Pomeau, P. Manneville: Commun. Math. Phys. **74**, 189 (1980)
12. I. Procaccia, H. Schuster: Phys. Rev. A **28**, 1210 (1983)
13. T. Geisel, J. Nierwetberg, A. Zacherl: Phys. Rev. Lett. **54**, 616 (1985)
14. J. D. Meiss, E. Ott: Phys. Rev. Lett. **55**, 2741 (1985) and Physica **20** D, 387 (1986)
15. C. F. F. Karney: Physica **8** D, 360 (1983); B. V. Chirikov, D. L. Shepelyanski: Physica **13** D, 395 (1984); P. Grassberger, H. Kantz: Phys. Lett. **113** A, 167 (1985)
16. e.g. T. Geisel: in Physics of Superionic Conductors, ed. M.B. Salamon, Topics in Current Physics, Vol. 15, p. 201. Berlin: Springer 1979, and H. Risken: The Fokker-Plank Equation, Springer Series in Synergetics, Vol. 18. Berlin: Springer 1984
17. see e.g. R.S. MacKay: Lect. Notes in Phys. **247**, 390 (1986)
18. I. C. Percival: AIP Conference Proceedings **57**, 1179 (1980); Aubry, S.: in Solitons and Condensed Matter Physics, eds. A.R. Bishop, T. Schneider, p. 264. Berlin: Springer 1978
19. B. Bagchi, R. Zwanzig, M. C. Marchetti: Phys. Rev. A **31**, 892 (1985)
20. T. Geisel, J. Nierwetberg: Z. Phys. B **56**, 59 (1984)

21. See e.g. E. W. Montroll, M. F. Shlesinger: in Nonequilibrium Phenomena II, eds. J. L. Lebowitz, E. W. Montroll (North Holland, Amsterdam 1984), p. 1
22. R. S. MacKay, J. D. Meiss, I. C. Percival: Phys. Rev. Lett. **52**, 697 (1984) and Physica **13** D, 55 (1984);
    D. Bensimon, L.P. Kadanoff: Physics **13** D, 82 (1984)
23. L. P. Kadanoff: Phys. Rev. Lett. **47**, 1641 (1981);
    S. J. Shenker, L. P. Kadanoff: J. Stat. Phys. **27**, 631 (1982);
    R. S. MacKay: Physica **7** D, 283 (1983);
    J. M. Greene, R. S. MacKay, J. Stark: Physica **21** D, 267 (1986)
24. B.A. Huberman, M. Kerszberg: J. Phys. A **18**, L 331 (1985)
25. S. Grossmann, F. Wegner, K. H. Hoffmann: J. Physique Lett. **46**, L 575 (1985)
26. For a recent review see e.g. C. De Dominicis, M. Schreckenberg: in Heidelberg Colloquium on Glassy Dynamics, eds.: J. L. van Hemmen, I. Morgenstern, Lect. Notes in Physics Vol. **275** (Springer, Berlin 1987), p. 255
27. D. Weiss, M. L. Roukes, A. Menschig, P. Grambow, K. v. Klitzing, G. Weimann: Phys. Rev. Lett. **66**, 2790 (1991)
28. R. Fleischmann, T. Geisel, R. Ketzmerick: Phys. Rev. Lett.,**68**, 1367 (1992)
29. T. Geisel, R. Ketzmerick, G. Petschel: Phys. Rev. Lett. **66**, 1651 (1991)
30. T. Geisel, R. Ketzmerick, G. Petschel: Phys. Rev. Lett. **67**, 3635 (1991)
31. J. Wagenhuber, T. Geisel, P. Niebauer, G. Obermair: Phys. Rev. B, **45**, 4372 (1992)

Spatially Chaotic Structures

R. Schilling

Institut für Physik der Universität Basel, Klingelbergstrasse 82,
CH-4056 Basel
Permanent Address: Institut für Physik, Johannes-Gutenberg-Univ.,
Staudinger Weg 7, W-6500 Mainz, Fed. Rep. of Germany

1   Introduction

One of the major tendencies in physics is to study systems behaving in as regular
a way as possible. This has led to a preferential treatment of problems involving
only a few degrees of freedom and possessing space and time symmetries, because
then the mathematical description can be substantially simplified. But these
properties do *not* guarantee regular behaviour. Poincaré [1] already noticed that
the classical three-body problem may exhibit rather irregular solutions of the
equations of motion. It was mainly the seminal work by Kolmogorov, Arnold
and Moser (KAM-theory) on conservative systems and by Ruelle and Takens on
dissipative systems that initiated the developement of quite a new field of physics
called *chaos theory*.

The KAM-theory proves that most of the tori of an integrable Hamiltonian
survive a weakly nonintegrable perturbation. The motion on tori which are
destroyed either becomes periodic or chaotic [2], [3]. An impressive success of
this theory is the explanation of the gaps in the asteroid belt between Mars
and Jupiter, observed by Kirkwood in 1866 (see e.g.[4]). The work by Ruelle
and Takens [5] proves the existence of a new type of attractors called strange
attractors. This work has brought a new interpretation of turbulence and its
onset. We will not follow this historical aspect here, however. For more details
the reader is referred to the textbooks on chaos theory (e.g. [6] - [8]) and the
contributions by Schuster and Geisel in the present book. But let us stress
that the remarkable property of *temporal* chaos is its occurrence in *deterministic*
systems. The chaotic behaviour of these systems is not due to external noise
but is generated *intrinsically*. As an example, the orbits of the rather simple
standard map:

$$I_{n+1} = I_n + k \sin \Theta_n$$
$$\Theta_{n+1} = \Theta_n + I_{n+1}$$

which has been studied extensively by numerous authors, is depicted in Figure 1. Three different types of orbits occur: periodic (fixed points), quasiperiodic
(closed curves also called Moser-tori) and chaotic ones ("snow-like" regions).

The following questions may now be asked: Is the chaotic behaviour only
restricted to time evolution or may it also occur in space? Where can static

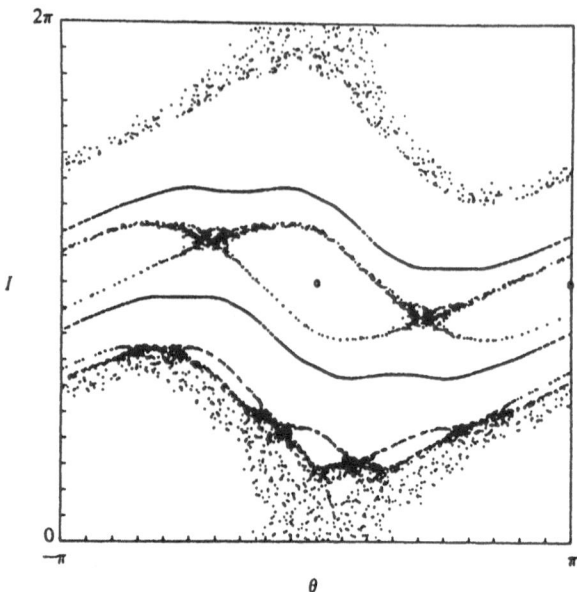

Figure 1: Numerical plot of five orbits of the standard map for k = 0.97 (after Greene [9]).

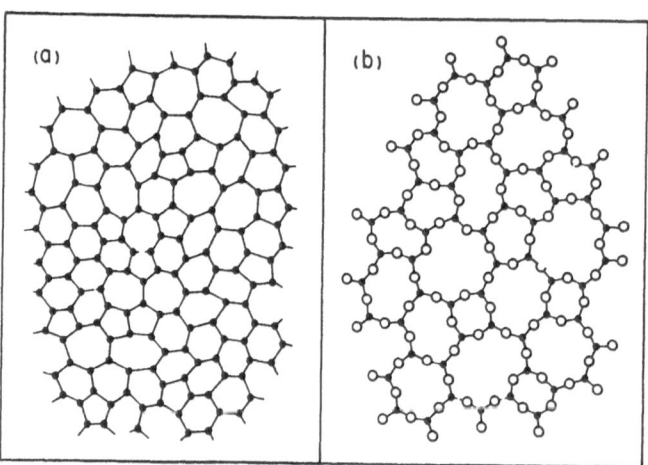

Figure 2: Schematic representation of a two-dimensional tri-coordinated structure (after Zallen [10]).

spatial chaos be found in nature? Concerning these questions we believe that amorphous, or more generally, disordered structures in solid-state science can be interpreted as "deterministic" (see below) spatial chaos. An example of an amorphous structure is shown in Figure 2. The positions of the atoms are *not* random. There is still some short-range order but no long-range correlations.

The obvious lack of translational and point symmetry makes the theoretical and experimental investigations of such structures difficult. Suppose the particle system can reasonably be described by a pair potential $v(|\vec{x}|)$. The total potential energy e.g. for N identical particles is then given by

$$V(\vec{x}_1, \vec{x}_2, \ldots, \vec{x}_N) = \frac{1}{2} \sum_{i \neq j} v(|\vec{x}_i - \vec{x}_j|) \quad . \tag{1}$$

From a theoretical point of view the main question is: Do there exist stationary and locally stable (metastable) configurations for V which are amorphous, or in other words, do the equations

$$\frac{\partial V}{\partial x_n^\alpha}(\vec{x}_1, \vec{x}_2, \ldots, \vec{x}_N) = 0 \quad , \quad n = 1, 2, \ldots, N \quad ; \quad \alpha = x, y, z \tag{2}$$

have solutions where the arrangement of the particles are spatially chaotic? The metastability requires that the dynamical matrix $\partial^2 V / \partial x_n^\alpha \partial x_m^\beta$ is positive semidefinite for these solutions. Most approaches to model amorphous materials (see e.g. [10], [11]) do not start on such a microscopic level, and those which investigate (2) usually apply numerical tools (see e.g. [12]). But these numerical calculations show the existence of many metastable amorphous configurations. In itself this fact is not less surprising than the occurrence of irregular temporal behaviour for nonlinear dynamical systems. Note that the function v in (1) does not depend on (i,j) explicitly as it does for most of the spin glass models [13]. This means that in our case the disorder is an *intrinsic* effect. Therefore we call this behaviour "deterministic".

In the last few years we have found simple models for the interaction energy V for which (2) can be solved exactly [14] - [16]. It is a characteristic feature of these models that (2) can be related to a nonlinear map with discrete "time". This ensures that spatially-chaotic configurations may be found with properties related to those of temporal chaos. On the other hand, these configurations show *glass-like* behaviour. Consequently there seems to exist a formal relationship between glassy behaviour and temporal chaos at least for these models. Before discussing the models, we will elucidate in the next section both the most important characteristics of temporal chaos and some of the typical properties of glassy materials [2].

Because most of the results presented here are already published, we concentrate on the presentation of the theoretical aspects and the physical ideas. The interested reader will find all technical details in our original papers.

---

[2] Although the term glass is mostly reserved for amorphous materials obtained from the melt (like vitreous silica) we use it in a loose way.

2 Properties of temporal chaos and glassy materials

Temporal chaos as well as glassy materials show a variety of properties. Only those which are important for later purposes will be discussed here.

2.1 Properties of temporal chaos

(i) *Sensitivity on initial conditions* is the most prominent feature of temporal chaos. Choosing two nearby initial conditions $\vec{x}(0)$ and $\vec{x}'(0)$ on different trajectories, we see that the corresponding trajectories are correlated over short times, but their distance diverges exponentially, i.e.

$$| \vec{x}'(t) - \vec{x}(t) | \ \sim \exp(\lambda t) \quad , \quad \lambda > 0 \qquad (3)$$

for large times t. λ is called the Lyapunov-exponent. Hence no long-time correlations exist.

(ii) The *embedding of the Bernoulli shift* is a second important property of chaotic systems [3], although it is generally hard to find. To explain the idea of embedding, let us consider a two-dimensional map T

$$\vec{z}_{n+1} = T(\vec{z}_n) \quad , \quad \vec{z}_n = (x_n, y_n) \in \mathbb{R}^2 \quad . \qquad (4)$$

Depending on T there exist symbols $\sigma \epsilon A$ which form an alphabet A, e.g. $\{-1, +1\}$ or $\{\alpha, \beta, \gamma\}$ etc. Let S be the space of all doubly-infinite sequences $\sigma = \{\ldots \sigma_{-2}\sigma_{-1}\sigma_0\sigma_1\sigma_2\sigma_3 \ldots\}$. Within S we can define a *symbolic dynamics* $s : S \to S$ as follows:

$$\sigma'_m = (s(\sigma))_m = \sigma_{m-1} \qquad (5)$$

corresponding to a shift of the subscripts, called Bernoulli shift. The embedding of the Bernoulli shift means that there exists a homeomorphism φ

$$\varphi : G \to S$$

such that the original dynamics T restricted to a certain region $G \subset \mathbb{R}^2$ of two-dimensional state space can be obtained by the symbolic dynamics as follows:

$$T |_G \ = \varphi^{-1} \circ s \circ \varphi \quad . \qquad (6)$$

This important result makes the occurrence of chaotic behaviour obvious . Suppose $A = \{-1, +1\}$ for simplicity. Choose $\sigma \in S$ such that -1 and $+1$ occur randomly, hence leading to a chaotic symbolic dynamics in S. Let $s^{(n)}$ be the n-th iterate of s. Because of the one-to-one correspondence (due to φ) between $\sigma^{(n)} = s^{(n)}(\sigma^{(0)})$ and $\vec{z}_n \in G$, the corresponding orbit $\{\vec{z}_n(\vec{z}_o)\}$ will be chaotic too. If σ is chosen periodic or quasiperiodic, the corresponding orbit $\{\vec{z}_n\}$ will be periodic and quasiperiodic , respectively. Thus one may say that the ergodic behaviour of the Bernoulli shift is the origin of chaotic behaviour of nonlinear (nonintegrable) dynamical systems.

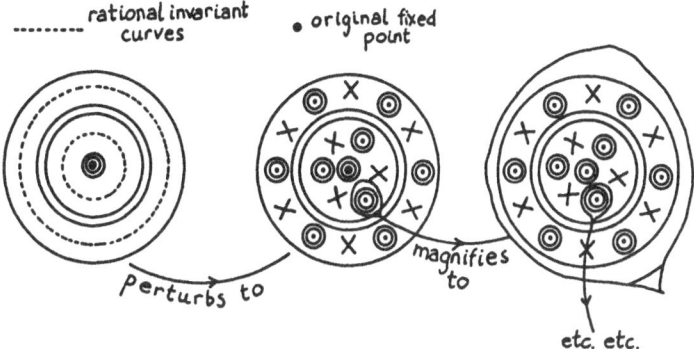

Figure 3: Illustration of self similarity for a two-dimensional map (after Berry [4]).

(iii) The final property we would like to mention is *self-similarity*. Figure 3 shows a schematic illustration. Integrable behaviour is illustrated on the left side. A perturbation leads to new elliptic and hyperbolic fixed points (middle). Magnifying one of the "islands" leads to a *similar* structure (right) as shown in the middle part of Figure 3. There exists a *hierarchy* of such islands. This hierarchical behaviour influences many of the physical quantities such as diffusion etc. (see also the contribution by Geisel in the present book).

2.2 Properties of glassy materials

(i) From a qualitative point of view the existence of *exponentially many metastable glassy configurations* is one of the crucial properties. This means that the potential energy of glass-forming systems is a complex landscape in configuration space (Figure 4). The "valleys" are the metastable configurations separated from each other by barriers. The height of the barriers may be of order 1 or even of order N, where N is the number of particles. Configurations corresponding to adjacent "valleys" differ only locally, and the rearrangement of one particle or a group of particles leads to transitions between adjacent "valleys". We will come back to this point later.

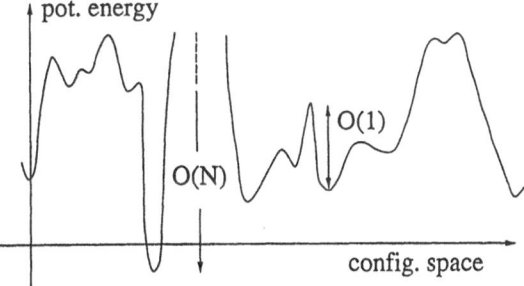

Figure 4: Schematic representation of the potential-energy landscape.

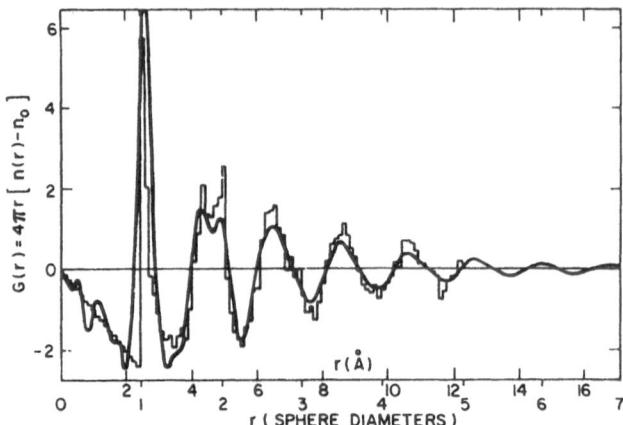

Figure 5: Reduced pair distribution function for amorphous $Ni_{76}P_{24}$ (solid line) and for Finney's dense random packing of hard spheres (thin line) (from Cargill [17]).

(ii) A quantitative characterization of the structure itself follows from the *pair distribution function* $G(r)$ or from the static structure factor $S(q)$. For amorphous structures $G(r)$ exhibits well pronounced first, second etc. neighbor peaks due to approximative crystalline short-range order, but less distinct higher-order peaks. For r to infinity $G(r)$ converges to a constant, proving the lack of long-range order (Figure 5). Similar behaviour is shown by $S(q)$.

(iii) Amorphous materials exhibit a number of *low-temperature anomalies* [18]. Experiments by Zeller and Pohl [19] have shown that the specific heat $c(T)$ and the thermal conductivity $\kappa(T)$ exhibit, respectively, an approximatly linear and quadratic temperature-dependence below 1 K. Actually many experimental data for $c(T)$ can be fitted by a power-law

$$c(T) \sim T^{1+\delta} \tag{7}$$

where δ varies between about -0.7 and 0.4. Figure 6 depicts two different experimental data for $c(T)$ with $\delta < 0$ and $\delta > 0$. The thermal conductivity and the other anomalies will not be discussed further.

(iv) So far we have only mentioned time-independent properties. Concerning *relaxation phenomena*, usually *nonexponential* behaviour is found near the glass transition temperature. For instance the density correlation function

$$\phi_q(t) = \frac{< \rho_q^*(t)\rho_q(0) >}{S(q)} \tag{8}$$

exhibits a Kohlrausch-Williams-Watts-law (KWW law):

$$\phi_q(t) \sim \exp[-(t/\tau)^\beta] \tag{9}$$

218

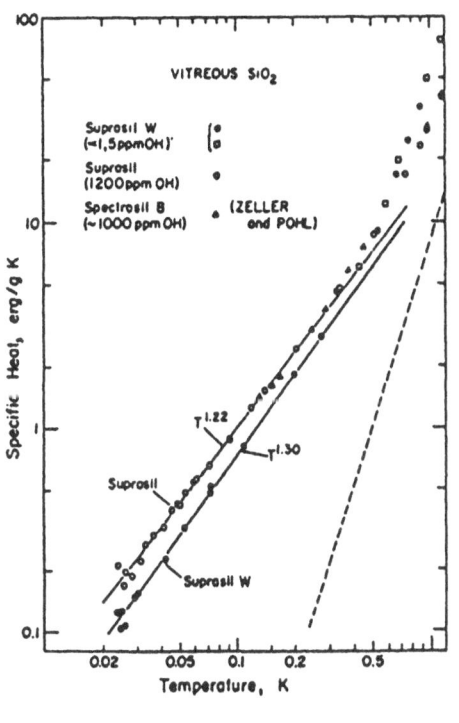

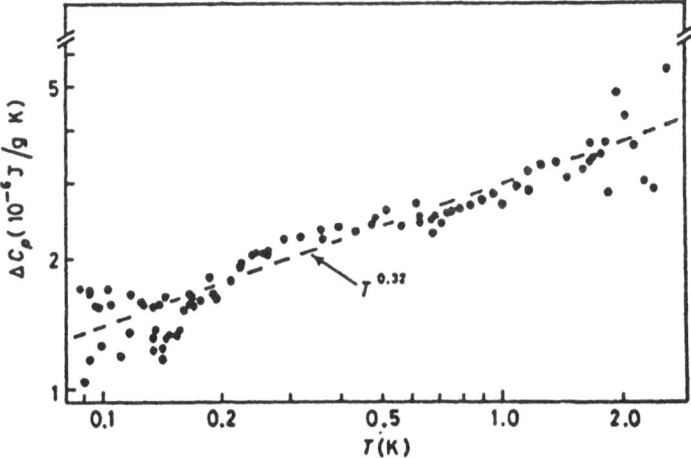

Figure 6: Specific heat for vitreous silica (top) (from [20]) and for TaS$_3$ (bottom) (from [21]). For the right Figure the phonon contribution is already subtracted.

with $\beta < 1$ (see e.g. [22]), also called stretched exponential. Here ρ_q is the Fourier transform of the density $\rho(\vec{x})$, $q = |\vec{q}|$, and $S(q)$ is the static structure factor. The KWW law is also found for other relaxation functions, e.g. the decay of the temperature increment in TaS_3 after a heat pulse shows a KWW law with β increasing from 0.3 at $T = 0.1$ K up to ~ 1 at $T \cong 0.5$ K [21].

There exist many different models explaining the stretched exponential behaviour. A concise overview of the literature and the mechanisms leading to a KWW law is given by Palmer [23]. Palmer discusses three properties of the free energy (or potential energy) surface which can lead to a stretched exponential. These are:

1. Many components, i.e. many metastable configurations exist

2. Hierarchical organization of components

3. Sparse network, i.e. many pathways in configuration space are effectively impassable due to microscopic constraints.

Quite a different approach is the mode-coupling theory of the liquid-glass transition [24], [25]. Although this theory does not consider the three properties in an explicit way, it nevertheless makes some new predictions concerning ergodicity-breaking and scaling behaviour. In particular a KWW law is found for ϕ_q.

(iv) Dwelling more upon these phenomena would require a more detailed discussion of the glass transition itself which is not our intention. But let us finally mention that the glass-forming process depends on the cooling rate. Very slow cooling results in more or less crystalline structures whereas high cooling rates, e.g. 10^6K per sec for metals, are necessary to produce amorphous structures. Cooling to zero temperature, the initial configuration (liquid) relaxes to one of the metastable configurations. The residual energy E_{res}, which is the energy of the relaxed configuration with respect to the ground-state energy, is an interesting quantity dependent on the cooling rate. This dependence itself reflects the properties of the complex potential-energy landscape.

3   One- and two-dimensional models

As already mentioned above, there exist models with a potential V such that (2) can be rewritten as a nonlinear map with discrete time. Aubry was the first to exploit this relationship in order to investigate one-dimensional incommensurate structures [26].

3.1   Connection between stationary configurations and nonlinear maps: One-dimensional models

Consider a chain of classical particles with interactions up to r-nearest neighbors:

$$V(\{x_n\}) = \sum_{n=1}^{N} \sum_{j=1}^{r} V_j(| x_{n+j} - x_n |) \tag{10}$$

where V_j is the j-th nearest neighbor potential. Note that V is *not* a sum over pair potentials and that V_j does *not* depend on n. Boundary conditions can be either periodic or free ends. Without restricting generality much, we assume

$$x_{n+1} > x_n \quad . \tag{11}$$

In this case (2) yields:

$$\sum_{j=1}^{r} [V_j'(x_{n+j} - x_n) - V_j'(x_n - x_{n-j})] = 0 \quad , \tag{12}$$

where $V_j'(x) = d/dx\, V_j(x)$ and boundary effects are ignored. Let us now assume that $V_r(x)$ is strictly convex or concave, i.e $V_r''(x)$ is strictly positive or negative, respectively. This allows us to solve (12) for x_{n+r}:

$$x_{n+r} = f(x_{n+r-1}, \dots, x_{n-r}) \tag{13}$$

with

$$f(x_{n+r-1}, \dots, x_{n-r}) = x_n + V_r'^{-1}\left(V_r'(x_n - x_{n-r}) - \sum_{j=1}^{r-1} [V_j'(x_{n+1} - x_n) - V_j'(x_n - x_{n-j})]\right). \tag{14}$$

Defining

$$(z^{(\nu)})_n = x_{n+r-\nu} \quad , \nu = 1, 2, \dots, 2r \tag{15}$$

$$z = (z^{(1)}, z^{(2)}, \dots, z^{(2r)})$$

(13) can be rewritten as a nonlinear map with discrete time n:

$$z_{n+1} = F(z_n) \tag{16}$$

where

$$F(z) = \begin{pmatrix} f(z^{(1)}, \ z^{(2)}, \quad \dots, \ z^{(2r)}) \\ z^{(1)} \\ z^{(2)} \\ \vdots \\ z^{(2r-1)} \end{pmatrix} . \tag{17}$$

This reveals the relationship between the solutions of (2) and the orbits of (16). The label n of the n-th particles is interpreted as discrete time of the map.

The energy (10) has two kinds of symmetry

$$V(\{x_n + a\}) = V(\{x_n\}) \quad \text{for all } a \in \mathbb{R} \tag{18}$$

and

$$V(\{x_{n+m}\}) = V(\{x_n\}) \quad \text{for all } m \in \mathbb{Z}, \tag{19}$$

assuming periodic boundary conditions. In analogy to usual classical mechanics, these symmetries imply the existence of a "constant of motion" which is an invariant I on the orbits of F:

$$I(x_{n+r-1}, \dots, x_{n-r}) = \sum_{l=1}^{r} \sum_{j=0}^{l-1} V_l'(x_{n+j} - x_{n+j-l}) \quad . \tag{20}$$

I is the *internal stress* in the chain. Using

$$y_n = x_{n+1} - x_n \tag{21}$$

$I(y_{n+r-2}, \ldots, y_{n-r}) = I = \text{const can be solved for } y_{n+r-2}:$

$$y_{n+r-2} = g(y_{n+r-3}, \ldots, y_{n-r}) \tag{22}$$

which again can be rewritten as:

$$\boldsymbol{u}_{n+1} = \boldsymbol{G}(\boldsymbol{u}_n) \tag{23}$$

with:

$$(u^{(\nu)})_n = y_{n+r-\nu-2}, \quad ; \nu = 1, 2, \ldots, 2r - 2 \tag{24}$$

and

$$\boldsymbol{G}(\boldsymbol{u}) = \begin{pmatrix} g(u^{(1)}, & u^{(2)}, & \cdots, & u^{(2r-2)}) \\ & u^{(1)} & & \\ & u^{(2)} & & \\ & \vdots & & \\ & u^{(2r-3)} & & \end{pmatrix}. \tag{25}$$

Thus the symmetries reduce the map to $(2r - 2)$ dimensions. If an external potential is added to (10), the symmetry (18) is broken. In this case the map $\boldsymbol{F}$ has to be investigated. Note that $\boldsymbol{F}$ and $\boldsymbol{G}$ are *nonlinear* if at least one of the potentials V_j are *anharmonic*.

If $\boldsymbol{F}$ and $\boldsymbol{G}$ are nonlinear, in general the three types of orbits shown in Figure 1 will occur. Due to the connection discussed above, the chaotic orbits correspond to irregular arrangements of the particles which we call *spatially-chaotic*. Their metastability remains to be proved.

3.2 Connection between stationary configurations and nonlinear maps: Two-dimensional models

For one-dimensional systems this connection is obvious. Both the label of the particle and the time are one-dimensional entities. In more than one dimension the situation is different. Let us restrict ourselves to two dimensions. Generalisation to $d \geq 3$ is straightforward. We consider the following type of model:

$$V(\{\vec{x}_n\}) = \sum_n \sum_\delta V_1(\vec{x}_{n+\delta} - \vec{x}_n) + \sum_n V_0(\vec{x}_n) \tag{26}$$

where $n \equiv (n_1, n_2) \in \Lambda \subset \mathbb{Z}_2$ is a *two*-dimensional label, and δ runs over the four nearest neighbor vectors of n. The first term describes the nearest-neighbor interactions between particles forming topologically a square lattice, and the second term is an external potential. In many cases $V_0(\vec{x})$ is assumed to be periodic. Using (26) Eq. (2) reads :

$$\sum_\delta [\frac{\partial}{\partial \vec{x}} V_1(\vec{x}_n - \vec{x}_{n-\delta}) - \frac{\partial}{\partial \vec{x}} V_1(\vec{x}_{n+\delta} - \vec{x}_n)] + \frac{\partial}{\partial \vec{x}} V_0(\vec{x}_n) = 0 \quad . \tag{27}$$

This equation is a "partial" *difference* equation, the solution of which needs the specification of $\vec{x}_n$ on the boundary of Λ. For instance if $\Lambda = \{(n_1, n_2) \mid 1 \leq n_j \leq N\}$, we could fix $\vec{x}_n$ for $(n_2 = 1), n_1 = 1, 2, \ldots, N$. In this case (27) must be completed by equations with (n_1, n_2) on the boundary of Λ. The resulting equations can again be rewritten as a $4N$-dimensional map:

$$
\begin{aligned}
v^1_{m+1} &= H^1(v^1_m, v^2_m) \quad , \quad m = 2, 3, \ldots, N; \alpha = 1, 2 \\
v^2_{m+1} &= H^2(v^1_m, v^{(2)}_m)
\end{aligned}
\tag{28}
$$

where

$$
v^\alpha_m = (x^\alpha_{1,m}, x^\alpha_{2,m}, \ldots, x^\alpha_{N,m}; x^\alpha_{1,m-1}, x^\alpha_{2,m-1}, \ldots, x^\alpha_{N,m-1}),
\tag{29}
$$

provided V_1 is strictly convex or concave.

Since we are interested in the thermodynamical limit $N \to \infty$, we would end up with an infinite-dimensional map. For such maps much less is to be found in the literature, in contrast to finite-dimensional maps.

This exposition for $d = 2$ (the same conclusion holds for $d \geq 3$) demonstrates the qualitative difference between $d = 1$ and $d \geq 2$ from a mathematical point of view. Nevertheless we will learn in the next chapter that one- and two-dimensional models have more in common than expected from this mathematical aspect.

4 Metastable and spatially-chaotic configurations

In this chapter we consider simple one- and two-dimensional models of type (10) and (26),respectively. In order that spatially-chaotic configurations occur, it is necessary that the interactions are:

1. anharmonic, leading to nonlinear terms in (12) and (27)

2. competing, leading to frustration like in spin glasses [13].

4.1 One-dimensional model

The simplest one-dimensional model of type (10) with competing interactions must contain at least nearest- and next-nearest neighbor couplings. We choose V_1 to be piecewise parabolic (Figure 7) with two parabola with the same second derivative patched together at c. This may result either in a single or a double well potential. Then V_1 is given by:

$$
V_1(y) = \frac{C_1}{2} \{[y - a_+ - a_- \sigma(y)]^2 - [c - a_+ - a_- \sigma(y)]^2\} \quad , \quad C_1 > 0
\tag{30}
$$

where

$$
\begin{aligned}
a_\pm &= \tfrac{1}{2}(a_2 \pm a_1) \\
\sigma(y) &= \operatorname{sgn}(y - c) \quad .
\end{aligned}
\tag{31}
$$

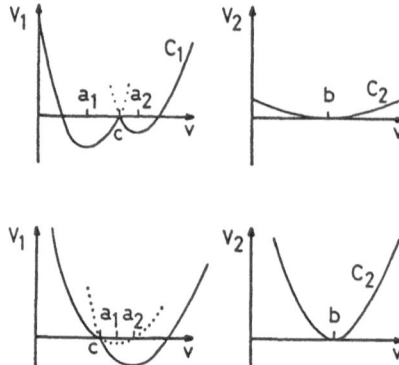

Figure 7: Examples for the nearest and next-nearest neighbor potentials.

The function $\sigma(y)$ divides the nearest-neighbor bonds into two classes, those with $y < c$ ($\sigma(y) = -1$) and those with $y > c$ ($\sigma(y) = +1$). The non-differentiability of V_1 at $y = c$ is not essential, since the cusp can be smoothed in a small neighborhood without changing the results significantly. For simplicity V_2 is chosen harmonic (Figure 7):

$$V_2(y) = \frac{C_2}{2}(y - b)^2, \quad C_2 > 0 \quad or \quad C_2 < 0 \quad . \tag{32}$$

For this potential, equation (12) can be solved exactly. The reader can find details in [15]. For vanishing internal stress $I = 0$ one finds for the nearest-neighbor bond lengths for the *infinite* chain:

$$y_n(\{\sigma_j\}) = A + B \sum_{j=-\infty}^{\infty} \eta^{|j|} \sigma_{n+j} \quad , \quad \sigma_j = \pm 1 \tag{33}$$

where A, B depend on C_1, C_2, a_1, a_2, c, b, and η depends only on the ratio of the coupling constants C_1/C_2. This result is only true if the self-consistency condition

$$\sigma(y_n(\{\sigma_j\})) = \sigma_n \tag{34}$$

holds. For this it is necessary and sufficient that $|\eta| < 1/3$ and that the geometrical parameters of V_1 and V_2 are in a certain range. Eq. (33) makes the connection between configurations of particles and sequences $\sigma = \{\sigma_j\}$ of symbols obvious, which is nothing but the embedding of the Bernoulli shift discussed in section (2.1.) The configurations (33) are metastable if $|\eta| < 1$, and their energy follows from an Ising-like Hamiltonian

$$E(\sigma) = J_0 \sum_{n \neq m} \eta^{|n-m|} \sigma_n \sigma_m - h \sum_n \sigma_n \tag{35}$$

J_0 and h depend on the model parameters.

224

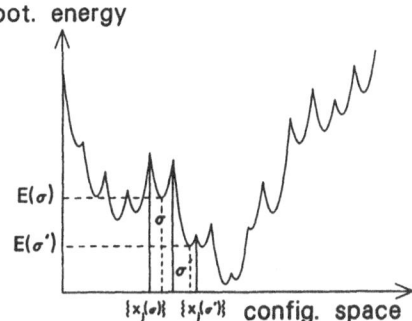

pot. energy

$\{x_j(\sigma)\}$ $\{x_j(\sigma')\}$ **config. space**

Figure 8: Illustration of results (33) and (35) in configuration space.

The one-to-one correspondence (for $|\eta| < 1/3$!) between metastable configurations and sequences of *pseudospins* of Ising-type prove that *spatially chaotic* configurations exist. Choosing $\sigma_j = \pm 1$ randomly, the corresponding configuration $\{y_n(\sigma)\}$ or $\{x_n(\{\sigma\})\}$ represents an irregular arrangement of particles. It is easy to show that the ground state is always periodic with period one ($\{\sigma_j = +1\}$ or $\{\sigma_j = -1\}$) or period two $\{\sigma_j = \pm(-1)^j\}$.

Figure 8 represents our result in the configuration space of the positions x_n. Exponentially many, i.e. 2^N valleys exist. Each valley is uniquely characterized by a sequence σ of pseudo-spins. The lowest point of each valley corresponds to a metastable configuration $\{x_n(\sigma)\}$ with energy $E(\sigma)$. Our model has the advantage that the *configurational* degrees of freedom, described by σ, are obtained exactly and can be separated from the vibrational ones (motion within a valley). On the other hand, the frequencies of these vibrational motions do not depend on σ, which is an artifact of the piecewise harmonic potential.

Thus, transitions between different metastable configurations are most simply described by transitions $\sigma \rightarrow \sigma'$ due to spin flips. Flipping the n-th spin corresponds to changing the bond length y_n from a value smaller than c to one larger than c or vice versa. The barriers $b_{n_1 n_2 \ldots n_\nu}^{(\nu)}(\sigma)$ for a ν-spin flip depend on ν, σ, and the positions n_i of the flips. For $\nu = 1$ one easily finds:

$$b_n^{(1)}(\sigma) = \frac{C_1}{2} \frac{1-\eta}{1+\eta}[y_n(\sigma) - c]^2 \qquad (36)$$

which is of order 1. For transitions between two metastable configurations involving ν-spin flips with $\nu \geq 2$, several paths in configurations space exist. There is at least one path with barrier of $O(\nu)$. For this path the ν spins are flipped simultaneously. But there also exist paths corresponding to a sequence of single spin flips where the maximum barrier on this path is of $O(1)$. This is different for higher-dimensional models.

4.2 Two-dimensional model

Although quasi one-dimensional systems do exist in nature (see e.g. [21]), it is important to investigate higher dimensional models. In this section we will study a layer of adsorbed atoms on a substrate (Figure 9). The potential energy V is given by

$$V(\{\vec{x}_j\}) = \frac{C}{4}\sum_{j,\delta}[\vec{x}_{j+\delta} - \vec{x}_j - a\vec{\mu}_\delta]^2 + \frac{\lambda C}{2}\sum_j[\vec{x}_j - a\vec{m}(\vec{x}_j)]^2 \quad , \quad \lambda > 0, C > 0 \quad (37)$$

with the *cell variable*:

$$m^\alpha(\vec{x}) = \mathrm{int}[\frac{x^\alpha}{a} + \frac{1}{2}] \quad , \alpha = 1, 2 \tag{38}$$

and $a\vec{\mu}_\delta$ the equilibrium positions of the nearest-neighbor bonds. Again these interactions are competing and anharmonic (piecewise harmonic). The solutions of (2) or (27) are as follows

$$\vec{x}_j(\{\vec{m}_k\}) = a\lambda \sum_k g_\lambda(\vec{j} - \vec{k})\vec{m}_k \tag{39}$$

where $g_\lambda(\vec{k})$ is a Green's function decaying exponentially for $|\vec{k}| \to \infty$. For details the reader is referred to [16]. (39) is a solution of (2) provided that the *fluctuations* of the nearest-neighbor distances are bounded by $r \in I\!N$, and that $\lambda > \lambda_c$. λ_c depends on r. This condition is quite analogous to $|\eta| < |\eta_c| = 1/3$ for the one-dimensional model. The energy of the configurations (39) is given by

$$E(\{\vec{m}_j\}) = \frac{1}{2}\sum_{k,l} K(\vec{k} - \vec{l})(\vec{m}_k - \vec{m}_l)^2 \tag{40}$$

with

$$K(\vec{j}) = \frac{1}{2}\lambda^2 a^2(g_\lambda(\vec{j}) - \frac{1}{\lambda}\delta_{j,0}) \quad . \tag{41}$$

The most important similarity to our previous model is the relationship between configurations $\{\vec{x}_j\}$ and sequences of symbols $\{\vec{m}_k\}$. If one is willing to introduce

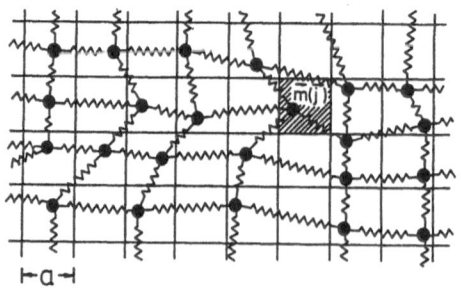

Figure 9: Top view of the adsorbed layer. The hatched region indicates that cell $\vec{m}(\vec{j})$ which is occupied by particle $\vec{j} = (j_1, j_2)$, $j_\alpha \in Z\!\!Z$. a is the period of the substrate.

a *two-dimensional*, discrete time (j_1, j_2), the result (39) can again be interpreted as the embedding of the Bernoulli-shift. In any case, the freedom to choose $\{\vec{m}_k\}$ for $\lambda > \lambda_c(r)$ guarantees the existence of spatially chaotic configurations in two dimensions. The discussion of the complex landscape, the configurational and vibrational degrees of freedom, the barriers etc. in the last section applies to the two-dimensional model as well. The main difference concerns the alphabet A of the symbols. For the former case it is just $\{-1, 1\}$, whereas for the adsorbed layer it is a subset of integers (see discussion above).

4.3 Three-dimensional models

We have not investigated a three-dimensional model yet. One could think of a generalization of the two-dimensional model, where $\vec{x}_j$ are the coordinates of *internal* degrees of freedom, e.g. the orientational vectors of polar molecules interacting with each other harmonically. In addition, these molecules may feel a crystal field with several local minima which could be approximated again by piecewise harmonic potentials. The results for the metastable configurations and their energy are of the same type as for the two-dimensional model but with three-dimensional Green's functions. The alphabet is related to the number of local minima of the crystal field. Thus, a similar relationship between metastable configurations and sequences of symbols exists as for the one- and two-dimensional models.

Let us finally mention a generalization of the previous class of two- and three-dimensional models to systems with a *periodic* substrate potential or crystal field and *pair potential* between the species. For example it has been shown by Stillinger [27] that the metastable configurations for an adsorbed layer with pair potential can be described by a lattice-gas model.

5 Glassy properties of spatially-chaotic structures.

The exact solutions we found in the last chapter for the metastable configurations allow to study their properties. Almost all solutions are spatially chaotic in the same sense as almost all real numbers of the unit interval are *normal* numbers. [3]

To make the connection between spatially-chaotic configurations and glassy structures, the symbols $\{\sigma_j\}$ and $\{\vec{m}_j\}$ describing, respectively, the configurational degrees of freedom of the one- and the two- dimensional model are assumed to be *quenched*. The disorder in such a frozen state is characterized by a probability distribution $P_{qu}(\{\sigma_j\})$ and $P_{qu}(\{\vec{m}_j\})$. An initial state at time $t = 0$ in thermal equilibrium at temperature T_{qu} subjected to a cooling procedure could yield P_{qu} in principle. But the analytical description of such nonequilibrium pro-

---

[3]The binary representation $\{\tau_i\}, \tau_i = 0, 1, i \geq 1$ of a normal number has the property that any subsequence $\tau_{i_1} \tau_{i_2} \ldots \tau_{i_n}$ occurs with *equal* probability 2^{-n} [28].

cesses is rather difficult. Therefore we assume an idealized quenching process for which the particle positions will freeze into that cell $S(\{\sigma_j\})$ and $S(\{\vec{m}_j\})$ of the configuration space where the system has been at $t = 0$. For instance, for the two-dimensional model it is:

$$P_{qu}(\{\vec{m}_j\}) = \frac{1}{Z_{conf}} \int_{S(\{\vec{m}_j\})} \prod_n d^2 x_n \exp[-\beta_{qu} V(\{\vec{x}_j\})], \quad \beta_{qu} = \frac{1}{k_B T_{qu}} \quad (42)$$

where Z_{conf} is the configurational part of the partition function

$$Z_{conf} = \int \prod_n d^2 x_n \exp[-\beta_{qu} V(\{\vec{x}_j\})] \quad . \quad (43)$$

Such an idealized freezing procedure obviously yields maximum disorder since crystalline short-range correlations can not be built up further. Although similar procedures can be performed for the one-dimensional model, a cruder approach is chosen. We assume that σ_i and σ_j $(i \neq j)$ are *uncorrelated* implying that

$$P_{qu}(\{\sigma_j\}) = \prod_j p(\sigma_j) \quad . \quad (44)$$

Since σ_j takes two values, $p \equiv p(\sigma_j = +1)$ has to be specified only. Now we can discuss the physical properties of the spatially chaotic configurations characterized by P_{qu}.

5.1 One-dimensional model

(i) Complex energy landscape

We have already discussed the complex energy landscape in configuration space. In the following we investigate the properties (ii) - (v) introduced in section 2.2.

(ii) Pair distribution function $G(r)$

A j-th nearest neighbor distance is given by

$$r_{n,j}(\boldsymbol{\sigma}) = \sum_{k=n}^{n+j-1} y_k(\boldsymbol{\sigma}) \quad . \quad (45)$$

Consequently we obtain for a realization $\{y_n(\sigma)\}$:

$$G(r) = \frac{1}{N} \sum_{n=1}^{N} \sum_j \delta(r - r_{n,j}(\boldsymbol{\sigma})) \quad . \quad (46)$$

In the thermodynamic limit the system is self-averaging such that $G(r)$ can be obtained by averaging over $P_{(qu)}(\boldsymbol{\sigma})$

$$G(r) = \sum_\sigma [\sum_j \delta(r - r_{0,j}(\boldsymbol{\sigma}))] P_{qu}(\boldsymbol{\sigma}) \quad . \quad (47)$$

For $P_{qu}(\boldsymbol{\sigma})$ from (44) with $p = p(\sigma_j = +1) = \frac{1}{2}$ a histogram for $G(r)$ with

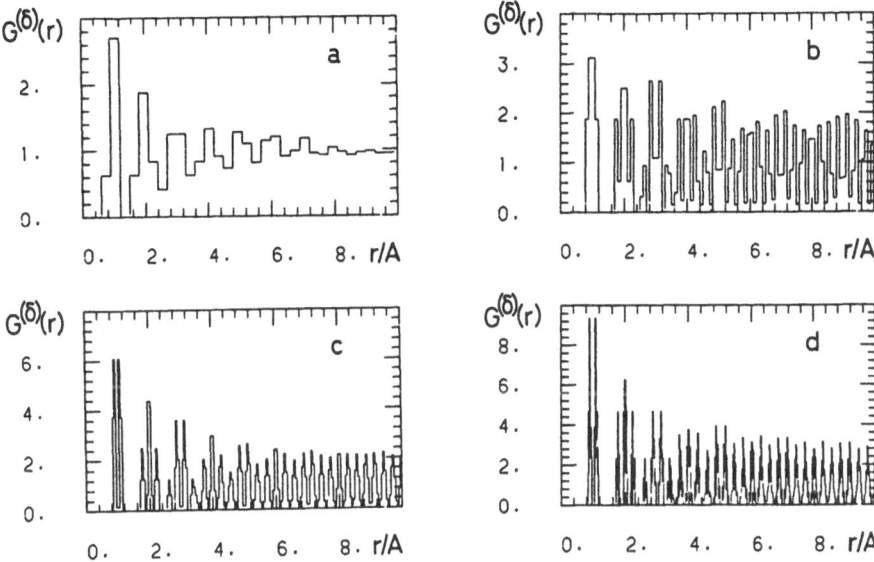

Figure 10: Histogram for the pair distribution function for $p = \frac{1}{2}, \eta = 0.2, \delta = 0.3A$ (a), $0.1A$ (b), $0.05A$ (c) and $0.02A$ (d). A is the mean lattice constant.

resolution δ is shown in Figure 10 . Well-pronounced j-nearest neighbor peaks for $j \leq 8$ exist. But for large r, $G^{(\delta)}(r)$ converges to 1 which has been proved exactly [29]. Thus short-range but *no* long-range order exists in accordance with glassy structures (cf. Figure 5). For high resolution, i.e. $\delta/A \ll 1$, a rather detailed structure occurs within the first, second etc. neighbor peaks. There are at least two reasons for this. First of all, in higher dimensions the averaging over the angle leads to a smoothing of the peaks which is absent for our model in $d = 1$, and second the nearest-neighbor distances $y_n(\sigma)$ form a *Cantor set* with an infinite number of gaps, and their distribution exhibits a hierarchical structure [15] contributing to the bizarre shape of $G(r)$.

(iii) Low-temperature specific heat

The low-temperature anomalies found for glassy structures below 1K have been attributed to two-level-systems (TLS) [30], [31]. These TLS are *configurational* excitations where an atom or a group af atoms rearranges locally. It was argued that the corresponding density of states $n(E)$ is constant for small energies leading to a specific heat

$$c_{TLS}(T) = k_B \int_0^\infty dE n(E) \frac{(\beta E)^2}{4 \cosh^2 \frac{\beta E}{2}} \sim T \quad , \quad \beta = 1/k_B T \qquad (48)$$

which is linear in T.

For the one-dimensional model there exists a *hierarchy* of two-level systems generated by rearrangements of the particles. The members of this hierarchy are classified by ν, the number of spin flips associated with a rearrangement.

Since the total length $L = \sum_{n=1}^{N-1} v_n$ depends on $\sum_{i=1}^{N-1} \sigma_i$, spin flips changing the "magnetization" $\sum_{i=1}^{N-1} \sigma_i$ will induce a change of L. Because at $T < 1K$ the transitions occur via tunneling, such rearrangements are excluded due to their macroscopic mass. The simplest TLS for which $\sum \sigma_i$ is unchanged corresponds to transitions where two adjacent spins of opposite sign flip:

$$\cdots - + - (+-) - - + \cdots \longleftrightarrow \cdots - + - (-+) - - + \cdots \quad .$$

Their classical excitation energy Δ, also called asymmetry, follows immediately from (35):

$$\Delta_n(\boldsymbol{\sigma}) = \frac{\varepsilon_0}{2\eta} \sum_{j=1}^{\infty} \eta^j (\sigma_{n+j} - \sigma_{n-1-j}) \quad , \varepsilon_0 = 8|J_0\eta|(1-\eta) \quad . \tag{49}$$

As mentioned before one has to take quantum corrections into account. For realistic data we found that for $0.05K \le T \le 1K$ these corrections either cancel or can be neglected [14]. Accordingly the density of states is given by:

$$n(E) = \frac{1}{N} \sum_{n=1}^{N} \delta(E - \Delta_n(\boldsymbol{\sigma})) \tag{50}$$

which becomes

$$n(E) = \sum_{\sigma} P_{qu}(\boldsymbol{\sigma})\delta(E - \Delta_0(\boldsymbol{\sigma})) \tag{51}$$

in the thermodynamical limit. Rewriting $\Delta_0(\boldsymbol{\sigma})$ as follows

$$\Delta_0(\boldsymbol{\sigma}) = \frac{\varepsilon_0}{2}(\sigma_1 - \sigma_{-2}) + \eta\Delta_0(\boldsymbol{\sigma'}) \quad ,$$

where σ_1, σ_{-2} are missing in $\boldsymbol{\sigma'}$ and exploiting that the σ_j's are uncorrelated, a functional equation can easily be derived for $n(E)$. One finds:

$$n(|\eta|E) = \frac{1}{|\eta|}\{(2p^2 - 2p + 1)n(E) + p(1-p)[n(E - \frac{\varepsilon_0}{|\eta|}) + n(E + \frac{\varepsilon_0}{|\eta|})]\} \quad . \tag{52}$$

Taking into account that $n(E) \equiv 0$ for $|E| > \varepsilon_0/(1 - |\eta|)$ we obtain

$$n(|\eta|E) = \frac{1}{|\eta|}(2p^2 - 2p + 1)n(E) \quad , \tag{53}$$

for $|E| \le \varepsilon_0 1 - 2|\eta|/|\eta|(1-|\eta|)$. This scaling relation reflects the *self-similarity* of the density of states which is shown in Figure 11. Here we stress an important point: Although the spatially chaotic configurations are *not* fractals from the geometrical point of view, they possess fractal properties concerning distribution functions.

From (48) and (52) a scaling relation can be derived for $c(T)$ resulting in a power-law [4]

---

[4]This power-law describes only the average behaviour of $c(T)$. For $|\eta| \ll 1/3$ the gaps in the density of states (cf. Fig.11) becomes larger. This yields a modulation of the power law due to Schottky anomalies. It is interesting that such modulations seem to occur also in Figure 6b. For $\approx 0.2 < |\eta| < 1/3$ these Schottky anomalies overlap such that a true power-law occurs.

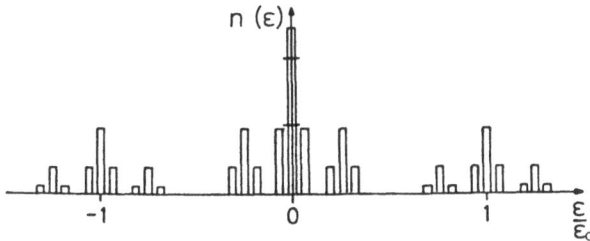

Figure 11: Density of states for $p = \frac{1}{3}, |\eta| = \frac{1}{4}$ and finite energy resolution.

$$c(T) \sim T^{\tilde{d}} \tag{54}$$

for low temperatures $k_B T \ll \varepsilon_0$ with a fractional exponent

$$\tilde{d} = \ln(2p^2 - 2p + 1)/\ln|\eta| \leq \ln 2/\ln 3 \cong 0.62 \quad . \tag{55}$$

This exponent is *not universal*. It depends both on p, the degree of disorder, and on $\eta = \eta(C_1/C_2)$. Such fractional exponents are really observed for quasi one-dimensional systems [21] as well as for three-dimensional ones [32], (cf. Figure 6) demonstrating that $n(E)$ is not constant at low energies and that it may reveal a self-similar structure. But we stress that this agreement between experimental and our theoretical results may be accidental.

(iv) Static and dynamical correlations functions in equilibrium

Recently we also investigated the dynamics of the chain. Integrating the equations of motion [5]

$$m\ddot{y}_n + \frac{\partial V}{\partial y_n} = 0 \quad , \quad n = 1, 2, \dots, N \tag{56}$$

numerically (N=5000), time averages were calculated. Here only a short account is presented, details will be given in [33]. The time-dependence of the configurations can be described by $\sigma_j(t) \equiv \sigma(y_j(t))$, the *dynamics of the symbols*, not to be confused with the symbolic dynamics. In order to test ergodicity, we calculated e.g. the correlation functions $1/N \sum_{j=1}^{N} \overline{\sigma_{j+n}(t)\sigma_j(t)}$ as a function of the temperature $T = (m/(k_B N) \sum_{j=1}^{N} \overline{(\dot{y}_j(t))^2}$, where the bar denotes time averaging.

These quantities can be compared with the corresponding canonical averages which are obtained as follows:

$$< f(\sigma) >= \frac{1}{Z_{conf}} \int \prod_n dy_n f(\sigma(y_1), \sigma(y_2), \dots, \sigma(y_N)) \exp[-\beta V(\{y_n\})]. \tag{57}$$

For an arbitrary spin function $f(\sigma)$, this can be rewritten as follows

---

[5] Note that for the dynamics y_n is interpreted as a coordinate, e.g. as the displacement of the particles from a periodic arrangement.

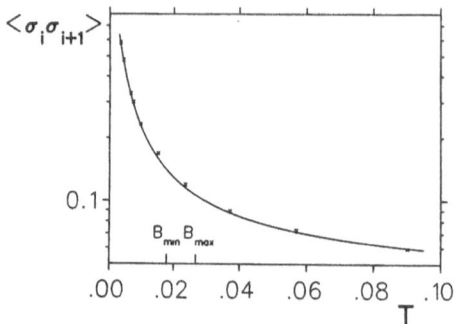

Figure 12: Temperature-dependence of the nearest-neighbor correlation function for $\eta = 0.048$ and $h = 0$ (cf.(35)). Numerical result (crosses), analytical result (solid line). B_{min} and B_{max} denote the minimum and maximum barrier height. T is measured in units of $7.25 \cdot 10^2 C_1 [kg/sec^2] K$.

$$< f(\sigma) > = \frac{1}{Z_{conf}} \sum_\sigma \int_{S(\sigma)} \prod_n dy_n f(\sigma(y_1), \ldots, \sigma(y_N)) exp[-\beta V(\{y_n\})] \quad , \quad (58)$$

where $S(\sigma)$ is that part of configuration space characterized by σ. For $k_B T$ smaller than the minimum barrier height B_{min} mainly those $\{y_n\}$ contribute which do not differ much from the local minima $\{y_n(\sigma)\}$. Expanding V around $\{y_n(\sigma)\}$ we obtain from (58)

$$< f(\sigma) > \cong \frac{1}{Z_\sigma} \sum_\sigma f(\sigma_1, \ldots, \sigma_N) e^{-\beta E(\sigma)} \tag{59}$$

with

$$Z_\sigma = \sum_\sigma e^{-\beta E(\sigma)} \tag{60}$$

and $E(\sigma)$ given by (35). For higher temperatures, (59) can still be used, but with $E(\sigma)$ replaced by an effective spin Hamiltonian with temperature-dependent coupling constants [33].

For $n = 1$, the numerical result for the time-averaged correlation function is compared with the canonical average $< \sigma_{j+n}\sigma_j >$ (Figure 12). Over the full temperature range a good agreement is found between both quantities. This was also the case for the magnetization $< \sigma_j >$, the 2-spin correlation function with $n = 2$, and a 3-spin correlation function. So far we may conclude that the chain is *ergodic*, at least for these spin functions and for the investigated temperature region. Even more interesting with respect to glassy behaviour is the time-dependent correlation function. We have restricted ourselves to the autocorrelation function $C(t) = 1/N \sum_{j=1}^{N} \overline{\sigma_j(t' + t)\sigma_j(t')}$. The result is shown for various temperatures in Figure 13. The log-lin-plot exhibits a *nonexponential* relaxation typical for glassy systems. The numerical data can be fitted by a stretched exponential for intermediate times. From this fit the relaxation time $\tau(T)$ and the exponent $\beta(T)$ are determined. Of course, it is hard to estimate

232

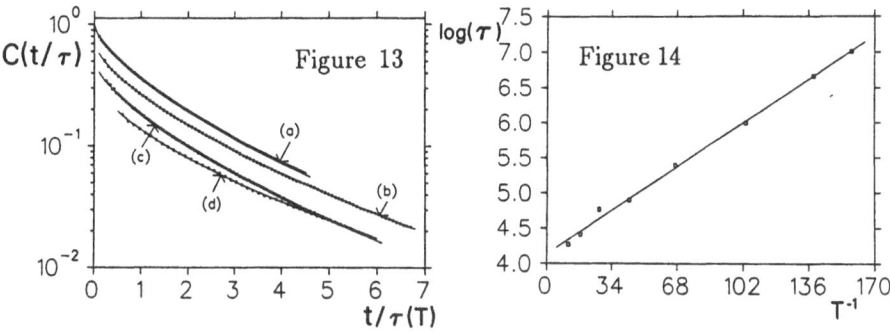

Figure 13: Time-dependent autocorrelation function for $\eta=0.0483$, h=0, T= 0.0097 (a), 0.0232 (b), 0.0365 (c), and 0.0569 (d). Numerical results (symbols) fit with stretched exponential (solid line). Units of T as in Figure 12.

Figure 14: Relaxation time as function of temperature for $\eta = 0.0483$ and $h = 0$. Units of T as in Fig. 12. Numerical results (squares) and Arrhenius fit (solid lines).

the significance of these fits, because the range for $t/\tau(T)$ is at most one order of magnitude. Whether the stretched exponential behaviour extends to much larger times is unclear, because for larger times the relative fluctuations of the data are so large that no conclusive statement can be made. This could be improved by averaging over many more sets of initial configurations or by considering larger systems which for the moment is beyond our computational possibilities. The exponent $\beta(T)$ decreases roughly with increasing temperature and is between 0.62 and 0.75 which is in a reasonable range for structural glasses [23]. $\tau(T)$ exhibits an Arrhenius behaviour as demonstrated in Figure 14.

An analytical calculation of the time-dependent correlation functions starting from the microscopic equations of motion seems rather unrealistic. But for temperatures small compared to B_{min} the dynamics of the spins σ_j may be described by kinetic equations analogous to Glaubers's kinetic Ising model [34], i.e. the dynamics is reduced to a master equation for $P(\sigma,t)$, the probability for a spin configuration at time t. Our numerical simulations show that single spin flips are generic. The related transition rates $W_i(\sigma \to \sigma')$ $(\sigma'_j \equiv \sigma_j$ for $j \neq i$ and $\sigma'_i = -\sigma_i)$ can be obtained with (36) from transition state theory. For $h = 0$ and $\eta \ll 1$ we have found

$$W_i(\sigma \to \sigma') = \alpha(T)[1 + \delta(T)\sigma_{i-1}\sigma_{i+1}][1 - \frac{1}{2}\gamma(T)\sigma_i(\sigma_{i-1} + \sigma_{i+1})] \qquad (61)$$

with

$$\alpha(T) = \alpha_0 e^{-\beta|J_0|} \cosh^2 2\beta J_0\eta \quad , \quad \delta(T) = \tanh^2 2\beta J_0\eta \qquad (62a)$$

$$\gamma(T) = \tanh 4\beta \mid J_0 \mid \eta \quad , \qquad (62b)$$

where α_0 sets the time scale. A microscopic justification of this approach will be given in [33]. (61) is the most general transition rate involving nearest-neighbor spins only and satisfying detailed balance. For $\delta(T) = 0$ the correlation functions can be calculated exactly. In this case, the autocorrelation function fits very well with a stretched exponential in an intermediate time range [35].Here again we must stress that the range of time where a stretched exponential was fitted is only a bit more than one order of magnitude. But using $\delta = -1$ and $\gamma = 0$, a stretched exponential with $\beta = 1/2$ was found *exactly* at low temperatures for $t \to \infty$ [36], [37]. For $\delta \neq 0$ and $\gamma \neq 0$ approximate analytical calculations could also be fitted by a stretched exponential but again for an intermediate time range only [35], [38]. The *qualitative* agreement between our numerical and these analytical results for a one-dimensional kinetic Ising model for intermediate times may support the ansatz that the chain of particles with interactions (30) and (31) is a realization of a kinetic Ising model, at least at low temperatures. Whether the decay of the autocorrelation function is a stretched exponential even on a larger time scale is not quite clear.

(v) Nonequilibrium phenomena: residual energy

For a system with complex potential energy landscape an intial equilibrium state will relax to one of the metastable configurations, when it is cooled to zero temperature. The energy E_{res} of the relaxed configuration with respect to the ground-state energy depends on the cooling rate γ. For spin glass models, Monte-Carlo simulations give evidence for a power-law dependence

$$E_{res} \sim \gamma^\alpha \quad , \quad \alpha > 0 \qquad (63)$$

and a logarithmic law

$$E_{res} \sim (-\ln\gamma)^{-\zeta} \quad , \quad \zeta > 0 \qquad (64)$$

for $\gamma \to 0$ in two and three dimensions, respectively [39]. On the other hand it was argued that the logarithmic behaviour should hold for all dimensions [40]. The arguments used in ref.[40] assume a *broad* distribution of asymmetries near the ground state, but this is not true for structural glasses. There the ground state is ordered (crystalline), and the low-lying *configurational* excitations have a gap. Therefore we do not expect a logarithmic law but a power-law, as can be concluded from ref.[40] assuming a sharp distribution of asymmetries at low energies.

To test this conjecture we simulated a cooling process by adding a frictional term $\gamma \dot{y}_n$ to the equation of motion (56), where γ plays the role of the cooling rate. Our first finding was an evidence for a power-law [41]. In the meantime we were able to investigate much smaller γ-values. The result for $e_{res}(\gamma)$, the residual energy per particle, is depicted in Figure 15. We believe that these results confirm the power-law behaviour for small γ. [6] A detailed description of our results is presented in [42]. We have also tried to model the freezing process for the chain by a kinetic Ising model with time-dependent temperature. Using the usual Glauber dynamics ($\delta = 0$) with $\alpha(T) = \alpha_0 e^{-\beta \varepsilon}$ and taking into account only nearest-neighbor interactions, an asymptotic analysis yields [43]

---

[6]For small η our numerical simulation shows that the asymptotic region is shifted to smaller γ values. Therefore the range where a power-law can be fitted for such η is rather small.

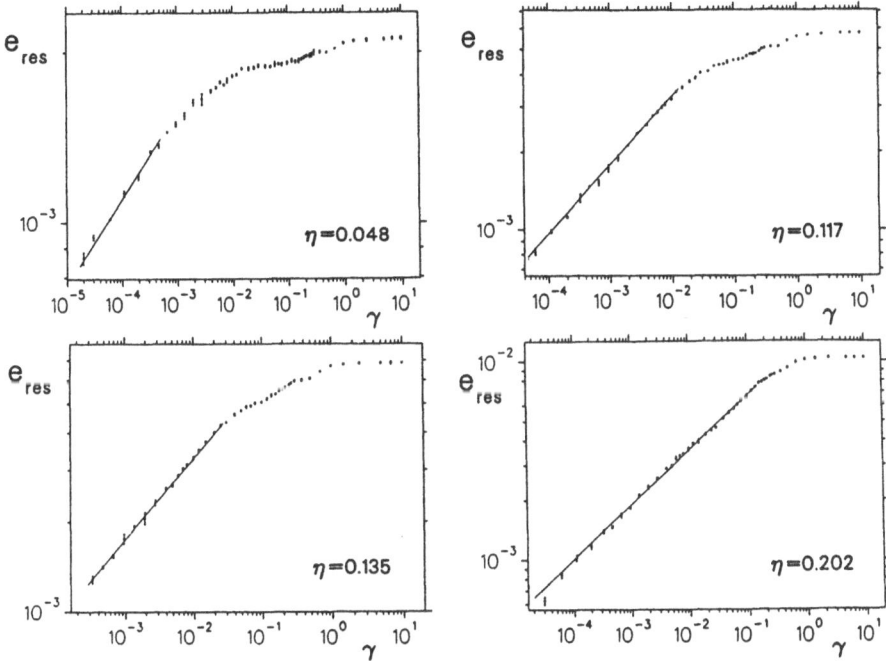

Figure 15: Residual energy as function of γ for different η. Solid lines are fits with a power-law.

$$e_{res} \sim \gamma^{\alpha} \tag{65}$$

with

$$\alpha = \frac{\delta}{2(1+\delta)} \quad , \quad \delta = \frac{8|J_0\eta|}{\varepsilon} \quad . \tag{66}$$

It is interesting to note that α equals the inverse dynamical critical exponent at the trivial fixed point at $T = 0$. This relationship was already found for an Ising model with alternating nearest-neighbor interactions [44].

For several η-values the exponent α from (66) agrees with the numerical value within 10 - 20 percent. An improvement of our analysis, i.e. taking $\delta \neq 0$ into account, leads to a better agreement for $\eta \leq 0.15$ [42]. In any case the η-dependence of α seems to be incorrect, at least for $\eta > 0.15$.

5.2   Two-dimensional model

For the two-dimensional model only very few results exist. Here we will discuss merely the TLS and their specific heat. Details can be found in ref.[16].

The simplest TLS are the configurational excitations where an atom jumps into one of the adjacent cells (Figure 16). Using (40), (41) and $\vec{m}_n \rightarrow \vec{m}'_n = \vec{m}_n + \vec{e}, \vec{m}_j = \vec{m}_j$ for all $j \neq n$, the energy difference $\Delta_{n,e}$ can easily be calculated:

$$\Delta_{n,e} = \frac{1}{2}\lambda a^2(\Delta g_\lambda)(0) + \lambda a^2 \sum_k \Delta g_\lambda(\vec{n} - \vec{k})(\vec{m}_k \cdot \vec{e}) \tag{67}$$

235

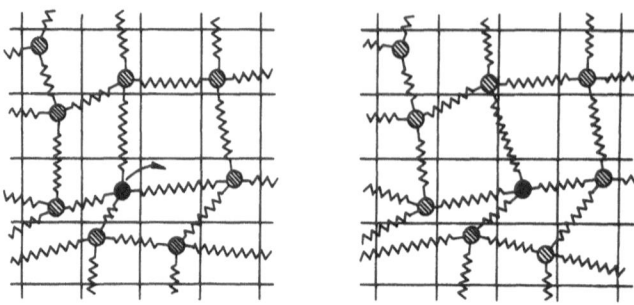

Figure 16: A configurational excitation forming a TLS: the black particle moves into a neighboring cell.

where $\Delta g_\lambda(\vec{j}) = \sum_\delta [g_\lambda(\vec{j} + \vec{\delta}) - g_\lambda(\vec{j})]$ is the discrete Laplacian and $\vec{e}$ is one of the nearest neighbors of $\vec{m} \in \mathbb{Z}^2$.

The distribution function for Δ depends sensitively on the *type* and the *degree* of disorder. For a special type of disorder the same results are found as for the one-dimensional model [16]. If on the other hand we use the reduced distribution function $P_{qu}(\{\vec{m}_j\})$ given by (42), the density of states will depend on the "quench-temperature" T_{qu}. For $Ca^2/2 \ll k_B T_{qu}$, i.e. for large disorder, P_{qu} is approximately given by

$$P_{qu}(\{\vec{m}_j\}) \cong \frac{1}{Z_{qu}} \exp\{-\beta_{qu} \frac{Ca^2}{2} \sum_j \sum_\alpha [(m^\alpha_{j+e_1} - m^\alpha_j - \vec{e}_\alpha \cdot \vec{\mu}_{e_1})^2$$
$$+ (m^\alpha_{j+e_2} - m^\alpha_j - \vec{e}_\alpha \cdot \vec{\mu}_{e_2})^2]\} \qquad (68)$$

where Z_{qu} is the corresponding partition function, $\vec{e}_1 = (1,0)$ and $\vec{e}_2 = (0,1)$. Using (68) we obtained the density of states shown in Figure 17. Assuming $Ca^2/2 = 0.1$ eV, the maximum energy in Figure 17 is about 4K. For low T_{qu} the density of states is rather bizarre like for the one-dimensional model (cf. Figure 11). Whether there exists a strict self-similarity is unclear. Increasing T_{qu} more and more $n(E)$ becomes smoother. For $T_{qu} = 5Ca^2$ (Figure 17d) the density of states is really constant on the scale of 1K and smaller, yielding a linear specific heat. But notice again that the density of states, and therefore the specific heat too, depends sensitively on the degree of disorder. This is a point usually passed over in the literature. In our opinion it is important and therefore worthwhile to try to classify different types and degrees of disorder, and to study their influence on physical quantities such as the specific heat.

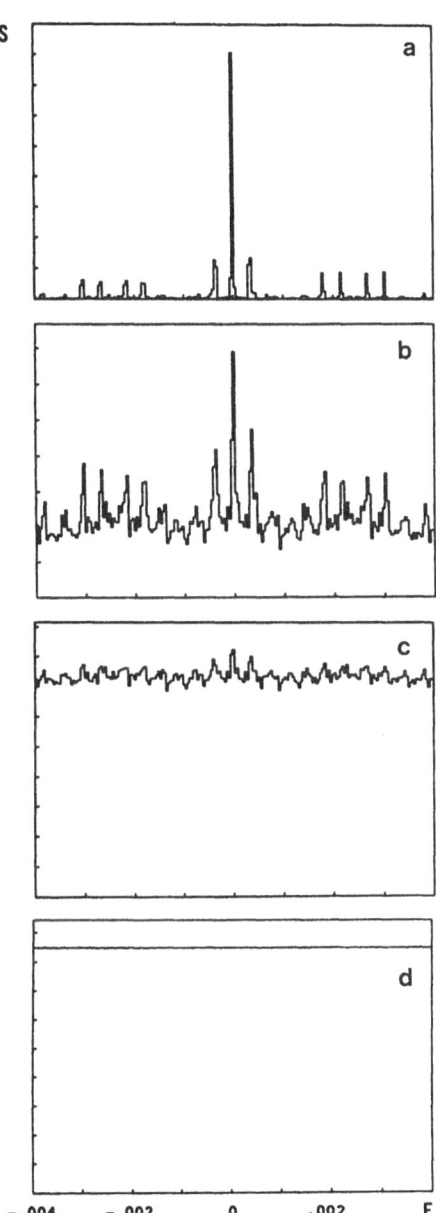

Figure 17: The density of states as a function of energy E for $\lambda = 10$. E is given in units of $Ca^2/2$ and $n(E)$ in arbitrary units. $T_{qu} = 0.5Ca^2$ (a),$0.7Ca^2$ (b), Ca^2 (c) and $5Ca^2$ (d).

6 Summary and Conclusion

The main purpose of this contribution was to investigate the existence and properties of spatial chaos. We have presented two types of models which exhibit *spatial chaos*. Its manifestation is the irregular arrangement of the particles in metastable configurations. The most important features of these models are the *anharmonicity* and the *competition* of the interactions. For the one-dimensional model a natural relationship exists to the *temporal* chaos generated by nonlinear dynamical systems with discrete time. This relationship exists also for the two-dimensional model if one is willing to use two-dimensional time.

Approximating the potential by piecewise harmonic interactions, the connection between the metastable configurations and sequences of symbols was found exactly. It is this connection which not only guarantees chaotic behaviour, but also allows us to reduce the physical description to the level of symbols. For both models the choice of the symbols is obvious.

The second concern has been to explore the properties of the spatially chaotic configurations with respect to amorphous materials. Certainly one may call the arrangement of atoms in amorphous structures chaotic. But the main question is: Do amorphous materials have something in common with the typical properties of chaos found in nonlinear *dynamical* systems? The answer found is positive. First of all the one-to-one correspondence between metastable configurations and sequences of symbols, in analogy to the embedding of the Bernoulli shift for nonlinear dynamical systems, proves that there exist exponentially many metastable configurations representing valleys of a complex energy landscape in configuration space. This is the basic property of glass-forming systems. Next, the existence of short-range order but no long-range correlations is true for both. The self-similarity and its hierarchical organisation as shown by the orbits of dynamical systems was found e.g. for the density of states of the two-level-systems, leading to a power-law behaviour for the specific heat at low temperatures. The corresponding exponent is fractional, is smaller than one, and depends on the coupling constants and the *degree of disorder*, i.e. the exponent is non-universal. It is interesting to note that such a fractional exponent was found recently in experiments on quasi one-dimensional systems. Moreover, we found that physical quantities, e.g. the specific heat, may also depend sensitively on the *type of disorder*. This point was mainly discussed for the two-dimensional model. More extensive theoretical and experimental studies are desirable to explore a possible classification of disorder and its influence on physical quantities.

The connection of the metastable configurations to sequences of symbols also turned out to be crucial for the dynamical features. Disregarding the vibrational motions, being only harmonic for our type of models, the time-dependence of the configurational degrees of freedom is described by the *dynamics of symbols*. We have found some evidence that this dynamics can be described by a kinetic Ising model in the case of one dimension. The numerically calculated autocorrelation function exhibits *nonlinear* relaxation characteristic for glassy systems. Although the data can be fitted by a Kohlrausch law for intermediate times, it is not clear whether such a stretched exponential behaviour is valid over a longer time range.

We also used the kinetic Ising model to describe the simulation of a quench process with finite cooling rate. The numerical results show a power law for the residual energy as function of the cooling rate γ for $\gamma \to 0$. A power-law is also found analytically for the kinetic Ising-model. Although the numerical and analytical exponents agree reasonably for $\eta \leq 0.15$, their η-dependence seems to be different for larger η. The resolution of this discrepancy will be one of our next steps, as well as the simulation of the two-dimensional model. This will be particularly interesting because the discrete Gaussian model (40) exhibits a phase transition, in contrast to the one-dimensional model.

In conclusion we may say that *simple* microscopic models for spatial chaos exist which are also able to describe some of the properties of glassy materials, at least on a qualitative level.

Acknowledgement

I am very grateful to W. Kob, P. Reichert and W. Uhler who have contributed much to the work presented here. In particular I am indebted to W. Kob for helpful discussions as well as for the preparation of most of the figures.

The financial support by the Swiss National Science Foundation is also gratefully acknowledged.

References

[1] H. Poincaré: *Les Methodes Nouvelles de la Mécanique Céleste*, Gauthier-Villars, Paris (1892)

[2] V.I. Arnold, A. Avez: *Ergodic Problems in Classical Mechanics*, Benjamin, New York (1968)

[3] J. Moser: *Stable and Random Motions in Dynamical Systems*, Princeton University Press, Princeton, N.J. (1973)

[4] M.V. Berry: In *Topics in Nonlinear Dynamics*. AIP Conference Proceedings, ed. S. Jorna, AIP, New York (1978)

[5] D. Ruelle, F.Takens: Comm. Math. Phys. **20**,167 (1971)

[6] A. J. Lichtenberg, M.A. Liebermann: *Regular and Stochastic Motion*, Springer, New York (1983)

[7] H.G. Schuster: *Deterministic Chaos. An Introduction*, Physik-Verlag, Weinheim (1984)

[8] P. Berge, Y. Pomeau, Ch. Vidal: *Order in Chaos*, John Wiley & Sons (1984)

[9] J. Greene: J. Math. Phys. **20**,1183 (1979)

[10] R. Zallen: *The Physics of Amorphous Solids*, John Wiley & Sons (1983)

[11] J. Ziman: *Models of Disorder*, Cambridge University Press (1979)

[12] F.H. Stillinger, R.A. La Violette: Phys. Rev. **B34**, 5136 (1986)

[13] K. Binder, A.P. Young: Rev. Mod. Phys., **58**, 801 (1986)

[14] R. Schilling: Phys. Rev. Lett. **53**, 2258 (1984)

[15] P. Reichert, R. Schilling: Phys. Rev. **B32**, 5731 (1985)

[16] W. Uhler, R. Schilling: Phys. Rev. **B37**, 5787 (1988)

[17] G.S. Cargill: Solid State Physics **30**, 227 (1975)

[18] W.A. Phillips (ed.): *Amorphous Solids: Low-Temperature Properties*, Springer (1981)

[19] R.C. Zeller, R.O. Pohl: Phys. Rev. **B4**, 2029 (1971)

[20] J.C. Lasjaunias, A. Ravex, M. Vandorpe, S. Hunklinger: Solid State Commun. **17**, 1045 (1975)

[21] K. Biljakovic, J.C. Lasjaunias, P. Monceau, E. Levy: Europhysics Lett. **8**, 771 (1989)

[22] J. Jäckle: *Models of the glass transition*, Reports on Prog. in Phys., **49**,171 (1986)

[23] R.G. Palmer: In *Heidelberg Kolloquium on Glassy Dynamics*, Lecture Notes in Physics, Vol.275 eds. J.L. van Hemmen, I. Morgenstern, Springer (1986)

[24] W. Götze: In *Liquid and Amorphous Metals*, Proc. of the Sixth Int. Conf. on Liquid and Amorphous Metals, R. Oldenbourg Verlag, München (1986)

[25] W. Götze: In *Amorphous and Liquid Materials* Nato ASI Series, eds. E. Lüscher, G. Fritsch, G. Jacucci, Martinus Nijhoff Publishers, Dordrecht (1987)

[26] S. Aubry: In *Solitons and Condensed Matter* eds. A.R. Bishop, T. Schneider, Solid State Sciences (Springer) **8**, 264 (1978)

[27] F.H. Stillinger: J. Chem. Phys. **88**, 380 (1987)

[28] I. Niven: *Irrational Numbers*, The Carus Mathematical Monographs 11, Wiley, New York (1956)

[29] P. Reichert,R. Schilling: J. Math. Phys. **26**, 1165 (1985)

[30] P.W. Anderson, B.I. Halperin, C.M. Varma: Philos. Mag. **25**, 1 (1972)

[31] W.A. Phillips: J. Low Temp. Phys. **7**, 351 (1972)

[32] J.C. Lasjaunias, A. Ravex: J. Phys. **F13**, L101 (1983)

[33] W. Kob, R. Schilling: Phys. Rev. **A42**, 2191 (1990)

[34] R.J. Glauber: J. Math. Phys. **4**, 294 (1963)

[35] J. Budimir, J.L. Skinner: J. Chem. Phys. **82**, 5232 (1985)

[36] J.L. Skinner: J. Chem Phys. **79**, 1955 (1983)

[37] H. Spohn: Comm. Math. Phys. **125**, 3 (1989)

[38] H.V. Bauer, K. Schulten, W. Nadler: Phys. Rev. **B38**, 445 (1988)

[39] G.S. Grest, C.M.Soukoulis, K. Levin: Phys. Rev. Lett. **56**, 1148 (1986)

[40] D.A. Huse, D.S. Fisher: Phys. Rev. Lett. **57**, 2203 (1986)

[41] W. Kob, R. Schilling: Z. Phys. **B68**, 245 (1987)

[42] W. Kob, R. Schilling: J. Phys. **A23**, 4673 (1990)

[43] R. Schilling: J. Stat. Phys., **53**, 1227 (1988)

[44] R.B. Stinchcombe, J.Jäckle, S. Cornell: preprint (1989)